MW01128748

# ADSL/VDSL PRINCIPLES:

*A Practical and Precise Study of Asymmetric Digital Subscriber Lines and Very High Speed Digital Subscriber Lines*

Dr. Dennis J. Rauschmayer

# *ADSL/VDSL Principles*

By Dr. Dennis J. Rauschmayer

Published by:
Macmillan Technical Publishing
201 West 103rd Street
Indianapolis, IN 46290 USA



Printed in the United States of America 1 2 3 4 5 6 7 8 9 0

02 01 00 99 4 3 2 1

Library of Congress Cataloging-in-Publication Data

Library of Congress Cataloging Number: 96-80467

ISBN: 1-57870-015-9

Universal Product Code: 619472001592

Warning and Disclaimer

This book is designed to provide information about Digital Subscriber Line technology. Every effort has been made to make this book as complete and as accurate as possible, but no warranty or fitness is implied.

The information is provided on an "as is" basis. The authors and Macmillan Technical Publishing shall have neither liability nor responsibility to any person or entity with respect to any loss or damages arising from the information contained in this book or from the use of the disks or programs that may accompany it.

## Trademark Acknowledgments

All terms mentioned in this book that are known to be trademarks or service marks have been appropriately capitalized. Macmillan Technical Publishing cannot attest to the accuracy of this information. Use of a term in this book should not be regarded as affecting the validity of any trademark or service mark.

DEVELOPMENT EDITOR
*Brad Miser*
*Christopher Cleveland*

PROJECT EDITOR
*Dayna Isley*

COPY EDITORS
*Geneil Breeze*
*June Waldman*

TECHNICAL REVIEWER
*Jacky Chow*

TEAM COORDINATOR
*Amy Lewis*

MANUFACTURING COORDINATOR
*Brook Farling*

BOOK DESIGNER
*Anne Jones*

COVER DESIGNER
*Sandra Schroeder*

COVER PRODUCTION
*Aren Howell*

DIRECTOR OF PRODUCTION
*Larry Klein*

PRODUCTION MANAGER
*Laurie Casey*

GRAPHICS IMAGE SPECIALIST
*Oliver Jackson*

PRODUCTION TEAM
*Louis Porter, Jr.*

INDEXER
*Bront Davis*

PROOFREADER
*Mary Ellen Stephenson*

## *Acknowledgments*

Writing this book was a great experience. Along the way, interactions with many of the people working on the standardization, design, and development of high-speed communications over copper increased my knowledge of the subject matter tenfold. I tip my hat to these engineers. This field truly has a bright future.

I must extend my thanks to the many folks who reviewed and contributed creative ideas to this text, including Chuck Stanski, Jeremy Nightingale, Debbie Wisniewski, and Jack Waller. A special thanks also goes out to Technical Editor Jacky Chow and Development Editor Brad Miser for playing an integral part in the development of this manuscript, as well as to the crew at Macmillan: Chris Cleveland, David Gibson, Tracy Hughes, and Dayna Isley.

Personally, I am indebted to many for the support I've received over the past year, including the fine folks at Pulsecom, especially Gerhard Pilcher, and my family—my sisters Eileen, Kathy, Joan, Nancy, Mary, Carol, and Linda; my brothers Joe and Rich; and most especially my mother.


## *About the Author*

**Dr. Dennis J. Rauschmayer** received a B.S. (summa cum laude) degree in electrical engineering in 1992 from Wilkes University, an M.S.E.E. degree from The George Washington University in 1994, and a Ph.D. in electrical engineering from Temple University in 1997.

From 1992 to 1996, Dr. Rauschmayer was a design engineer with Pulsecom, where he designed voice-band line cards with automatic gain control and adaptive hybrid equalization to optimize performance against voice-band data modems and on-hook data transmission devices. Since 1996 he has been employed as principle scientist, broadband copper access, at Pulsecom. His work includes research, design, and development of high-speed copper access devices as well as network design of systems utilizing xDSL access devices.

Dr. Rauschmayer is an active member of ANSI Working Group T1E1.4, ITU-T, and the ADSL Forum, making numerous contributions assisting in the effort to standardize ADSL and VDSL technologies.

## *Dedication*

To Jen

*—for every single smile—*

# Overview

# Table of Contents

CHAPTER 1

# *Introduction and Motivation*

In this chapter:

- Overview of this book
- Basic discussion of the outside plant
- Motivation for higher bit rates
- Introduction to DSL technologies

This book discusses networks capable of delivering high-speed data to all end users. Specifically, the book discusses having widespread users connected to the network using traditional telephone lines with access rates in excess of 1 Mbps. Digital Subscriber Line (DSL) technologies, a family of relatively new and exciting high-speed access technologies using ordinary telephone lines, are especially well suited for the access portion or "on ramp" of such a network.

The principles of two specific DSL technologies, *Asymmetric Digital Subscriber Line* (ADSL) technology and *Very High Speed Digital Subscriber Line* (VDSL) technology, are the primary focus of this book, along with networks to which they allow access. ADSL provides an end user with data rates up to 8 Mbps in one direction and up to 800 Kbps in the other. VDSL provides even higher speeds as well as asymmetric or symmetric operation. This book presents in-depth discussions of ADSL and VDSL as well as how networks are designed to use them. Background discussion and theory are given for the issues faced by DSL technologies and the fundamental performance limits that can be achieved. Most of the discussions in this book are explained in terms suitable for both technical and nontechnical readers. In addition, detailed theory and extensive equations are provided for those who want to dive more deeply into the subject matter.

The book is logically divided into three main sections. The first section deals with the background necessary to understand the challenges and limits of DSL technologies. This section includes Chapter 1 through Chapter 5. Chapter 1 provides a basic introduction and motivation explaining why DSL technologies are important to widespread network access. Chapter 2 discusses physical and electrical characteristics of *twisted pairs* (telephone wires), including electrical and physical properties, and it discusses the basics of *crosstalk*. Chapter 3 continues this discussion by presenting tools used to analyze any type of twisted pair. Chapter 4 focuses on several different types of DSL technologies and the *Power Spectral Densities* (PSDs) used by each. From this, the crosstalk results of Chapter 2 are expanded to show mathematical models of how different technologies interact with one another. Chapter 5 takes this one step farther by combining these effects with the tools from Chapter 3 to predict the theoretical limits of performance of ADSL and VDSL.

The second part of the book deals with ADSL and consists of Chapter 6 through Chapter 8. Chapter 6 discusses, in general terms, the different blocks often present in DSL transmitters and receivers; the material in Chapter 6 is applicable to ADSL and VDSL as well as many other types of communications systems. Chapter 7 is dedicated to different types of ADSL transmitter and receiver systems. The specific implementations of the blocks discussed in Chapter 6 are explained. Real-world insight and issues are provided. Finally, Chapter 8 discusses how ADSL is integrated into *wide area networks* (WANs). Higher layer protocol issues are discussed as well as several network implementations.

The final part of the book deals with VDSL and consists of Chapter 9. There, candidate VDSL implementations (which are still under development) are discussed as well as some issues above and beyond ADSL. Finally, a brief discussion of possible VDSL WAN networks is provided.

## Telephone and Data Systems

In the modern world, a telephone in every home and every building—and sometimes in every room—is the norm. This is relevant to DSL technology in that in each instance the telephone is connected to a pair of wires (or copper pair) that run out of the building and somewhere within the next 20,000 feet or so connect into telephone switching equipment. (This copper pair is also known as the *subscriber line.*) Thus, every home and building already has a physical means through which data can pass.

### Coexistence

Figure 1.1 shows a typical section of the telephone system. The copper loops, or subscriber lines, could be serving a group of homes, an apartment building, an office building, or more generically, a subscriber. Common devices at the subscriber end of the subscriber line include telephones, fax machines, and voice band modems. Generically, these will be referred to as *Plain Old Telephone Service* (POTS) devices.

**Figure 1.1** A typical example of subscriber lines and a Central Office (CO) housing a telephony switch.

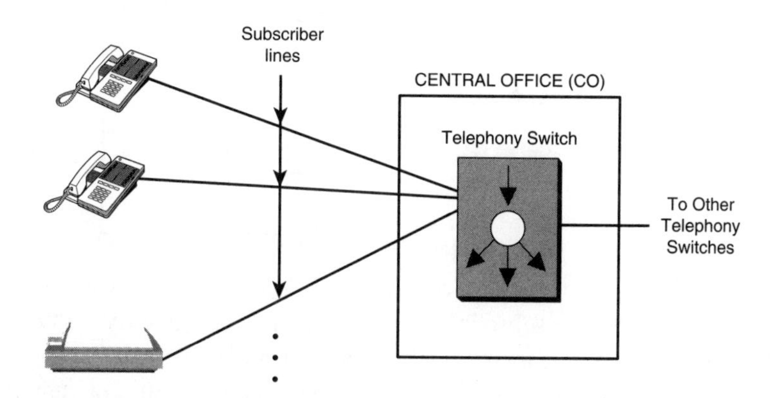


The telephone switching equipment allows one end user to call another end user. Although the links between the telephone switches are typically high-bandwidth optical fibers, the links must be capable of serving many users simultaneously, and so each end user can only use a small portion of the data bandwidth. Furthermore, the switching equipment is designed to handle only small amounts of bandwidth per end user. Typically, the bandwidth available for POTS spans from about 200 Hz to around 3200 Hz and has a capability of carrying a maximum of 64 Kbps.

Now consider switching devices and terminal devices designed and optimized to handle large amounts of data rather than only voice signals for each end user. Further, consider that this new equipment should use the same subscriber lines that the telephone system previously used. Finally, add the criteria that this new system and the original telephone system both use the same subscriber line simultaneously without interfering with one another. This is the basis of DSL technology and is shown in Figure 1.2.

Figure 1.2 A conventional telephone system coexisting with a data overlay network.

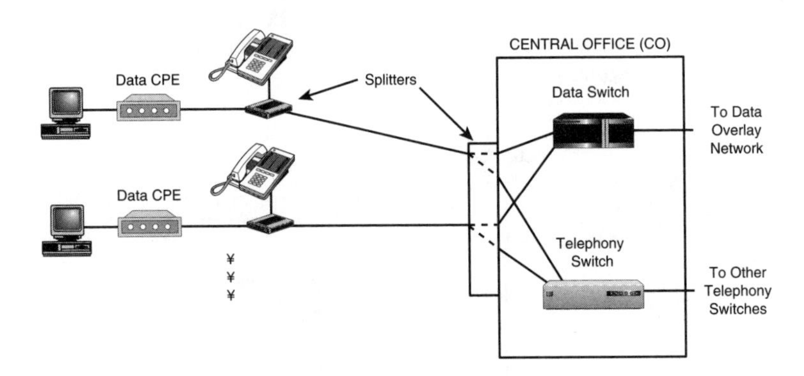


In Figure 1.2, the data network is called a *data overlay network* because it essentially overlays the POTS signals on the subscriber lines. Both types of signals, the POTS signals and the data signals, are routed to their respective end terminals at the subscriber location and also to their respective switching equipment at the switching center, or central office (CO). A device called a *POTS splitter* is responsible for splitting and recombining the two types of signals at both ends of the subscriber line. Typical terminal equipment at the subscriber end for the data network are computers and set-top boxes.

## *Subscriber Lines*

In all, about 560 million subscriber lines exist in the world (about 156 million are in the United States). Other common names for a subscriber line include a *copper pair*, *subscriber loop* or simply *loop*, and *twisted pair*. Although most lines have typically been used for voice or low-speed data, a typical loop can support much higher data rates depending on its type, length, and environment. Physical and electrical characteristics of twisted pairs are discussed in detail in Chapter 2.

Asymmetric Digital Subscriber Lines (ADSL) and Very High Speed Digital Subscriber Lines (VDSL), the main DSLs discussed in this book, are methods of transporting data over twisted pairs. As you will see, both technologies require considerable processing power, which normally increases cost and complexity. Much of this is due to the fact that twisted pairs, although capable, were never designed to carry high bit rates because they originally were designed and deployed for relatively low data content voice signals.

Given the differences between POTS and DSL technologies, a natural question might be, "Why try to piggyback a brand-new technology on old lines rather than install fiber or high-speed wireless connections to homes?" The simple answer is that the copper is already there. It is an *installed base*. Installing a new base of any kind is costly and takes considerable time, so why not use what is already in place?

Copper loops provide one of the only means to deliver high bit rate data to many subscribers. Even though the switches and the POTS splitters must still be added, this is far less of a burden than installing a whole new infrastructure.

## *Motivation for Higher Bit Rates*

Demand for new technologies needs a driver. Several drivers exist for ADSL, including the Internet and World Wide Web (WWW), video services, consumer services, and entertainment services. Synergy with other technological advances has also helped motivate for ADSL. The most noteworthy of these is the widespread deployment of personal computers (PCs) and an increasing awareness in the general public of how to use PCs. The following sections discuss motivations for ADSL and VDSL.

### *Growth of the PC*

PCs are common in the home. In 1996 alone, more than 71 million PCs were shipped to retail distributors. PCs are becoming user friendly enough that no detailed technical knowledge is required to use many of the available software applications. The number of PC households is expected to continue to grow every year for the foreseeable future. Figure 1.3 shows the projected growth of PC households until the year 2000.

Figure 1.3 The growth of personal computers (PCs) worldwide.

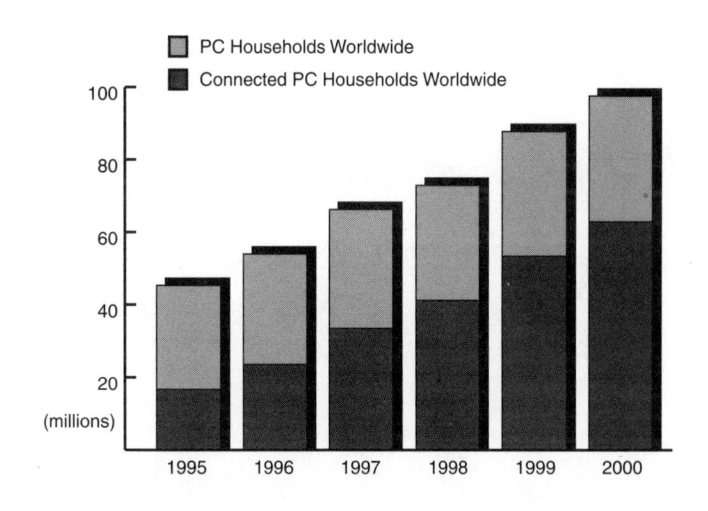

## *Bulletin Board Services*

Bulletin board services (BBSs) became popular in the 1980s. Although BBS systems do not presently provide impetus for higher bit rates, they are worthy of discussion because they are the predecessors to modern Internet service providers. Typically, a PC could dial into a BBS through a 300 to 2400 bps modem and access information (such as online encyclopedias and dictionaries) or enter into chat conversations with other users. Most early BBS systems had limited graphics so that 2400 bps was adequate.

## *Internet*

The Internet gained popularity in the late 1980s. At that time, Internet users were primarily universities and some businesses. Most users had a technical background, and there was little widespread interest in the Internet outside the scientific community. Email over the Internet as well as newsgroups started to spread interest into other disciplines. As more users started to regularly access the Internet, more diverse information and topics became available, creating an increased demand for wider access to the Internet.

## *World Wide Web*

Originally, the Internet was primarily text based. The World Wide Web (WWW) changed this by allowing both text and graphics to be sent and received. This increased the need for higher access rates because graphics typically require larger data transfers than text. The WWW had a simple and attractive graphical user interface, requiring only the use of a mouse and little actual typing. The size of the WWW is growing dramatically. Figure 1.4 illustrates the growth of the number of WWW hosts over the past several years. With more than 16 million existing sites, the WWW's diversity in subject matter is astounding.

Figure 1.4 The growth of Internet hosts (courtesy of Network Wizards, http://www.nw.com/).

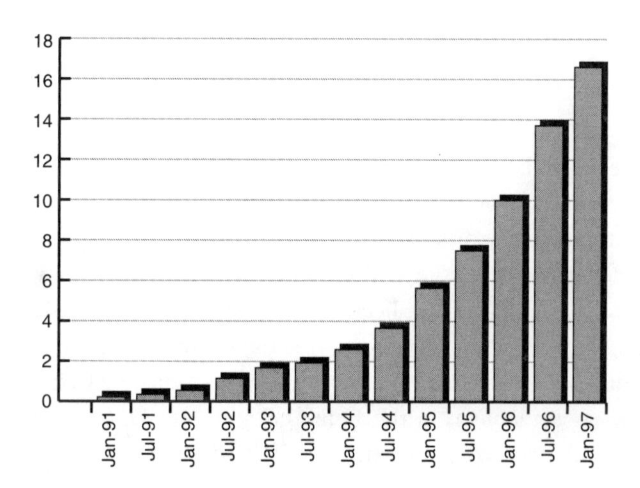


The size of a typical WWW page varies depending on the content. Some pages now contain multimedia, including text, graphics, audio, and video. Few high-quality graphics, even in compressed form (JPEG) are under 40 kilobits (kb) in size. Most audio clips run in the hundreds of kilobits range and most video clips in the several megabit range. As consumers continue to use and enjoy the WWW, the demand for higher bandwidth will grow because users do not want to wait for data transfers to take place and will demand faster response.

New WWW technology allows for real-time video and audio as well as three-dimensional graphics. As connection speeds increase, Web pages can become even more sophisticated. Games, shopping, entertainment, and educational applications will demand and use even more bandwidth as it becomes available, further increasing the demand for solutions providing higher bandwidths.

## *Internet Service Providers*

Internet service providers (ISPs) and network service providers (NSPs) have become common in the 1990s. A typical NSP allows dial-up access to the WWW and Internet and provides other services such as email, educational tools, banking, and entertainment. Usually, a consumer connects to an NSP through a voice band modem over the Public Switched Telephone Network (PSTN). Some higher speed access technologies might also be available to connect to NSPs including ISDN (128 Kbps), T1 (1.544 Mbps), and fractional T1 (n × 64 Kbps), but these are usually expensive and geared toward businesses rather than residential subscribers.

The popularity of ISPs and NSPs has grown tremendously in the 1990s. The number of subscribers to these services has grown to more than 20 million and is predicted to further increase.

## Video on Demand

*Video on demand* (VOD) is another application for ADSL and VDSL technologies. A typical VOD channel requires around 3 Mbps of throughput. ADSL is one of the first technologies allowing the delivery of VOD over the telephone line, and VDSL could allow the delivery of multiple channels. In the United States, cable systems are widely accessible, and VOD might not be as popular an application for DSL technology as Internet access. However, it does give telephone companies (telcos) options they wouldn't otherwise have, which is important in a business and marketing sense. In other parts of the world, cable TV systems are not as available as in the United States, and DSL technologies could play a large role in delivering video to the home.

## Small Office/Home Office

The small office/home office (SOHO) provides another good opportunity for DSL technology. In this application, DSL technology can be used to connect the SOHO to a corporate LAN, the Internet, or both. DSL technology would allow a telecommuter to access a LAN with similar perceived speed and efficiency as if she were at the corporate office. This could create a large demand for higher speed access. The number of telecommuters in the United States is also projected to grow into the year 2000 as shown in Figure 1.5.

A typical corporate LAN might have a 10 Mbps backbone connecting PCs and servers. Linking to this at a rate of several megabits per second via ADSL, although not optimal, would still provide an efficient means of connection, and the performance difference would most likely go unnoticed by most users. At VDSL speeds, even higher access rates are achievable.

Figure 1.5 The increasing number of telecommuters in the United States will continue into the year 2000.

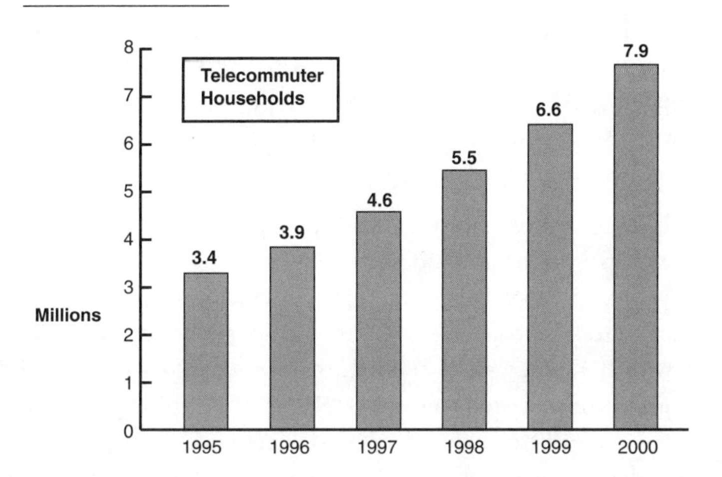


## *Modem Evolution*

Voice band modems have evolved quite a bit in the 1990s. In a period of about four years, the speed of voice band modems increased by an order of magnitude. Figure 1.6 shows the trends in voice band modems for the PC since 1990 and predictions until the year 2000.

Figure 1.6 Personal computer modem shipment.

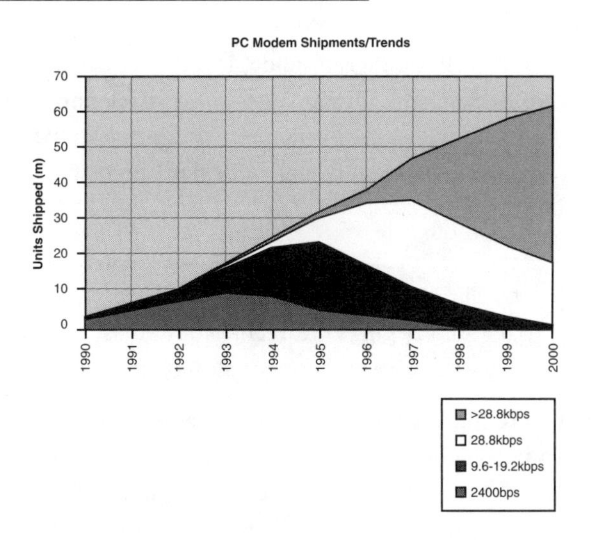

The rise and fall of lower speed voice band modems during the 1990s, including 2400 bps, 9.6 Kbps, and 28.8 Kbps modems, is apparent.

Modems with speeds greater than 28.8 Kbps include higher speed voice band modems such as 56 Kbps modems, Integrated Services Digital Network (ISDN) modems, different types of DSL modems, cable modems, and other types, such as wireless modems. Most likely, all these will be deployed in the future, although the types of modems dominating the deployment remain to be seen.

ISDN modems have struggled since the 1980s to achieve wide-scale deployment. ISDN modems use twisted pairs and can send and receive data up to 144 Kbps. Standardization discrepancies have made ISDN provisioning and installation difficult, thus impeding widespread use. Other factors have kept ISDN out of main stream usage, including cost and incompatibility with POTS service. Generally, the cost of ISDN was too high, and a POTS line cannot share the same twisted pair as ISDN. ISDN is typically switched through the telephone network and not through a data overlay network. Even with its faults, over the last decade, ISDN has been one of the only means of achieving data rates significantly greater than voice band modems.

## *Why DSL Technology?*

With other broadband technologies in the running, several key attributes make DSL networks an attractive choice for high-bandwidth access. Although a reader of this book is likely to be aware of these technologies, a quick review of some key attributes will be beneficial.

As discussed earlier, xDSL technology can be deployed over existing twisted pairs. This installed base is widespread, and the total investment that telephone companies have placed in it over the years is immeasurable. Not only does using this copper further utilize this inherent vestment, but it also allows the delivery of new, high-speed service without investing in a brand-new infrastructure.

Most new technologies require considerable time even before feasibility trials can be done. For example, a wireless satellite system would require designing, building, and deploying the satellites before a trial could begin. DSL technology trials, on the other hand, can be much simpler due to the installed base of copper along with existing switching centers.

Many applications require high bandwidth, and new applications are sometimes created when high bandwidth is available. Applications in medicine, real estate, law, business, and many other fields can be supported with DSL technologies. Furthermore, DSLs can be deployed to many locations needing this bandwidth in a cost-efficient manner by using the existing infrastructure.

With the growing popularity of online services, the PSTN is being overrun with traffic. (This currently is a heated debate between Internet service providers who claim that the PSTN is in no danger of being overrun by data calls and the telephone companies who claim the opposite.) Originally, the PSTN was designed to handle voice traffic having patterns different from data traffic. For example, most voice calls are short in duration, whereas a data call to an Internet service provider might last several hours. The statistical models used to design the PSTN did not adequately consider the data call model. The data overlay network involved in xDSL technology will relieve the telephone switching system of some data traffic, avoiding major upgrades to the infrastructure.

## *The Market Demand*

The market demand for a technology plays a large role in advancing the technology itself. This is because companies need to make a profit and often will only undertake projects that are predicted to generate income over in the future. The detailed justifications of a business case for DSL technologies are beyond the scope of this book; however, some of the more general arguments in favor of the DSL technology, such as the advantages of xDSL technology over competing options like cable modems and wireless data services, are worthy of discussion. The competing technologies are discussed later in this chapter.

Figure 1.7 depicts the predicted rollout of DSL modems over the next several years.

Figure 1.7 The predicted rollout volumes of xDSL modems through 2002.

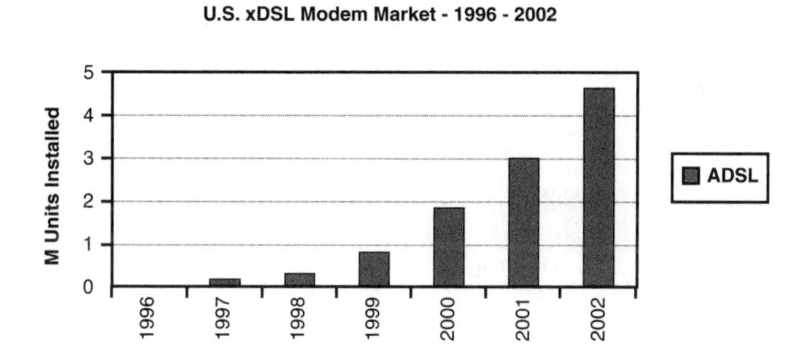

The most significant growth is predicted from 1999 through 2002. In all, more than 10 million DSL units are predicted to be installed by the year 2002. If each had a cost of $300.00, probably a conservative estimate, xDSL would be a more than $3 billion industry over the four years! This is a good business case indeed.

## *The DSLs*

This section discusses the various types of DSLs that have evolved over the past few years. Some of the technologies are driven by companies, whereas others are supported by the industry as a whole, including service providers and vendors. For these, standards bodies are defining or have defined specifications for implementation. Because of all the acronyms used for different DSL technologies, this section uses the generic term *xDSL* to generally refer to DSLs. Table 1.1 shows a summary of many xDSL technologies.

Table 1.1 Capabilities of Various DSL Technologies

| DSL | Upstream Rates | Downstream Rates | Symmetric/ Asymmetric | POTS Coexistence | Standardization |
|---|---|---|---|---|---|
| ADSL | 100–800 Kbps | 1–8 Mbps | Asymmetric | Yes | Yes |
| HDSL | Fixed 784, 1544, 2048 Kbps | Fixed 784, 1544, 2048 Kbps | Symmetric | No | Yes |
| HDSL2 | Fixed 1544, 2048 Kbps | Fixed 1544, 2048 Kbps | Symmetric | No | Yes |
| IDSL | Fixed 128 Kbps | Fixed 128 Kbps | Symmetric | No | No[1] |
| MDSL | 64–200 Kbps | 800–2000 Kbps | Asymmetric | Yes | No |
| SDSL | 384 Kbps | 384 Kbps | Symmetric | Yes | No |
| VDSL | Up to 25 Mbps | Up to 25 Mbps | Both | Yes | Under development |

[1]IDSL itself has no standard, but is similar to ISDN, for which a standard does exist.

## ADSL

As stated earlier, Asymmetric Digital Subscriber Lines (ADSL) technology is one of the main topics of this book. ADSL is characterized by a different data rate from the service provider to the customer (the downstream direction) as compared to the data rate from the customer to the service provider (the upstream direction). This difference is the origin of the name Asymmetric. Data rates are typically about 10 times as high in the downstream direction and range anywhere from around 1 Mbps to around 8 Mbps—the maximum rate can depend on the length and condition of the loop determined during initialization or on a maximum provisioned rate.

For example, consider a loop capable of supporting 4 Mbps in the downstream direction and 440 Kbps in the upstream direction. A modem having the capability to operate at 6 Mbps downstream will realize during initialization and training that 6 Mbps cannot be supported on the particular loop and will negotiate the rate downward to 4 Mbps. This is an example of a rate determined by the length and condition of the loop. The service provider might, however, have provisioned the modem to run even slower than this (for example, at a maximum of 2 Mbps). (It is not yet clear whether network providers such as telcos will provision modems to limit the data rate, but it is expected to be a desired option. A telco might want to do this to provide various tiers of service with, of course, higher pricing for higher bandwidth capable tiers.) In this case, even though the modem could operate at the higher speed of 4 Mbps, it will limit the downstream bit rate to 2 Mbps.

ADSL is typically geared to support applications such as Internet access, LAN bridging, and VOD. ADSL has been in many technology trials since the early to mid-1990s.

## HDSL

High Bit Rate Digital Subscriber Line (HDSL) technology provides symmetric bit rates. Typically, HDSL modems are designed to run at fixed speeds. Some common operating speeds are 768 Kbps, 1.544 Mbps, and 2.048 Mbps. The latter two values correspond to T1 and E1 lines, respectively. HDSL usually needs two separate twisted pairs, and both must be full duplex. Applications for HDSL are growing, but the original driver was the replacement of T1 and E1 lines because HDSL had a longer range than these older technologies. HDSL uses 2B1Q modulation and cannot coexist on the same twisted pair with POTS.

**Note**

The code 2B1Q is a four-level baseband code; 2B1Q stands for two binary, one quatenary signal. That is, each symbol is one of four possible symbols and conveys two bits of data.

## *HDSL2*

HDSL2 is a follow-on technology to HDSL. Essentially, it is meant to transmit T1 and E1 rate signals over a single twisted pair. Currently, HDSL2 is undergoing the standardization process. POTS cannot be run on the same twisted pair as HDSL2.

## *IDSL*

IDSL is one of the new abbreviations to come along, and it stands for ISDN Digital Subscriber Line. IDSL uses the same 2B1Q line code technology as used in ISDN. IDSL is a symmetric service and cannot share a twisted pair with POTS. Initial rates for IDSL include 56 Kbps, 64 Kbps, 128 Kbps, and 144 Kbps.

## *MDSL*

Medium Bit Rate Digital Subscriber Line (MDSL) technology is basically a watered down version of ADSL. MDSL is an asymmetric technology running at slower speeds than ADSL—typically 800 Kbps to 1 Mbps downstream and 100 Kbps upstream. MDSL was proposed to reduce the cost and complexity of a DSL modem while still providing reasonably high data rates. Applications for MDSL include Internet access and LAN bridging. MDSL is currently under study but is not being standardized.

## *RDSL and RADSL*

RDSL and RADSL can stand for two different things. The first is Rate Adaptive ADSL, meaning an ADSL line that can automatically select an optimum operating rate for a twisted pair. Originally, ADSL was not thought of as adaptive in rate. This name was coined when vendors began proposing rate adaptive functionality. This name is seldom used because ADSL itself is usually assumed to be rate adaptive.

The second technology denoted by RDSL and RADSL is Reverse ADSL. Reverse ADSL was proposed as an ADSL system in which the asymmetry of data rate is reversed. Thus, the upstream direction runs faster than the downstream direction. The proposed main application for Reverse ADSL was for subscribers wanting to provide WWW servers. Typically, a WWW server provides more data than it receives, motivating the need for a higher upstream data rate and a slower downstream data rate. Because of spectrum incompatibilities, Reverse ADSL has received little attention by vendors or standards bodies.

### SDSL

Symmetric Digital Subscriber Line technology is the little brother to HDSL. As the name implies, data rates in the upstream and downstream direction are symmetric. Typical data rates are 384 Kbps and 768 Kbps. SDSL can be used to provide fractional T1 services or LAN and Internet access. No standard currently exists for SDSL.

### VDSL

Very High Speed Digital Subscriber Line (VDSL) technology is the other major topic of this book. VDSL can be viewed as ADSL's big brother. VDSL can run with either symmetric or asymmetric rates. The highest symmetric rate being proposed for VDSL is 26 Mbps in each direction. Other typical VDSL rates being proposed are 13 Mbps symmetric, 52 Mbps downstream and 6.4 Mbps upstream, 26 Mbps downstream and 3.2 Mbps upstream, and finally, 13 Mbps downstream and 1.6 Mbps upstream. VDSL can coexist with POTS on a twisted pair.

The maximum possible length of a VDSL loop is much less than for that of an ADSL loop. VDSL is planned for use as the last link in fiber in the loop (FITL) and fiber to the curb (FTTC) networks. FITL and FTTC networks have small access nodes, are remote from the CO, and are fed by high bandwidth fiber. A typical node might reside on a curb in a neighborhood and serve 10 to 50 homes. Thus, the loop lengths from the node to the subscriber tend to be shorter than loops from the CO to the subscriber. Although the applications are not yet clear, VDSL could be used as the last link in a full-service multimedia network delivering voice, video, and data to a subscriber.

## Other High-Speed Data Options

xDSL is not the only option available to provide high-bandwidth solutions. Several others offer formidable competition to xDSL.

### Cable Modems

Cable television delivered via coaxial cable (coax) is available throughout most of the United States and much of Europe. Coax can handle high bandwidths, and cable modems take advantage of this. Cable systems are traditionally a broadcast medium with one line serving many subscribers. Because of this, the subscribers must share the bandwidth of the coax cable. Also, the upstream direction from the subscriber back to the coax distribution point traditionally has not been used to transfer information and, in many areas, does not support transmission. Reconditioning the infrastructure to support upstream transmission is costly and time consuming, and deployment in a geographical area cannot be done until the entire infrastructure in that area is upgraded. Despite this, cable modem standards are under development, and cable modems will continue to be a significant competitor to xDSL.

### Wireless Technologies

Several wireless technologies also have the capability to distribute data to subscribers. Direct Broadcast Satellite (DBS) relies on geostationary satellites to supply data—primarily programming in broadcast form—to subscribers. Multichannel Multipoint Distribution Systems (MMDS) can send 33 television channels in analog form or around 100 channels in digital form to subscribers. These technologies require a satellite dish receiver and a set-top box at the subscriber location. Although primarily geared for broadcast at the present time, these technologies might become players in the high-speed interactive data world in the future.

## End-to-End System Diagram

Many companies throughout the world are developing products based on ADSL and VDSL. Specialized chipsets are available for ADSL and should be available for VDSL in 1999. To better discuss all the aspects of ADSL and VDSL systems, a generic model on which to base the discussion will be helpful. Figure 1.8 shows this model for an ADSL access network, and the various components, including hardware equipment, line codes, and protocols are discussed in the following sections.

Figure 1.8 An end-to-end system diagram of an ADSL or VDSL network.

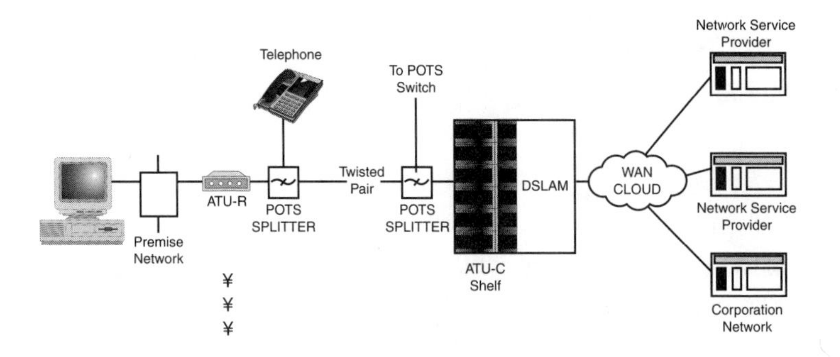


### Customer Interface

Many possible options are available for the customer interface. Common interfaces include 10BaseT and a 25.6 Mbps ATM interface (sometimes simply called an ATM-25 interface). These are more common types of high-speed interfaces to the PC. Also for a PC interface, an internal card sometimes called a *network interface card* (NIC) is used. Certain other interfaces, such as a V.35 interface to a set-top box and newer types of high-speed interfaces, such as Universal Serial Bus (USB), are also possibilities for the customer interface.

## The POTS Splitter

The POTS splitter allows ADSL or VDSL signals to coexist on a twisted pair with telephony signals. One POTS splitter is needed on each end of the twisted pair. In one direction, the POTS splitter combines the two kinds of signals, whereas in the other direction it properly splits the signals.

The POTS splitter is basically a three-port device containing a bidirectional high-pass filter and a bidirectional low-pass filter. The POTS splitter might be fully or partially integrated into the ATU-R and ATU-C.

## The Twisted Pair

The twisted pair is made up of two copper conductors twisted around one another; these are usually less than 18,000 feet in length for ADSL and less than 4,500 feet in length for VDSL. (In both cases, the loop length varies over a wide range up to the maximum. The actual length dictates what data rates can be supported; longer loops support only lower data rates.) The physical and electrical properties of twisted pairs are discussed in detail in Chapter 2.

## xTU-R

The xTU-R modem resides at a customer's location (house or business). The xTU-R might be an external modem, a card plugging into a PC, or, in some cases, part of a bigger piece of network equipment such as a router. For ADSL, this modem is called the ATU-R and for VDSL is called the VTU-R.

## xTU-C

The xTU-C is the counterpart to the xTU-R and resides at the central office. Generally, the xTU-C is part of the network access equipment. The xTU-C normally consists of cards plugging into an access shelf. More than one modem can exist on each card, or the modem functions can be distributed among several cards, but an xTU-C connects only to one xTU-R at a time. For ADSL, this modem is called the ATU-C, and for VDSL, it is called the VTU-O (the O represents the fact that the VDSL node normally resides at an optical node).

**Note**

Note that the xTU-C may reside outside the central office and the data traffic backhauled over high-speed links such as OC-3 optical fiber.

## *DSLAM*

DSLAM has become one of the more popular abbreviations in xDSL technology; it stands for Digital Subscriber Line Access Multiplexer. The DSLAM aggregates subscriber line traffic onto higher speed links for the backbone switching equipment.

## *NSP Network*

The network service provider (NSP) network is an important part in achieving of an overall service network. NSP is a generic term for an Internet service provider, entertainment provider, corporate network, or any other type of provider whose customers gain access to services via the xDSL network.

## *Line Codes*

The line code is the physical layer modulation scheme used by the xDSL technology. This book discusses line codes for both ADSL and VDSL technologies. For ADSL, discrete multitone modulation (DMT) is the standard line code. Carrierless amplitude and phase (CAP) modulation and quadrature amplitude modulation (QAM) are other line codes competing with DMT as ADSL modulation schemes. CAP and QAM produce similar signals under certain conditions and are lobbying for standardization. CAP- and DMT-based ADSL modems have been successful in field trials, and both, along with QAM, are discussed in detail in later chapters. CAP and QAM are often referred to as single carrier modulation schemes, whereas DMT is referred to as a multicarrier modulation scheme.

**Note**

The reader may note that referring to CAP as a single carrier modulation scheme contradicts the "C" in its acronym, which stands for carrierless. CAP is generated using digital filters instead of by modulating a carrier, thus the name carrierless. In the most widely used implementation, the digital filters generate sine and cosine waves, and the result is a signal very similar (and sometimes identical) to a carrier modulation scheme such as QAM.

A line code has not yet been standardized for VDSL. Codes in competition include single carrier systems based on CAP and QAM, multicarrier systems based on symmetric DMT (SDMT), and also a DMT technique called Zipper. All these line codes are discussed in later chapters.

**Note**

One proposal for a VDSL line code is actually a hybrid approach with a single carrier downstream scheme and a multicarrier upstream scheme based on Discrete Wavelet Multitones (DWMT).

### *Data Link Framing and Encapsulation*

Different types of data link framing and encapsulation techniques being standardized for ADSL include ATM-based approaches and frame-based approaches. VDSL framing is not as far along, although initial applications make ATM a strong candidate. Chapter 8 discusses framing and encapsulation methods for ADSL; Chapter 9 covers proposed framing and encapsulation methods for VDSL.

### *Backbone/ATM Cloud*

In discussing ADSL and VDSL systems, it is assumed that a network backbone exists linking the subscriber lines to network services. In many cases, *Asynchronous Transfer Mode* (ATM) will be the underlying transport mechanism in the backbone, although the protocols running on top of ATM may vary. For example, in some cases, protocols that encapsulate Internet protocol (IP) data might be discussed, whereas in other cases, tunnel protocols might be used. It is also possible to have other types of backbone WAN networks including an IP routed network or a frame relay network.

## *ADSL and VDSL Standardization*

Standardization is an important aspect in the xDSL marketplace. Service providers deploying equipment like to have the assurance that what they are purchasing from one will properly interoperate with what they might purchase from other vendors. The voice band modem market is an example of the benefits of standardization as almost any pair of off-the-shelf modems interoperate with one another.

### *ADSL Standardization*

Several different standardization organizations (sometimes referred to as standards bodies) play a role in ADSL specifications. The following sections discuss the various standards bodies involved.

## *Committee T1*

The original body studying ADSL was committee T1, which is an open group of industrial participants that create standards for the United States. The American National Standards Institute (ANSI) accredits committee T1. Several subcommittees exist within the committee. These subcommittees are dedicated to different areas of standardization and include the following:

- T1A1 is dedicated to performance and signal processing
- T1E1 is dedicated to interfaces, power, and protection for networks
- T1M1 is dedicated to internetwork operations, administration, maintenance, and provisioning
- T1P1 is dedicated to wireless/mobile services and systems
- T1S1 is dedicated to services architecture and signaling
- T1X1 is dedicated to digital hierarchy and synchronization

The subcommittee relevant to ADSL is T1E1, with xDSL falling under the interfaces category. T1E1 is further broken down into working groups. The working group assigned to DSL access (including ADSL) is T1E1.4. The scope of working group T1E1.4 includes the development of standards for the physical layer including transmission techniques and interface functionalities. The T1E1.4 working group developed the T1.413 standard for ADSL. This standard was published in 1995.

## *The ADSL Forum*

The ADSL Forum is a group of industry participants formed in 1994 to promote ADSL with both marketing and technical efforts. The marketing effort is primarily to educate the public on ADSL. On the technical front, committees exist to work in implementation issues not covered by T1E1.4 (primarily layers above the physical layer as well as the management and signaling planes). Several subgroups exist in the ADSL Forum including the following:

- A packet mode protocols group dedicated to defining ADSL functionality with a packet-based (non-ATM) access
- An ATM mode group dedicated to defining ADSL functionality within ATM networks
- A network migration group dedicated to defining the evolution of networks implementing ADSL

- A network management group dedicated to defining the management methods for ADSL including the management information base (MIB)
- A test group dedicated to defining test requirements for ADSL systems

The ADSL Forum publishes technical specifications in many of the areas listed previously. Further information and updates of work in progress are available at `http://www.adsl.com`.

## *The ITU*

In 1998, the International Telecommunications Union (ITU) joined the effort to standardize ADSL. This is the same standardization body that has published specifications for voice band modems including V.34. The ITU is working on many items related to ADSL including the following:

- **G.DMT**. a project based largely on T1.413
- **G.lite**. a project aimed at lower speed, reduced complexity ADSL applications
- **G.test**. a project aimed at specifying test specifications for xDSL technologies
- **G.OAM**. a project aimed at specifying operations, administration, and maintenance aspects of xDSL systems
- **G.HS**. a project aimed to define handshaking protocols to allow startup negotiation among xDSL modems

Many items being worked on by the ITU overlap work being done by T1E1.4 or the ADSL Forum. In most cases, cooperation between the groups exists to facilitate convergence of xDSL technology.

## *ETSI*

The European Technical Standards Institute (ETSI) is an open European body dedicated to determining and producing telecommunications standards. A technical body within ETSI known as TM is responsible for transmission and multiplexing. A subgroup of technical body TM known as TM6 is responsible for xDSL standards development including ADSL. TM6 is very much a European counterpart to T1E1.4. Cooperation between T1E1.4 and ETSI TM6 on ADSL standards development is common.

## *VDSL Standardization*

T1E1.4 and ETSI TM6 also study projects dedicated to VDSL. From 1996 to 1998, the bodies individually drafted technical requirements documents. In the future, it is expected that an actual standard will be developed, most likely with cooperation between the groups. It is likely in the future that other groups will begin work on aspects of VDSL as well.

# *Summary*

This chapter provided an introduction to xDSL technology as well as a guide to what can be found in subsequent chapters of this book. The chapter also discussed drivers for high-bandwidth access to various types of networks. It also reviewed the benefits of xDSL technology, as well as how xDSL technology uses the existing PSTN infrastructure.

The chapter also explained the different flavors of xDSL technology and key characteristics of each. The chapter explored the role and importance of standardization bodies in ADSL and VDSL technologies and described several bodies along with the area of ADSL and VDSL that they cover.

CHAPTER 2

# Twisted Pair Environment

In this chapter:

- History of twisted pairs
- Physical and electrical characteristics of twisted pairs
- Transmission line parameters
- Derivation of basic crosstalk equations

This chapter is devoted to the physical and electrical characteristics of twisted pairs. A twisted pair channel is noisy, lossy, and prone to crosstalk. ADSL and VDSL modulation schemes are designed to overcome and mitigate these impairments. Thorough understanding of ADSL and VDSL begins with thorough understanding of the environment in which they must exist.

This chapter discusses the physical and electrical properties of common twisted pairs. The transmission line characteristics and basic crosstalk equations in twisted pair bundles will be derived. This information becomes useful in subsequent chapters dealing with transmission performance. But first, an introduction to twisted pairs is in order.

## A Brief History of Twisted Pairs

The history of telephone wires is more than a century old and has roots in the wires originally used to carry information from the telephone's predecessor, the telegraph.

Samuel F. B. Morse constructed the first commercial telegraph line between Baltimore, Maryland, and Washington, D.C., in 1844. After failing to successfully install underground lines due to high leakage from the environment, Morse decided that an aerial approach would be better. Morse proceeded to install 700 wooden poles spaced about 300 feet apart for the 40-mile span. A 2-foot cross arm was nailed to the top of each pole, and a wire was run through a slot at the end of each cross arm. Thus, the groundwork was laid for the deployment of twisted pairs.

After the telephone began to gain popularity, the need for many more wires was necessary, and the telephone cable—consisting of many wires bound in a sheath—was developed. One of the biggest problems at first was crosstalk between wires, as the two-wire metallic circuit had not yet been employed. In New York City, in 1879, cable was laid across the Brooklyn Bridge, years before it was even open. In 1880, an electrician speaking for the Metropolitan Telephone and Telegraph Company reported:

> We have over the East River Bridge at the present time, four cables, 3800 feet long, each cable with seven conductors.... In using the cables and talking on one wire, you could hear whatever was said on another wire.[1]

In the following years, many schemes were employed to reduce crosstalk, most dealing with shielding the individual wires or providing grounding at intervals. Finally in 1881, a patent for the metallic twisted pair circuit, which worked well in suppressing crosstalk, was given to Alexander Graham Bell, who testified to the Honorable Commissioner of Patents about his experiments as follows:

> The methods by which I sought to accomplish this result, on the 26th of November, 1876 was to employ a metallic circuit, the direct and return wires which should be parallel to the disturbing wires and to one another.... The improvements in the arrangement that I devised in England were:
>
> To place the direct and return lines close together...and
>
> To twist the direct and return lines around one another so that they should be absolutely equidistant from the disturbing wires.[1]

Thus, the twisted pair was born. Since that time, the deployment of twisted pair lines has been extensive and includes many types and sizes of cable.

Several surveys have been conducted on deployed twisted pair subscriber loop statistics with some results including average lengths, impedances, and losses.[2,3]

**Note**

For an interesting history of all methods of mass communications in the United States, including the rise of the telegraph and telephone and the development of the twisted pair, I highly recommend visiting the communications exhibits at the Smithsonian American History Museum in Washington, D.C.

## Introduction to Twisted Pairs

Before getting into the technical and mathematical details regarding twisted pairs, let's review exactly what the term twisted pair means and why, in nontechnical terms, twisted pairs are used for communications.

As described in the previous section, a twisted pair is made up of two wires twisted around one another. The two wires are electrically driven and sensed differentially, and neither is connected to a common ground point. Because the wires are physically close and geometrically symmetric to one another, any outside signal influencing one of the wires influences the other wire in the same way. If a receiver senses only the difference between the two wires, the outside influence subtracts out. Historically, the two wires making up a twisted pair are called tip and ring.

**Note**

The names *tip* and *ring* come from connector cables used by operators in which one wire terminated at the tip of the connector, and one wire terminated in a ring around the connector. The names are still widely used today.

## Differential Mode and Common Mode Signals

An important concept to understand about twisted pairs is the difference between differential mode and common mode signals. A ***differential mode signal*** is applied between the two wires making up the twisted pair. A ***common mode signal*** is applied between the two individual wires of the twisted pair and ground. Differential mode signals are sometimes called *metallic signals*, and common mode signals are sometimes called *longitudinal signals*. Figure 2.1 illustrates differential mode signals and common mode signals.

Figure 2.1 Differential mode and common mode signals.

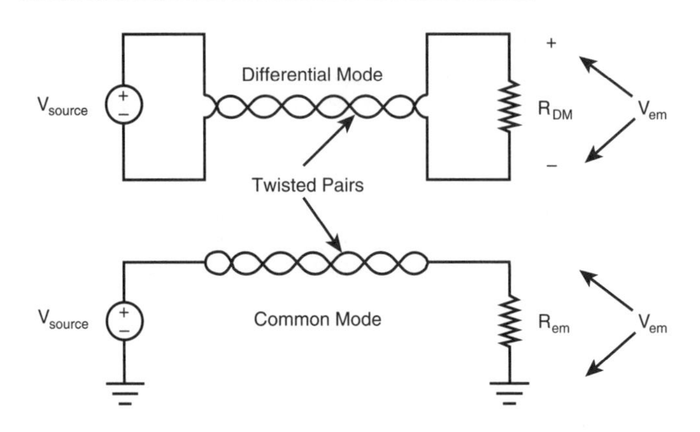


Data and voice signals on a twisted pair are always transmitted as differential signals. Common mode signals are typically due to outside influences such as 60 Hz induction from nearby power lines, RF interference from radio stations and other active operating bands, and disturbances from signals on other twisted pairs.

**Note**

In some systems, such as HDSL powering units, the common mode is used to transmit D.C. power for remote unit powering.

At the receiver of a twisted pair system, only the differential voltage of the twisted pair is processed for information. The common mode signal is rejected by transformers and well-balanced differential amplifiers, which effectively subtract the voltage on the individual wires of the twisted pair.

For example, consider the twisted pair shown in Figure 2.2.

Voltages of each wire are shown along the length of the pair with respect to ground. Note that the voltage on the twisted pair appears as a 1000 Hz, 1 V peak sine wave riding on a 60 Hz wave. The 60 Hz is in phase on each wire, and the 1000 Hz wave is 180 degrees out of phase on each wire. After the differential amplifier, which essentially subtracts the voltage on one wire from the voltage on the other, the 60 Hz voltage subtracts away and is no longer present, and only the 1000 Hz waveform remains (in this simple example, loss on the twisted pair is ignored). This is sometimes referred to as common mode rejection.

Figure 2.2 An example of common mode rejection on a twisted pair.

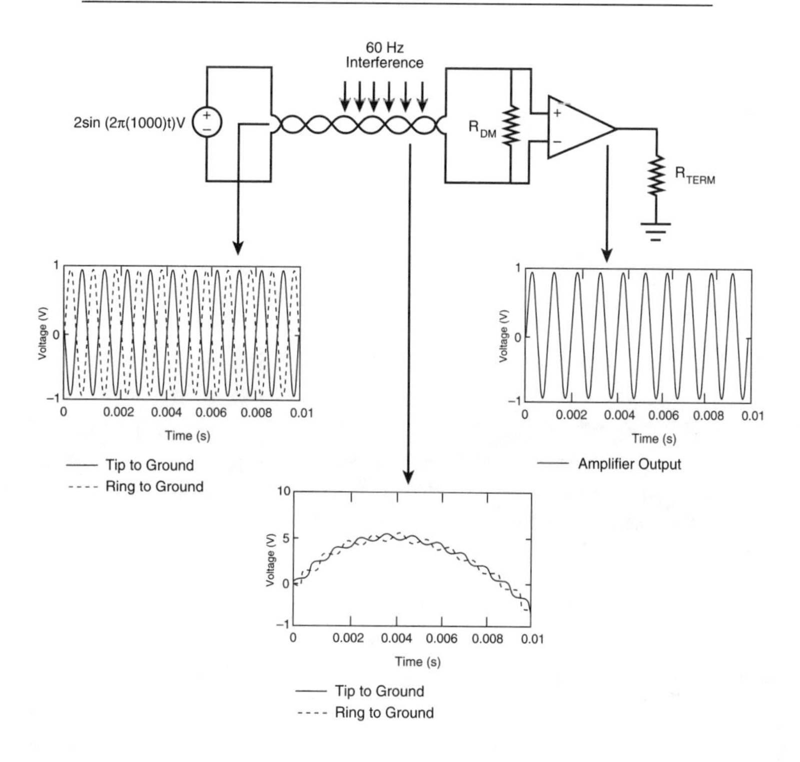


# *Two-Wire Full-Duplex Systems*

If you don't have experience with twisted pair applications, the use of two wires to send signals in both directions simultaneously, or full-duplex operation, might seem extraordinary. If fact, full duplex use of a twisted pair is common and simple. Figure 2.3 shows a basic scheme.

Here, the twisted pair part of the circuit is called the *two-wire path*, and the transmit and receive portion on each end is called the *four-wire path*. The conversion at each end of the twisted pair between two-wire and four-wire paths is called *the two-wire to four-wire hybrid*, or simply the *hybrid*. Signals arriving at point A pass through TX1 and on through the transformer to the twisted pair. Some portion of the signal at point B will enter the receive path of side 1. This signal is called the *echo signal*. The echo signal is canceled by a balance filter B1 and thus will not pass to point C. The original signal will arrive and pass through side 2's transformer and into the receive path of side 2. If side 2 is simultaneously transmitting a signal, any echo from that signal will be canceled by B2 and will not affect the signal proceeding to point D.

Figure 2.3 A full-duplex twisted pair circuit with two-wire to four-wire hybrids on each end.

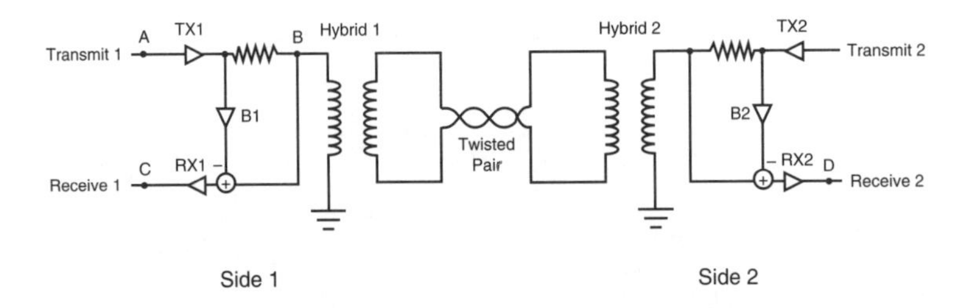


**Note**

Take note that different hybrid designs exist, including designs using transformers to cancel the anticipated echo, using analog or digital amplifiers to cancel the "average expected" echo, and using adaptive filters to measure the echo and adjust to do cancellation. Most high-speed data systems use the latter method.

Using a hybrid to achieve full-duplex communications on a twisted pair is typical of voice systems and some *echo canceled systems* including ISDN, HDSL, and some ADSL systems. Other systems, especially high bit rate systems, do not use echo cancellation because they often require complex echo cancellation filters and front-end analog-to-digital converters. Instead, these systems may use *frequency division multiplexing* (FDM) or *time division multiplexing* (TDM). Such systems are discussed in detail in later chapters.

## Physical Characteristics

With the massive number of twisted pairs deployed over many years, care must be taken not to mix generalizations of the characteristics of common twisted pairs with the characteristics of all deployed twisted pairs. In other words, some twisted pairs have characteristics different from those described in this chapter. This fact is especially important when troubleshooting "real world" issues and problems.

### Cable Bundles

Twisted pairs are usually deployed in *cable bundles* containing multiple twisted pair lines in a cable sheath. The number of pairs contained in the cable varies from as little as one pair to as many as 4,200 pairs. Common cables used in the subscriber loops contain 25 to 100 twisted pairs. A color coding scheme is used to identify specific twisted pairs within a bundle.[4,5]

Each wire in a twisted pair has an insulator coating. Older wires used a paper-based insulation—these are called *pulp-insulated twisted pairs.* Newer twisted pairs use various types of plastics as insulators. The plastics used include polyethylene, polypropylene, and PVC. These types of twisted pairs are generically referred to as *plastic-insulated cables* or PIC. Some cables might also be packed with gel to increase resistance to moisture.

Cable bundles normally have an outer metallic shield. This shield is grounded at the cable ends to help reduce interference from outside sources. The type of shield varies from cable to cable. Cables may also contain support structures such as coated steel armor and galvanized steel supporting strands.

Individual twisted pairs encountered in the outside plant normally do not have dedicated shields. The twisting along the length of a specific pair is constant, but the amount of twist varies between pairs from 2 to 6 inches in length. The varying twist length of neighboring twisted pairs helps further reduce crosstalk between the pairs.

## Gauges and Line Characteristics

Twisted pairs are most often characterized by their gauge using the American Wire Gauge (AWG) designations; the gauge is indicative of the diameter of the copper wires making up the twisted pair. Typical twisted pair gauges include #19, #22, #24, and #26 with diameters ranging from 0.03589 inches (0.912 mm) to 0.01594 inches (0.404 mm) with the smaller gauge number corresponding to the larger diameter. Gauges #24 and #26 are most common in DSL applications. Table 2.1 shows the preceding gauges along with the diameters and DC resistance per 1,000 feet of each.

Table 2.1 Common Twisted Pair Wire Gauges and Characteristics

| AWG | Diameter (in) | DC Resistance 20° (ohms/kft) |
|---|---|---|
| 19 | 0.03589 | 16.9 |
| 22 | 0.02535 | 33.8 |
| 24 | 0.02010 | 53.4 |
| 26 | 0.01594 | 85.8 |

Twisted pairs are also defined by categories specifying electrical tolerances of the pair. Installations for high-speed data are typically required to be category 3 or higher cables. Higher category numbers correspond to higher performance twisted pairs in terms of the loss and crosstalk properties of each. Table 2.2 lists the different cable categories, the approximate bit rates supported, and the target applications of each category.

Table 2.2 Twisted Pair Categories and Characteristics

| Category | Bit Rates | Applications |
|---|---|---|
| 1 | Unspecified | |
| 2 | 1 Mbps | Some low-speed data circuits (DDS) |
| 3 | 16 Mbps | 16 Mbps Used for 10BaseT and 4 Mbps Token Ring |
| 4 | 20 Mbps | 10BaseT and 16 Mbps Token Ring |
| 5 | 100 Mbps | 10/100BaseT, other high-speed copper technologies |

For DSL technologies, category 3 or higher twisted pair is preferred. As discussed in Chapter 1, reuse of the existing copper plant is a major driver for ADSL and VDSL. Existing cable is normally as good as or better than category 3 specifications.[6] Detailed specifications for typical wiring categories can be found in *Telecommunications Wiring for Commercial Buildings, A Practical Guide*.[7]

## Electrical Characteristics of Twisted Pairs

For any transmission line, the basic parameters of interest are resistance, inductance, capacitance, and admittance. These parameters are also called the *RLCG parameters*. In the voice band, the RLCG parameters are close to constant for #24 and #26 gauge twisted pairs. At frequencies above the voice band, resistance, inductance, and admittance vary with frequency. These three parameters also vary with gauge. Capacitance tends to be constant over both frequency and gauge.

Depending on the situation, one of two assumptions can be made when discussing line parameters. The first assumption ignores the effects of admittance and assumes that inductance and capacitance are constant with frequency. Also, resistance is assumed to be proportional to $f^{1/2}$ where f is frequency in Hz. This approximation is used when deriving near-end and far-end crosstalk models. Later in this chapter, and in Chapter 3, you will see that this approximation makes for a simple and accurate model for crosstalk on DSLs.

The second approximation is used for actual numeric calculations of twisted pair characteristics. This approximation will be used when finding the transfer function and insertion losses of a twisted pair as well as the characteristic impedance. This approximation was numerically determined for #24 and #26 twisted pairs.[8]

### Analytic RLCG Models

Analytic RLCG models have been accepted and used in literature for many years. The parameters are given in Eqtn. 2.1–2.4 and the constants in Table 2.3. The resulting RLCG parameters are with respect to 1,000 feet, or 1 kft, a common measure of distance for twisted pairs.

Eqtn. 2.1

$$R = R_0 f^{\frac{1}{2}}$$

Eqtn. 2.2

$$L = L_0$$

Eqtn. 2.3

$$C = C_0$$

Eqtn. 2.4

$$G = 0$$

Table 2.3 Parameters for the Analytic Twisted Pair Model

| Parameter | #24 Gauge | #26 Gauge |
|---|---|---|
| $R^0$ (ohms/kft) | 0.15 | 0.195 |
| $L^0$ (mH/kft) | 0.188 | 0.205 |
| $C^0$ (nF/kft) | 15.7 | 15.7 |

## *Numeric RLCG Models*

The numeric RLCG models for #24 and #26 gauge twisted pairs are valid over the frequency range from D.C. to 10 MHz. The model parameters were found by curve fitting to measured cable.[8] Eqtns. 2.5–2.8 are the general equations for all four parameters; the constant values applicable to #24 and #26 gauge pairs are given in Table 2.4. The resulting RLCG parameters in Table 2.4 are with respect to 1 km.

Eqtn. 2.5

$$R(f) = \sqrt[4]{r_{oc}^4 + a_c f^2}$$

Eqtn. 2.6

$$L(f) = \frac{l_o + l_\infty \left(\frac{f}{f_m}\right)^b}{1 + l_\infty \left(\frac{f}{f_m}\right)^b}$$

Eqtn. 2.7

$$G(f) = g_o f^{g_e}$$

Eqtn. 2.8

$$C(f) = c_\infty$$

Table 2.4 Parameters for the Numeric Twisted Pair Model

| Parameter | #24 Gauge | #26 Gauge |
|---|---|---|
| $r_{oc}$ (ohms/km) | 174.55888 | 286.17578 |
| $a_c$ (ohms$^4$/km$^4$Hz$^2$) | 0.053073481 | 0.14769620 |
| $l_o$ (H/km) | $6.1729593*10_{-6}$ | $675.36888*10_{-6}$ |
| $l_\infty$ (H/km) | $478.97099*10_{-6}$ | $488.95186*10_{-6}$ |
| $f_m$ (Hz) | 553760.63 | 806338.63 |
| b | 1.1529766 | 0.92930728 |
| $g_o$ (Siemen/Hz*km) | $0.23487476*10_{-12}$ | $4.3*10_{-8}$ |
| $g_e$ | 1.38 | 0.70 |
| $c_\infty$ (nF/km) | $50*10_{-9}$ | $49*10_{-9}$ |

The RLCG parameters for both the analytic models and the numeric models are graphed in Figures 2.4–2.7.

In Figure 2.7, only the numeric model is shown because the analytic admittance is assumed to be zero. As frequency increases past 100 kHz, note the closeness of the numeric resistance model to the analytic model dependent on $f^{1/2}$. This is largely due to skin effect of the copper wires at high frequency, which for a cylindrical conductor, is known to exhibit such a dependence on frequency.[9]

Figure 2.4 Resistance models for #24 and #26 gauge twisted pairs.

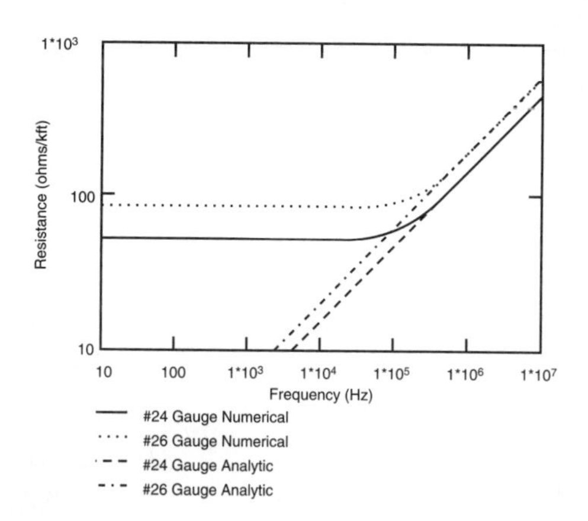


Figure 2.5 Inductance models for #24 and #26 gauge twisted pairs.

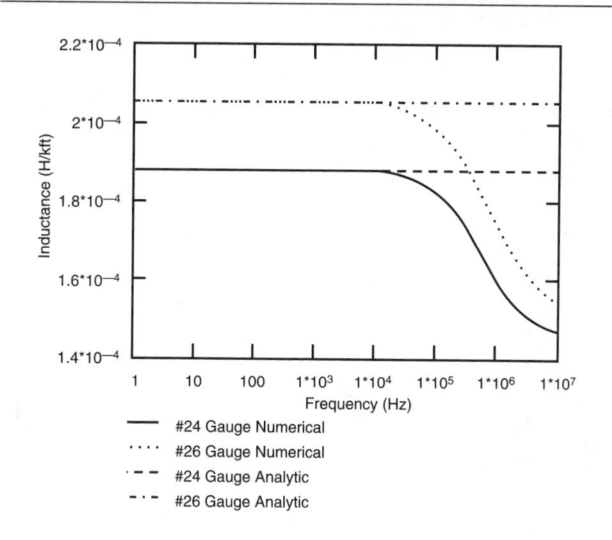

Figure 2.6 Capacitance models for #24 and #26 gauge twisted pairs.

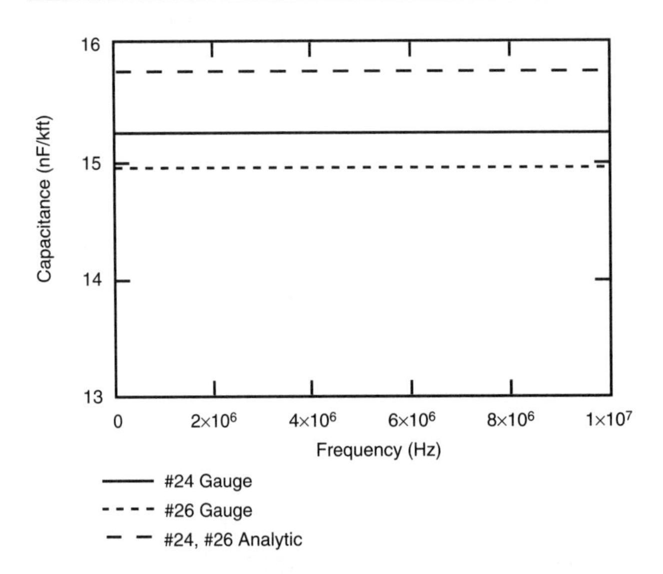


Figure 2.7 Admittance models for #24 and #26 gauge twisted pairs.

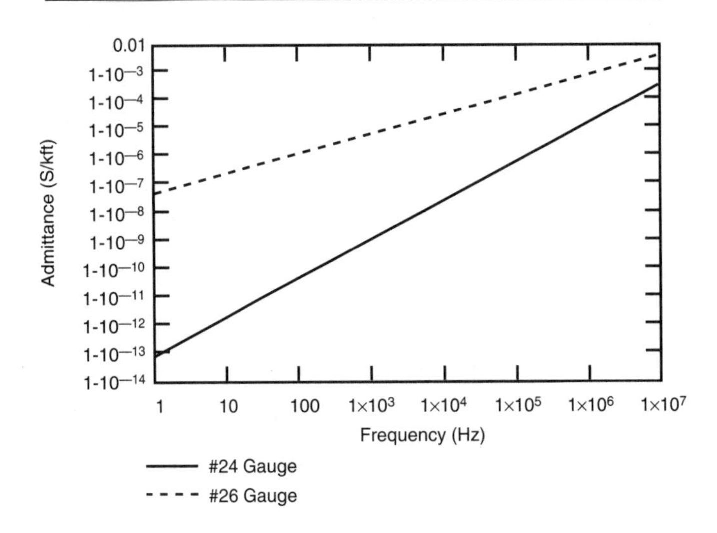


Note that in Figure 2.5, the inductance for both #24 and #26 wires varies by about 20 percent in the region above 100 kHz. Assuming that inductance is constant in the analytic model is certainly incorrect, but the choice to trade off error for simplicity is reasonable. The following sections show that the assumption of zero admittance in the analytic model, although not accurate according to the numeric model, is reasonable at frequencies above 100 kHz. At these frequencies, the admittance needs to compare to jωC where j is the imaginary unit vector and ω is the radian frequency variable (2πf).

## Transmission Line Parameters

The RLCG parameters of a transmission line define two common parameters useful in characterizing voltage and current along the line. These are the ***propagation constant*** (γ) and ***characteristic impedance*** ($Z_o(f)$). With these parameters, the voltage at some point z on a transmission line of length l with a load $Z_L(f)$ is given by Eqtn. 2.9. (The assumption is made that z=0 at the source voltage and increases positively until z=l at the termination load.)

Eqtn. 2.9

$$V(f,z)=V_0(f)\left(e^{-\gamma(f)z}+\Gamma_L(f)e^{\gamma(f)z}\right)$$

In Eqtn. 2.9, $V_0(f)$ is the source voltage, and $\Gamma_L(f)$ is the reflection coefficient of the line given by Eqtn. 2.10.

Eqtn. 2.10

$$\Gamma_L(f)=\frac{Z_L(f)-Z_o(f)}{Z_L(f)+Z_o(f)}$$

Note that for a known source voltage, termination impedance, and position, the voltage on the line is a function of only propagation constant and characteristic impedance.

Maximum power is delivered to a termination load on a transmission line when the characteristic impedance of the line and the load impedance are equal. This is desired in the case of a DSL because the termination load is the front of the channel receiver; delivering the maximum possible power to the receiver is necessary to optimize performance.

Under this condition, the expression for voltage on the line is simplified as the reflection coefficient in Eqtn. 2.10 becomes zero and the second exponential term in Eqtn. 2.9 drops out of the equation. When the load impedance is the same as the transmission line characteristic impedance, we say that the transmission line is matched. Hereafter, unless otherwise stated, it will be assumed that the twisted pair transmission lines under investigation are always matched, and the expression for voltage at some point z along the line is given by Eqtn. 2.11.

Eqtn. 2.11

$$V(f,z)=V_0(f)e^{-\gamma(f)z}$$

**Note**

Later in the chapter, you will see that this is a reasonable assumption over the frequencies of interest. Even when the characteristic impedance and load impedance are slightly different, the second term in Eqtn. 2.9 is negligible compared to the first term.

Knowledge of the propagation constant of a twisted pair is key to understanding the effect that a twisted pair has on signals. The propagation constant can further be broken down resulting in Eqtn. 2.12.

Eqtn. 2.12

$$\gamma = \alpha + j\beta = \sqrt{(R + j\omega L)(G + j\omega C)}$$

In Eqtn. 2.12, dependence of all quantities on frequency has been dropped only for notational simplicity. The terms $\alpha$ and $\beta$ are the real and imaginary parts of the propagation constant, respectively. Substituting these into Eqtn. 2.11 yields the following in Eqtn. 2.13.

Eqtn. 2.13

$$V(f,z) = V_0(f)e^{-(\alpha + j\beta)z} = V_0(f)e^{-\alpha z}e^{-j\beta z}$$

The transfer function between the source voltage and some point z on the twisted pair is given by H *(f)* as defined in Eqtn. 2.14.

Eqtn. 2.14

$$H(f) = \frac{V(f,z)}{V_0(f)} = e^{-\alpha z}e^{-j\beta z} = \left|e^{-\alpha z}\right| \angle \beta z$$

The last part of Eqtn. 2.14 is valid because the exponential involving $\beta$ has unity magnitude and only contributes to phase shifting the voltage. On the other hand, the term involving $\alpha$ affects only the magnitude of the voltage. Thus, from Eqtn. 2.14, you can see that the real part of the propagation constant determines by how much the voltage will change over a length of twisted pair and the imaginary part how much the phase will change.

## *Propagation Constant*

Later in this chapter, you will see that crosstalk between two twisted pairs becomes worse as frequency increases. Also, propagation loss of a twisted pair increases as frequency increases. In this section, an expression will be derived for the propagation constant of a twisted pair at these high frequencies. Two basic assumptions will aid in characterizing the propagation constant: first, $j\omega C >> G$, or equivalently, $G=0$; and second, $j\omega L > R$. Note that the second condition is less strict than the first. These approximations are valid for #24 and #26 twisted pairs at frequencies above 100 kHz.

Using the approximations set forth, the propagation constant of a twisted pair can be written as shown in Eqtn. 2.15.

Eqtn. 2.15

$$\gamma = \sqrt{(R + j\omega L)(G + j\omega C)} \approx \sqrt{j\omega C}\sqrt{(R + j\omega L)}$$

In Eqtn. 2.15, the first assumption was used to eliminate the admittance term. Factoring the equation yields the result in Eqtn. 2.16.

Eqtn. 2.16

$$\gamma = \sqrt{j\omega C}\sqrt{(R + j\omega L)} = j\omega\sqrt{LC}\sqrt{\left(\frac{R}{j\omega L} + 1\right)}$$

$$\approx j\omega\sqrt{LC}\left(\frac{R}{j\omega L} + 1\right)$$

$$= R\sqrt{\frac{C}{L}} + j\omega\sqrt{LC}$$

In this equation, the final approximation is based on the second assumption $(x+1)^{1/2}=(x+1)$ for small x.

In terms of $\alpha$ and $\beta$, the real and imaginary parts of the propagation constant, the result from Eqtn. 2.16 yields these terms as shown in Eqtn. 2.17 and Eqtn. 2.18.

Eqtn. 2.17

$$\alpha = R\sqrt{\frac{C}{L}} = \alpha_o f^{1/2}$$

Eqtn. 2.18

$$\beta = \omega\sqrt{LC} = \beta_o f$$

In Eqtns. 2.17 and 2.18, $\alpha_o$ and $\beta_o$ are introduced as constants. The final result for $\alpha$ comes from using the analytic approximation for the RLCG parameters. $\alpha$ is proportional to the square root of frequency because capacitance and inductance are both assumed constant, and R is assumed proportional to $f^{1/2}$ due to skin effect. The final result for $\beta$ shows that $\beta$ is proportional to frequency due to the $\omega(2\pi f)$ term.

Figures 2.8 and 2.9 show the real and imaginary part of propagation constant of a #24 and #26 gauge twisted pair. At frequencies above 100 kHz, both plots are consistent with Eqtn. 2.17 and Eqtn. 2.18. Note that the real part of the propagation constant, which was already established to have a direct effect on loss, increases with frequency and is higher for #26 wire than for #24.

Figure 2.8 Real part of the propagation constant for #24 and #26 twisted pairs.

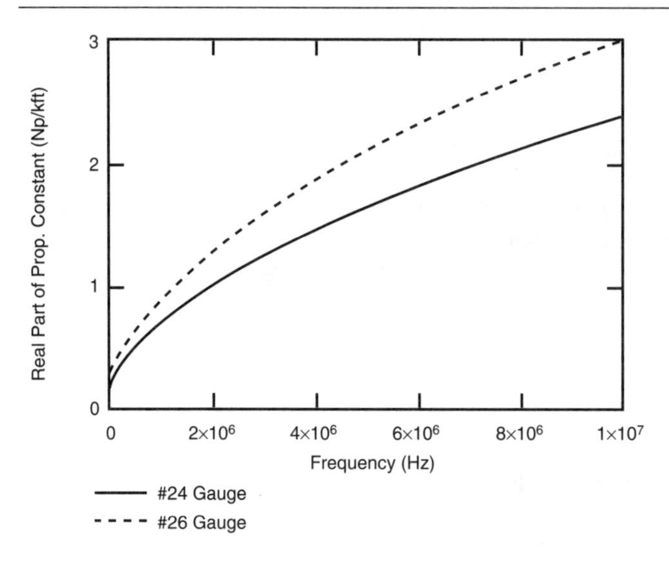


Figure 2.9 Imaginary part of the propagation constant for #24 and #26 twisted pairs.

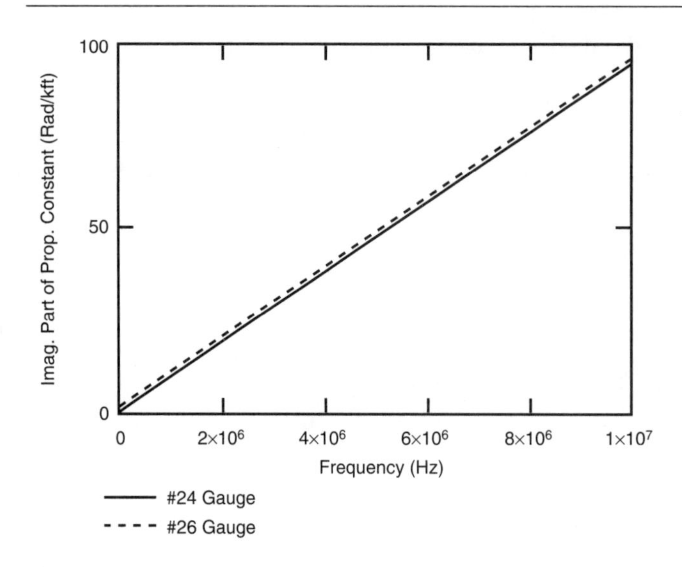


In Eqtn. 2.11, an expression for the transfer function of a twisted pair was given in terms of the propagation constant. Figures 2.10–2.13 show the transfer function magnitude and phase for both #24 and #26 gauge cables.

Figure 2.10 Transfer function magnitude for various #24 gauge twisted pairs.

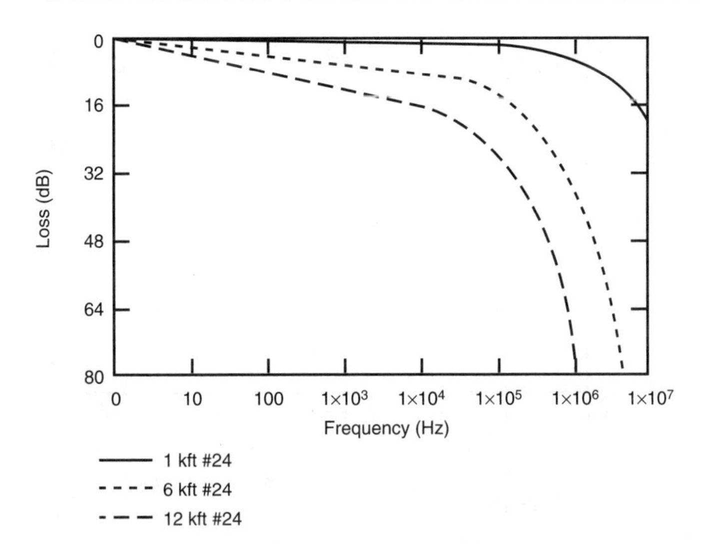


Figure 2.11 Transfer function phase for various #24 gauge twisted pairs.

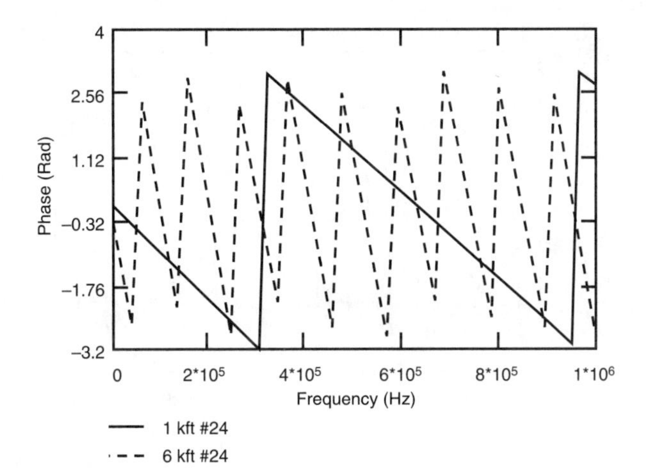

Figure 2.12 Transfer function magnitude for various #26 gauge twisted pairs.

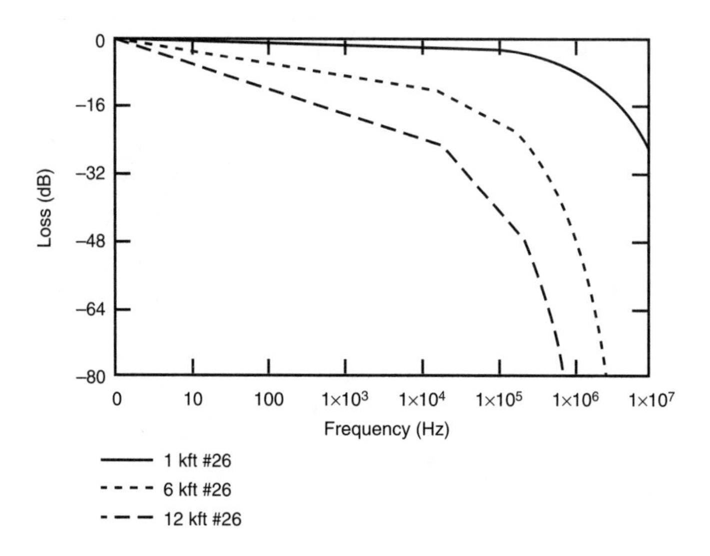


Figure 2.13 Transfer function phase for various #26 gauge twisted pairs.

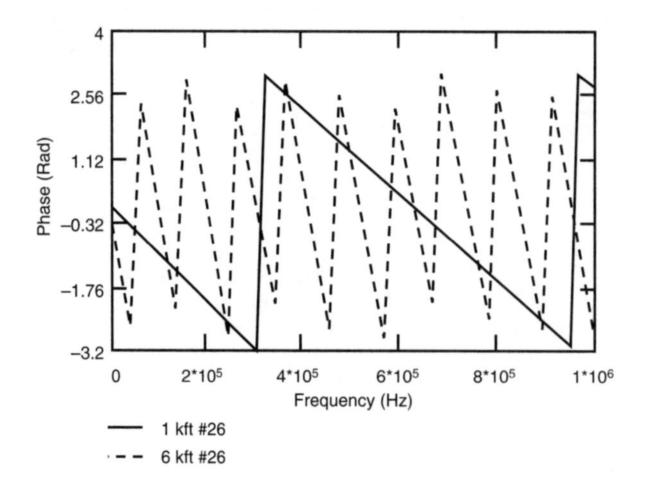


Figures 2.10–2.13 show various cable line lengths. Note that the loss of the cable increases as line length increases and that it also increases with higher frequencies. Intuitively, these behaviors make sense. Longer lines have more loss than shorter lines, and higher frequencies are attenuated more by a pair of wires. Note also in Figures 2.11 and 2.13, where

phase is shown for only a 1,000 ft. and 6,000 ft. loop, the greater phase slope of the longer line. This trend continues as the loop length is further increased.

## *Characteristic Impedance*

The preceding analysis assumes that the twisted pair was terminated in its characteristic impedance, $Z_o$. The characteristic impedance of any transmission line is defined in terms of the transmission line RLCG parameters and is given by Eqtn. 2.19.

Eqtn. 2.19

$$Z_o = \sqrt{\frac{R + j\omega L}{G + j\omega C}}$$

Note that the characteristic impedance does not depend on the length of the twisted pair. Figure 2.14 shows the characteristic impedance magnitude for a #24 and #26 gauge twisted pair. The respective phases are shown in Figure 2.15.

Figure 2.14 The characteristic impedance magnitude for #24 and #26 gauge twisted pairs.

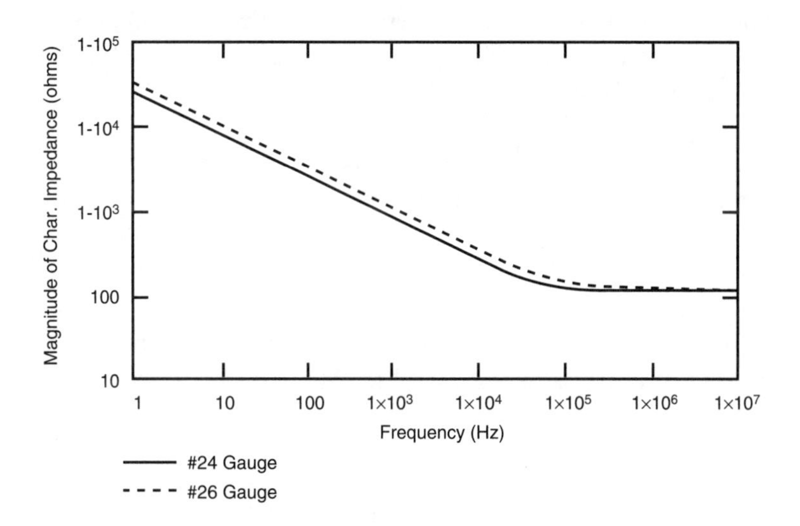

Figure 2.15 The characteristic impedance phase for #24 and #26 gauge twisted pairs.

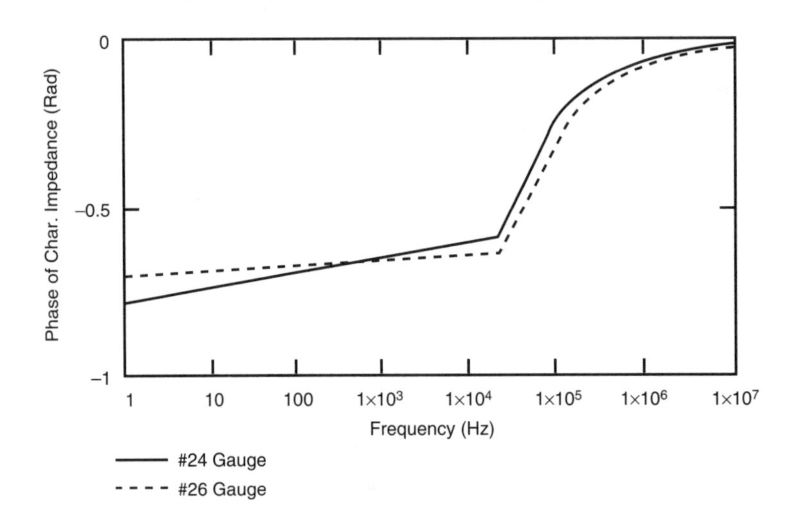


Although the characteristic impedances vary with frequency, a 100 ohm purely resistive impedance is often assumed. (In some cases, especially in Europe, 135 ohms is assumed.) The error of this assumption is negligible in many cases because the reflection coefficient further reduces the effect of small mismatches as indicated in Eqtn. 2.10.

**Note**
In an actual implementation, it might be reasonable to design a terminating impedance to better favor either a #24 gauge or a #26 gauge twisted pair.

## Crosstalk

This section is dedicated to defining crosstalk on twisted pairs and creating a mathematical model for it.

Crosstalk refers simply to disturbance on one twisted pair due to signals on another twisted pair. Usually, the twisted pairs are in the same cable or *binder group*. This text refers to the twisted pair being disturbed as the *disturbed pair* and to the twisted pair causing the disturbance as the *disturber* or *disturbing pair*. In cables with many pairs, it is common to have multiple disturbers. This also is addressed.

**Note**
If you have ever been on the telephone and faintly heard someone else's conversation on your line, you have experienced crosstalk firsthand.

## *Crosstalk Types*

Crosstalk is generally characterized as either *near-end crosstalk* (NEXT) or *far-end crosstalk* (FEXT). NEXT is characterized by the disturbing pair's source being local to the disturbed pair's receiver. In this case, the disturbing signal starts down the disturbing pair, couples into the disturbed pair, and then propagates back to the disturbed pair's receiver. FEXT is characterized by the disturbing pair's source being distant from disturbed pair's receiver. In this case, the disturbing signal propagates down the disturbing pair, crosstalks into the disturbed pair, and propagates the rest of the distance along the disturbed pair into the disturbed pair's receiver. Figure 2.16 illustrates NEXT and FEXT.

Figure 2.16 Near-end crosstalk (NEXT) and far-end crosstalk (FEXT).

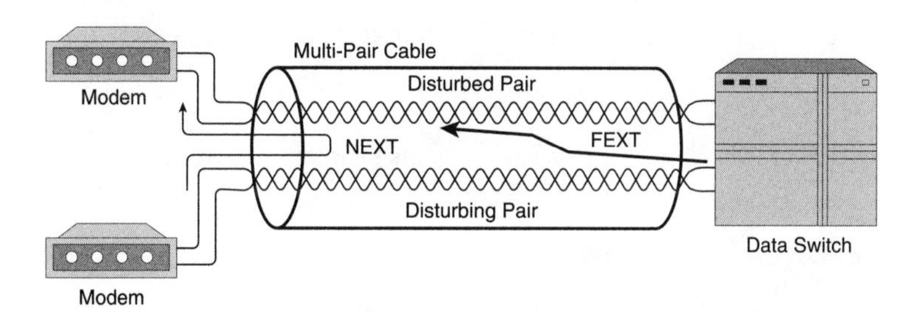


As shown in Figure 2.16, disturbing signals due to NEXT usually have to travel over less length of the twisted pair before entering a disturbed pair's receiver as compared to FEXT. For this reason, NEXT from a particular disturber is usually more damaging than FEXT. Note that either NEXT or FEXT or both might exist for a disturbed pair, and furthermore, more than one NEXT or FEXT disturber might exist.

NEXT or FEXT from a disturber whose line code is the same as that of the disturbed pair creates self-NEXT or self-FEXT, respectively. This type of disturbance is often present as similar services are often deployed from a central location. Later, you will see that ADSL and VDSL systems avoid self-NEXT, although self-FEXT is inevitable. Other common disturbers (both NEXT and FEXT) for ADSL systems include ISDN, HDSL, and T1 systems.

## *Unbalance Models*

The following sections break the derivation of a mathematical model for NEXT and FEXT into two parts. The first part derives a mathematical model for crosstalk between two very short twisted pair sections. The second part takes that generic result and applies it to integrals specific to either NEXT or FEXT. It is interesting to note that the models stem from work done just after the turn of the 20th century by George Campbell[10,11] and also from work done in the 1960s by Cravis and Carter.[12] Both works are considered classics in crosstalk modeling and referenced in many journal articles on crosstalk.

For two twisted pairs located near one another, capacitive and inductive components exist between each of the four wires and also between each wire and ground. Analysis shows that crosstalk occurs when an unbalance exists among the capacitive and inductive components. The following assumptions are made while deriving the crosstalk models:

- The total crosstalk is a sum of the crosstalk due to capacitive and inductive crosstalk.
- Secondary effects (or crosstalk due to crosstalk) can be ignored.

### *Capacitive Unbalance Model*

The capacitive unbalance analysis follows the derivation done by Miller.[13] Figure 2.17 shows a model of a portion of length of two twisted pair circuits in reasonably close proximity (for example, in a cable or binder group).

The four wires that compose the model each have some self-impedance, an admittance to one another, and a capacitance to ground. Furthermore, it is assumed that resistance between wires is negligible due to the plastic insulation. Thus, all admittances are purely capacitive (consistent with the analytic RLCG model).

Figure 2.17 Capacitive model of a short section of two twisted pairs.

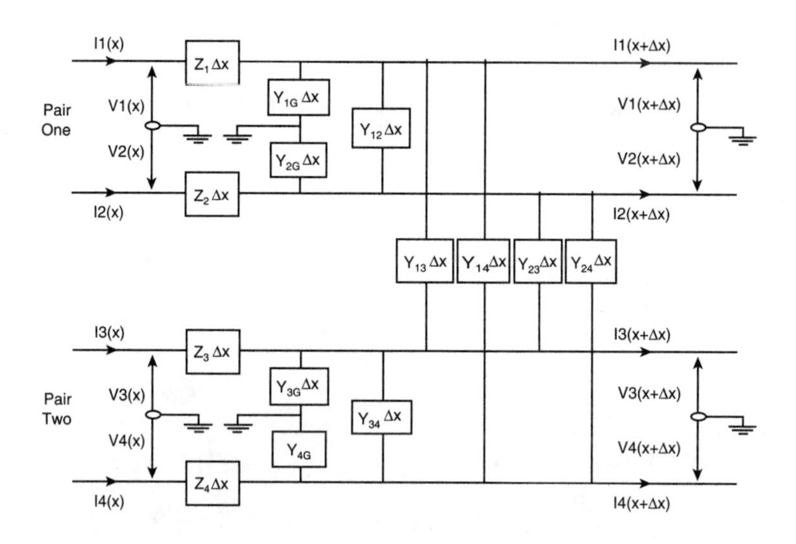


For convenience, you can write all capacitances as admittances (Y=jωC). You can then write equations for the voltage and current on each of the four wires at the location x+Δx using basic circuit theory as in Eqtns. 2.20–2.27.

Eqtn. 2.20

$$V_1(x+\Delta x) = V_1(x) - I_1(x)Z_1\Delta x$$

Eqtn. 2.21

$$V_2(x+\Delta x) = V_2(x) - I_2(x)Z_2\Delta x$$

Eqtn. 2.22

$$V_3(x+\Delta x) = V_3(x) - I_3(x)Z_3\Delta x$$

Eqtn. 2.23

$$V_4(x+\Delta x) = V_4(x) - I_4(x)Z_4\Delta x$$

Eqtn. 2.24

$$I_1(x+\Delta x) = I_1(x) - V_1(x)Y_{1G}\Delta x - [V_1(x) - V_2(x)]Y_{12}\Delta x$$
$$-[V_1(x) - V_3(x)]Y_{13}\Delta x - [V_1(x) - V_4(x)]Y_{14}\Delta x$$

Eqtn. 2.25

$$I_2(x+\Delta x) = I_2(x) - V_2(x)Y_{2G}\Delta x - [V_2(x) - V_1(x)]Y_{12}\Delta x$$
$$-[V_2(x) - V_3(x)]Y_{23}\Delta x - [V_2(x) - V_4(x)]Y_{24}\Delta x$$

Eqtn. 2.26

$$I_3(x+\Delta x) = I_3(x) - V_3(x)Y_{3G}\Delta x - [V_3(x) - V_1(x)]Y_{13}\Delta x$$
$$-[V_3(x) - V_2(x)]Y_{23}\Delta x - [V_3(x) - V_4(x)]Y_{34}\Delta x$$

Eqtn. 2.27

$$I_4(x+\Delta x) = I_4(x) - V_4(x)Y_{4G}\Delta x - [V_4(x) - V_1(x)]Y_{14}\Delta x$$
$$-[V_4(x) - V_2(x)]Y_{24}\Delta x - [V_4(x) - V_3(x)]Y_{34}\Delta x$$

In Eqtns. 2.20–2.27, you can divide through by $\Delta x$, and as $\Delta x$ approaches zero, the definition of the derivative can be invoked. The eight simultaneous equations can then be rearranged and placed in matrix form as shown in Eqtn. 2.28.

Eqtn. 2.28

$$\frac{d}{dx}\begin{bmatrix} V_1 \\ V_2 \\ V_3 \\ V_4 \\ I_1 \\ I_2 \\ I_3 \\ I_4 \end{bmatrix} = \begin{bmatrix} 0 & 0 & 0 & 0 & -Z_1 & 0 & 0 & 0 \\ 0 & 0 & 0 & 0 & 0 & -Z_2 & 0 & 0 \\ 0 & 0 & 0 & 0 & 0 & 0 & -Z_3 & 0 \\ 0 & 0 & 0 & 0 & 0 & 0 & 0 & -Z_4 \\ -\begin{pmatrix} Y_{1G}+Y_{12} \\ +Y_{13}+Y_{14} \end{pmatrix} & 0 & 0 & 0 & 0 & 0 & 0 & 0 \\ Y_{12} & -\begin{pmatrix} Y_{2G}+Y_{12} \\ +Y_{23}+Y_{24} \end{pmatrix} & 0 & 0 & 0 & 0 & 0 & 0 \\ Y_{13} & 0 & -\begin{pmatrix} Y_{3G}+Y_{13} \\ +Y_{23}+Y_{34} \end{pmatrix} & 0 & 0 & 0 & 0 & 0 \\ Y_{14} & 0 & 0 & -\begin{pmatrix} Y_{4G}+Y_{14} \\ +Y_{24}+Y_{34} \end{pmatrix} & 0 & 0 & 0 & 0 \end{bmatrix} \cdot \begin{bmatrix} V_1 \\ V_2 \\ V_3 \\ V_4 \\ I_1 \\ I_2 \\ I_3 \\ I_4 \end{bmatrix} = A \cdot S$$

In Eqtn. 2.28, for convenience of notation only, the dependence on x from all voltages and currents has been dropped, and S and A have been introduced to denote the 8×8 impedance matrix and the 8×1 voltage and current vector, respectively.

At this point, it is helpful to transform the voltages and currents on the four lines to metallic and longitudinal voltages and currents. A simple set of linear transformations shown in Eqtns. 2.29–2.36 accomplishes this.

Eqtn. 2.29

$$V_{1M} = V_1 - V_2$$

Eqtn. 2.30

$$V_{2M} = V_3 - V_4$$

Eqtn. 2.31

$$V_{1L} = \frac{V_1 + V_2}{2}$$

Eqtn. 2.32

$$V_{2L} = \frac{V_3 + V_4}{2}$$

Eqtn. 2.33

$$I_{1M} = \frac{I_1 - I_2}{2}$$

Eqtn. 2.34

$$I_{2M} = \frac{I_3 - I_4}{2}$$

Eqtn. 2.35

$$I_{1L} = I_1 + I_2$$

Eqtn. 2.36

$$I_{2L} = I_3 + I_4$$

In these equations, the first subscript $_{(1 \text{ or } 2)}$ identifies the twisted pair, and the second subscript $_{(M \text{ or } L)}$ indicates metallic or longitudinal signal type, respectively. Thus for example, $V_{1M}$ would represent the metallic voltage on twisted pair 1.

To transform Eqtn. 2.28 to be in terms of metallic and longitudinal signals, Eqtns. 2.30–2.36 must be solved for the original voltages and currents. By letting S be a column vector of the metallic and longitudinal voltages and currents, it can easily be shown that Eqtn. 2.37 is true.

Eqtn. 2.37

$$S = T \cdot \begin{bmatrix} V_{1M} \\ V_{1L} \\ V_{2M} \\ V_{2L} \\ I_{1M} \\ I_{1L} \\ I_{2M} \\ I_{2L} \end{bmatrix} = T \cdot S'$$

In Eqtn. 2.37, S is the column vector of voltage and currents, and T is a transformation matrix given by Eqtn. 2.38.

Eqtn. 2.38

$$T = \begin{bmatrix} \frac{1}{2} & 1 & 0 & 0 & 0 & 0 & 0 & 0 \\ \frac{-1}{2} & 1 & 0 & 0 & 0 & 0 & 0 & 0 \\ 0 & 0 & \frac{1}{2} & 1 & 0 & 0 & 0 & 0 \\ 0 & 0 & \frac{-1}{2} & 1 & 0 & 0 & 0 & 0 \\ 0 & 0 & 0 & 0 & 1 & \frac{1}{2} & 0 & 0 \\ 0 & 0 & 0 & 0 & -1 & \frac{1}{2} & 0 & 0 \\ 0 & 0 & 0 & 0 & 0 & 0 & 1 & \frac{1}{2} \\ 0 & 0 & 0 & 0 & 0 & 0 & -1 & \frac{1}{2} \end{bmatrix}$$

You can then transform the system of equations given in Eqtn. 2.28 as in Eqtn. 2.39.

Eqtn. 2.39

$$\frac{d}{dx} S' = T^{-1} A T S'$$

The result of this transformation shows that the derivatives with respect to x of the metallic and longitudinal voltages depend only on the metallic and longitudinal currents, and the derivative with respect to x of the metallic and longitudinal currents depends only on the metallic and longitudinal voltages. Concentration will be placed only on the latter, simplifying the system of equations to four equations and four unknowns. Thus, the simplified set of equations in terms of metallic and longitudinal signals is shown in Eqtn. 2.40.

Eqtn. 2.40

$$\frac{d}{dx}\begin{bmatrix} I_{1M} \\ I_{1L} \\ I_{2M} \\ I_{2L} \end{bmatrix} = \frac{-j \cdot \omega}{4}\begin{bmatrix} a_{11} & a_{12} & a_{13} & a_{14} \\ a_{21} & a_{22} & a_{23} & a_{24} \\ a_{31} & a_{32} & a_{33} & a_{34} \\ a_{41} & a_{42} & a_{43} & a_{4} \end{bmatrix} \cdot \begin{bmatrix} V_{1M} \\ V_{1L} \\ V_{2M} \\ V_{2L} \end{bmatrix}$$

The elements of the four-by-four matrix are given by Eqtns. 2.41–2.50.

Eqtn. 2.41

$$a_{11} = C_{1G} + C_{2G} + 4C_{12} + C_{13} + C_{14} + C_{23} + C_{24}$$

Eqtn. 2.42

$$a_{21} = a_{12} = 2C_{1G} - 2C_{2G} + 2C_{13} + 2C_{14} - 2C_{23} - 2C_{24}$$

Eqtn. 2.43

$$a_{31} = a_{31} = -C_{13} + C_{14} + C_{23} - C_{24}$$

Eqtn. 2.44

$$a_{41} = a_{14} = -2C_{13} - 2C_{14} + 2C_{23} + 2C_{24}$$

Eqtn. 2.45

$$a_{22} = 4C_{1G} + 4C_{2G} + 4C_{13} + 4C_{14} + 4C_{23} + 4C_{24}$$

Eqtn. 2.46

$$a_{23} = a_{32} = -2C_{13} + 2C_{14} - 2C_{23} + 2C_{24}$$

Eqtn. 2.47

$$a_{24} = a_{42} = -4C_{13} - 4C_{14} - 4C_{23} - 4C_{24}$$

Eqtn. 2.48

$$a_{33} = C_{3G} + C_{4G} + C_{13} + C_{14} + C_{23} + C_{24} + 4 \cdot C_{34}$$

Eqtn. 2.49

$$a_{34} = a_{43} = 2C_{3G} - 2C_{4G} + 2C_{13} - 2C_{14} + 2C_{23} - 2C_{24}$$

Eqtn. 2.50

$$a_{44} = 4C_{3G} + 4C_{4G} + 4C_{13} + 4C_{14} + 4C_{23} + 4C_{24}$$

In these equations, capacitance has been used instead of admittance to further stress that the impedances between all wires is capacitive. For crosstalk between twisted pairs, you should be most concerned with the coupling from the metallic signal of the disturbing pair into the metallic signal of the disturbed pair. The parameter $a_{31}$ from Eqtn. 2.43 defines this coupling, metallic voltage in the disturbing pair ($V_{1M}$) to metallic current in the disturbed pair ($I_{2M}$). Parameter $a_{31}$ is referred to as the *capacitance unbalance* of a twisted pair. For notational purposes, $a_{31}$ will be renamed as $C_{M1M2}$. Thus, crosstalk resulting from capacitance unbalance is given by Eqtn. 2.51.

Eqtn. 2.51

$$I_{2M} = \frac{-d}{dx}\frac{j\omega}{4}C_{M1M2}V_{1M}$$

## *Inductive Unbalance Model*

Crosstalk due to mutual inductance unbalance is a bit easier to derive than capacitive unbalance. Consider a small section of two twisted pairs as shown in Figure 2.18.

Figure 2.18 Mutual inductance model of a short section of two twisted pairs.

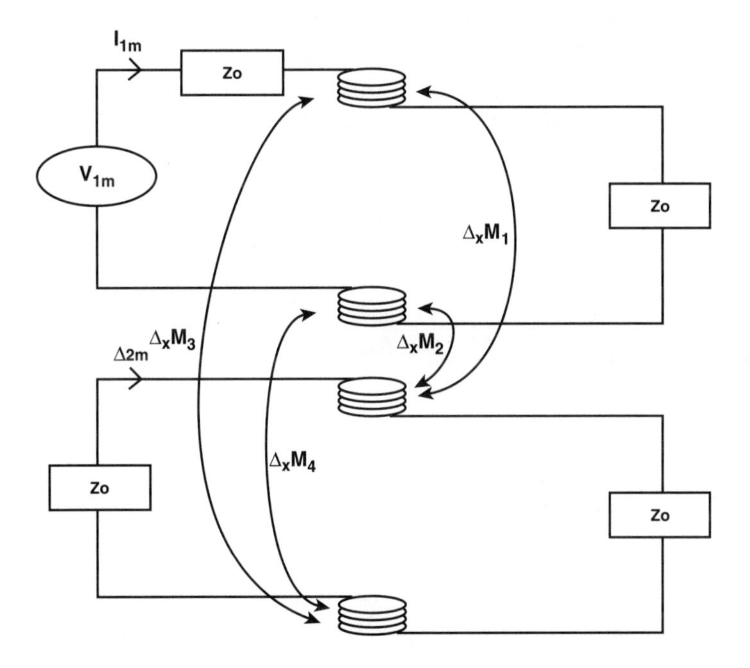


Mutual inductance exists pair-wise between each of the wires in the two twisted pairs. The effective termination at each end of the two pairs is the characteristic impedance. No crosstalk occurs if the couplings are balanced with respect to one another. For example, if $M_1$ and $M_3$ produce exactly opposite currents in the disturbed pair and $M_2$ and $M_4$ do the same, then all the coupled currents will cancel one another, and no net crosstalk will be present on the disturbed pair.

To mathematically analyze Figure 2.18, it is helpful to draw the equivalent circuit shown in Figure 2.19.

Figure 2.19 Equivalent circuit for the mutual inductance unbalance of a short section of two twisted pairs.

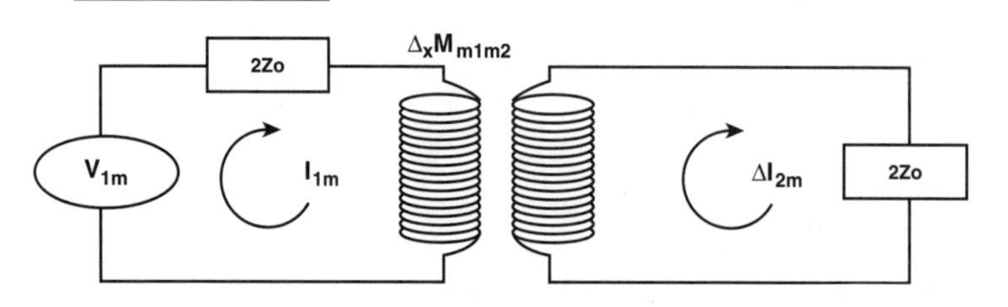


The mutual inductance denoted by $M_{M1M2}$ is given by Eqtn. 2.52.

Eqtn. 2.52

$$M_{M1M2} = M_1 + M_2 + M_3 + M_4$$

Note that $M_{M1M2}$ will be zero when the mutual inductances are balanced. In Figure 2.19, the current $\Delta I_{2M}$ is the current generated on the disturbed twisted pair on the small section shown.

The circuit in Figure 2.19 can be analyzed by writing the following loop equations (Eqtn. 2.53 and Eqtn. 2.54) for $I_{1M}$ and $\Delta I_{2M}$:

Eqtn. 2.53

$$2ZI_{1M} - M\Delta xI_{2M} = V_{1M}$$

Eqtn. 2.54

$$-M\Delta xI_{1M} + 2ZI_{2M} = 0$$

These equations can then be solved simultaneously, and the resulting $\Delta I_{2M}$ is in Eqtn. 2.55.

Eqtn. 2.55

$$I_{2M} = \frac{j\omega M \Delta x V_{1M}}{4Z^2 - M^2\omega^2\Delta x^2}$$

Because M is usually very small and Δx is a very small portion of line, the second term in the denominator is small with respect to the first term and can be dropped. Also, the definition of derivative can be invoked as Δx approaches zero, so Eqtn. 2.55 can be written as Eqtn. 2.56.

Eqtn. 2.56

$$\frac{dI_{2M}}{dx} \approx \frac{j\omega M V_{1M}}{4Z^2}$$

Thus, the result of inductive coupling here gives a similar result to the earlier result for capacitive coupling in that the current induced on a disturbed line due to a voltage on a disturbing line is proportional to the frequency. (This assumes that the characteristic impedance is constant with frequency as was shown earlier in this chapter.)

## *General Unbalance Expression*

From the capacitive and inductive unbalance expressions, the total crosstalk on a small portion of twisted pair can be written as in Eqtn. 2.57.

Eqtn. 2.57

$$\frac{dI_{2M}}{dx} = \left(\frac{j\omega M}{4Z^2} + \frac{j\omega C_{M1M2}}{4}\right)V_{1M}$$

If the characteristic impedance of the line is a constant over frequency, then the current in Eqtn. 2.57 is proportional to frequency for a given voltage. With this in mind, all the constants of Eqtn. 2.57 can be grouped together to form a new unbalance constant, $Q_{M1M2}$ such that Eqtn. 2.58 holds.

Eqtn. 2.58

$$\frac{dI_{2M}}{dx} = \left(\frac{j\omega M}{4Z^2} + \frac{j\omega C_{M1M2}}{4}\right)V_{1M} = j\omega Q_{M1M2} V_{1M}$$

## *Near-end Crosstalk (NEXT)*

Now that a generic expression has been obtained for the crosstalk from a very short segment of a disturbing pair to a very short segment of a disturbed pair, attention can be turned to the specific cases of NEXT and FEXT. Consider a signal V traveling along a disturbing pair. At some distance x1 along a disturbing pair, it couples into a disturbed pair due to unbalance and then travels back to the receiver of the disturbed pair as shown in Figure 2.20.

Figure 2.20 An example of near-end crosstalk due to only one unbalance point (x1).

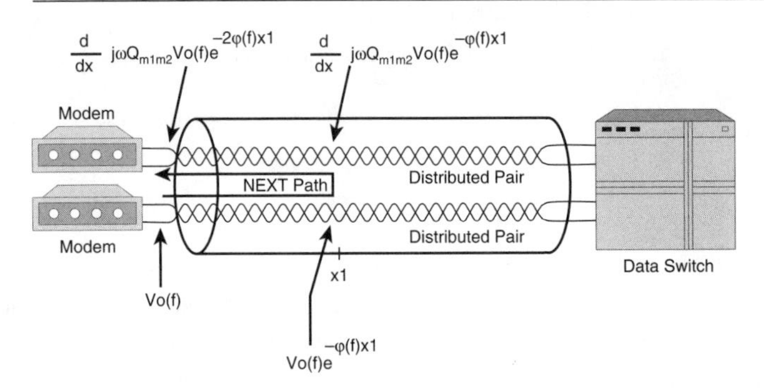


If the signal exciting the disturbing pair is $V_o$, then from Eqtn. 2.11 the signal on the disturbing pair at x1 is as shown in Eqtn. 2.59.

Eqtn. 2.59

$$V_1(f, x1) = V_o(f)e^{-\gamma(f)x1}$$

This signal couples due to unbalance onto the disturbed pair at x1. From the unbalance term of Eqtn. 2.58, the resulting current on the disturbed pair is shown by Eqtn. 2.60.

Eqtn. 2.60

$$I_2(f, x1) = \frac{d}{dx} j\omega Q_{M1M2} V_1(f, x1) = \frac{d}{dx} j\omega Q_{M1M2} V_o(f)e^{-\gamma(f)x1}$$

To reach the disturbed pair's receiver, the signal must propagate the distance x1 along the disturbed pair. This introduces further attenuation of $e_{ms\gamma x1}$. The current arriving at the disturbed pair's receiver can thus be written as in Eqtn. 2.61.

Eqtn. 2.61

$$I_2(f,0) = I_2(f, x1)e^{-\gamma(f)x1} = \frac{d}{dx} j\omega Q_{M1M2} V_o(f)e^{-2\gamma(f)x1}$$

The expression in Eqtn. 2.61 gives the current that NEXT will produce if only one unbalance point existed along the twisted pairs. Usually, unbalance will exist at many points along the pairs, and furthermore, the unbalance coefficient, $Q_{M1M2}$ might be different at each point. To find the total current at the disturbed pair's receiver due to all the unbalance points along the pairs, you can let the unbalance coefficient be a function of position on the line, $Q_{M1M2}$, and integrate along the length of the line. The total current is then given by Eqtn. 2.62.

Eqtn. 2.62

$$I_2(f) = \int_0^L j\omega Q_{M1M2}(x) V_o(f) e^{-2\gamma(f)x1} dx$$

For performance analysis, the concern is more with the power due to NEXT than with the current. Given a current and a termination impedance, $Z_{term}$, power can be found by Eqtn. 2.63.

Eqtn. 2.63

$$P(f) = I(f) I^*(f) Z_{term} \quad P(f) = I(f) I^*(f) Z_{term}$$

In Eqtn. 2.63, * is the complex conjugate operator. For the NEXT expression, still in integral form, explicit calculation of the current, and therefore power, would be impossible because knowledge of $Q_{M1M2}$ at every point x on the pairs is not known. Assuming that the termination impedance is equal to the line's characteristic impedance, from Eqtn. 2.62, the NEXT power at the disturbed receiver can be written as in Eqtn. 2.64.

Eqtn. 2.64

$$E[P(f)] = E\left[|Z_o| I_2(f) I_2(f)^*\right]$$
$$= E\left[|Z_o| \int_0^L j\omega Q_{M1M2}(x) V_o(f) e^{-2\gamma(f)x} dx \int_0^L - j\omega Q_{M1M2}(y) V_o(f) e^{-2\gamma^*(f)y} dy\right]$$

The operator E* is used here to denote the expected value. All terms that do not involve x or y can be pulled outside the integral so that Eqtn. 2.64 simplifies to Eqtn. 2.65.

Eqtn. 2.65

$$E[P(f)] = E\left[|Z_o| I_2(f) I_2(f)^*\right]$$
$$= E\left[|Z_o| \omega^2 V_o^2(f) \int_0^L \int_0^L Q_{M1M2}(x) Q_{M1M2}(y) e^{-2\gamma(f)x} e^{-2\gamma^*(f)y} dxdy\right]$$

The double integration in Eqtn. 2.65 can be simplified by assuming that the unbalance at some point x1 on the twisted pair is independent of the unbalance at another point x2. Using this assumption leads to the expression in Eqtn. 2.66.

Eqtn. 2.66

$$E[Q_{M1M2}(x)Q_{M1M2}(y)] = k \cdot \delta(x - y)$$

Here, k is the second moment of the random variable $Q_{M1M2}$, and δ is the khroniker delta function such that Eqtn. 2.67 is true.

Eqtn. 2.67

$$\delta(x) = \begin{cases} 1 & \text{if } x = 0 \\ 0 & \text{otherwise} \end{cases}$$

Because $Q_{M1M2}$ is the only random variable in Eqtn. 2.65, the expectation can be taken inside, and the first integration performed such that the result in Eqtn. 2.68 is obtained.

Eqtn. 2.68

$$\begin{aligned} E[P(f)] &= |Z_o|\omega^2 V_o^2(f)\int_0^L\int_0^L E[Q_{M1M2}(x)Q_{M1M2}(y)]e^{-2\gamma(f)x}e^{-2\gamma^*(f)y}dxdy \\ &= |Z_o|\omega^2 V_o^2(f)\int_0^L\int_0^L k\delta(x-y)e^{-2\gamma(f)x}e^{-2\gamma^*(f)y}dxdy \\ &= |Z_o|\omega^2 V_o^2(f)\int_0^L ke^{-2\gamma(f)x}e^{-2\gamma^*(f)x}dx \end{aligned}$$

The sifting property was used to obtain the last line of Eqtn. 2.68.

If the exponential terms in Eqtn. 2.68 are combined and the propagation constant expressed in terms of α and β, Eqtn. 2.69 results.

Eqtn. 2.69

$$\begin{aligned} E[P(f)] &= |Z_o|\omega^2 V_o^2(f)\int_0^L ke^{-2\gamma(f)x}e^{-2\gamma^*(f)x}dx \\ &= |Z_o|\omega^2 V_o^2(f)\int_0^L ke^{-2x[(\alpha(f)+j\beta(f))+(\alpha(f)-j\beta(f))]}dx \\ &= |Z_o|\omega^2 V_o^2(f)\int_0^L ke^{-4x\alpha(f)}dx \end{aligned}$$

The final integration can now be carried out. The result is in Eqtn. 2.70.

Eqtn. 2.70

$$
\begin{aligned}
E[P(f)] &= |Z_o|\omega^2 V_o^2(f)\int_0^L ke^{-4x\alpha(f)}dx \\
&= |Z_o|\omega^2 V_o^2(f)k\frac{e^{-4x\alpha(f)}}{-4\alpha(f)}\Big|_0^L \\
&= \frac{|Z_o|\omega^2 V_o^2(f)k}{-4\alpha(f)}\left(e^{-4L\alpha(f)}-1\right)
\end{aligned}
$$

For long lines, the exponential in the result will be small compared to 1, and Eqtn. 2.70 can be simplified. In addition, Eqtn. 2.70 can be divided by the power of the disturbing transmitter ($|V_o|^2/Z_o$) to obtain the power transfer function between the disturbing transmitter and the disturbed receiver, denoted as $H_{NEXT}$ as in Eqtn. 2.71.

Eqtn. 2.71

$$
H_{NEXT}(f) = \frac{|Z_o|^2\omega^2 k}{4\alpha(f)} = \frac{k' f^2}{\alpha_o f^{1/2}} = K_{NEXT} f^{3/2}
$$

In Eqtn. 2.71, $K_{NEXT}$ is a constant of proportionality encompassing all the constants in the first two terms of Eqtn 2.71. Thus, for a disturber with a *flat power spectral density* (PSD), you have an expression showing that NEXT is proportional to $f^{3/2}$. Eqtn. 2.71 is sometimes called a *crosstalk loss function*. It is more generic than a general NEXT power equation because it does not depend on the PSD of the disturber. Instead, it can be multiplied by any PSD to give the expected NEXT power that the disturber would cause.

Note that NEXT does not depend on the length of the twisted pair. This is because the interference due to NEXT, although a sum of crosstalk at every point along the pair, will be dominated by the unbalance points nearest to the receiver. The crosstalk from the nearer unbalance points will dominate the crosstalk from distant unbalance points due simply to the length of twisted pair that each must propagate through.

Note also that $K_{NEXT}$ is a random variable. For many different disturbers into a disturbed pair, $K_{NEXT}$ will tend toward a Gaussian distribution due to the central limit theorem.[14,15,16]

## *Far-end Crosstalk (FEXT)*

FEXT will be derived using a similar method as used in the NEXT derivation. Figure 2.21 shows an example of FEXT due to an unbalance point at x1 only.

Figure 2.21 An example of far-end crosstalk due to only one unbalance point (x1).

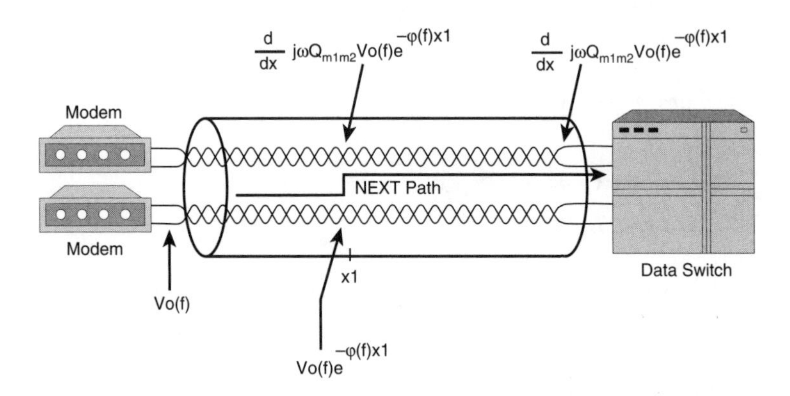


The signal expressions on the disturbing pair are the same as for the NEXT case, and the current at x1 on the disturbed pair is also the same as in the NEXT derivation. Repeating the expression for this current from Eqtn. 2.60 yields Eqtn. 2.72.

Eqtn. 2.72

$$I_2(f, x1) = \frac{d}{dx} j\omega Q_{M1M2} V_1(f, x1) = \frac{d}{dx} j\omega Q_{M1M2} V_o(f) e^{-\gamma(f)x1}$$

To reach the disturbed pair's receiver, the signal must propagate over the remainder of the disturbed pair (a distance of L-x1). Upon reaching the disturbed pair's receiver, the resulting expression for current on the disturbed line is shown in Eqtn. 2.73.

Eqtn. 2.73

$$\begin{aligned} I_2(f, L) &= \frac{d}{dx} j\omega Q_{M1M2} V_o(f) e^{-\gamma(f)x1} e^{-\gamma(f)(L-x1)} \\ &= \frac{d}{dx} j\omega Q_{M1M2} V_o(f) e^{-\gamma(f)L} \end{aligned}$$

Just as in the NEXT case, the total crosstalk signal on the disturbed pair will be a summation of crosstalk due to unbalance points all along the twisted pairs and not just the one at x1. Eqtn. 2.73 can be integrated to find this expression. Moving one step farther, one can write an expression for the total FEXT power similar to Eqtn. 2.64 as in the following Eqtn. 2.74.

Eqtn. 2.74

$$E[P(f)] = E\left[|Z_o| I_2(f) I_2(f)^*\right]$$
$$= E\left[|Z_o| \int_0^L j\omega Q_{M1M2}(x) V_o(f) e^{-\gamma(f)L} dx \int_0^L -j\omega Q_{M1M2}(y) V_o(f) e^{-\gamma^*(f)L} dy\right]$$
$$= |Z_o| \omega^2 V_o^2(f) e^{-\gamma(f)L} e^{-\gamma^*(f)L} \int_0^L \int_0^L E[Q_{M1M2}(x) Q_{M1M2}(y)] dxdy$$

The relationship given in Eqtn. 2.66 can again be used to help carry out the integral in Eqtn. 2.74 so that the expression for FEXT simplifies to Eqtn. 2.75.

Eqtn. 2.75

$$E[P(f)] = |Z_o| \omega^2 V_o^2(f) e^{-2\alpha(f)L} \int_0^L k dx = |Z_o| \omega^2 V_o^2(f) e^{-2\alpha(f)L} L$$

The constants in Eqtn. 2.75 can be grouped together in a single constant, and the average FEXT transfer function can be written as Eqtn. 2.76.

Eqtn. 2.76

$$H_{FEXT}(f) = K_{FEXT} f^2 e^{-2\alpha(f)L} L$$

Note that $K_{FEXT}$, like $K_{NEXT}$ is also a random variable. When multiple disturbers exist, $K_{FEXT}$ can be expected to tend toward a Gaussian random variable due to the central limit theorem. Note also that FEXT, unlike NEXT, is a function of the length of the twisted pair. This fact is intuitively satisfying because FEXT disturbances due to any unbalance point along the twisted pairs must propagate the entire length of the pair to disturb the receiver. (At this point, the reader should realize that both NEXT and FEXT increase with frequency, and thus a twisted pair designed for low frequency signals presents challenges when high frequency signals are involved.)

## *Summary*

This chapter was dedicated to better understanding twisted pair physical and electrical properties. The primary line constants, RLCG, of a twisted pair were discussed. Numeric and analytic models of a twisted pair's propagation constant and characteristic impedance were derived. The attenuation of various twisted pair cables was also illustrated.

Crosstalk between twisted pairs was discussed in some detail. The basic cause of crosstalk, unbalance between neighboring pairs, was derived from basic circuit theory. From the unbalance expressions, expressions for near-end crosstalk (NEXT) power and far-end crosstalk (FEXT) power were derived.

Understanding the challenges that a twisted pair environment poses to an ADSL or a VDSL environment will help the reader understand the motivation for some of the modulation and coding schemes presented in later chapters. Also, the models developed here will be used directly in later chapters to predict and bound the performance of certain ADSL and VDSL examples.

## *Exercises*

1. For #24 and #26 gauge twisted pairs, plot over frequency the characteristic impedance and propagation constant using the analytic and numeric models.
2. Find the loss at 500 kHz of a 10 kft, #26 gauge twisted pair using both the analytic and numeric models.
3. For a #26 loop, plot Eqtn. 2.70 for various lengths. At what length does the approximation of Eqtn. 2.71 become valid?
4. Graph the NEXT and FEXT loss functions on the same set of axes for different lengths of #24 and #26 twisted pair. At what length does NEXT become dominant over FEXT?
5. Plot the loss of a #26 twisted pair at 500 kHz against the length of the pair. At what length is the loss equal to the NEXT loss at 500 kHz?

## *Endnotes*

1. Rhodes, Frederick Leland, *Beginnings of Telephony.* New York: Harper and Brothers, 1929.
2. Manhire, L. M., "Physical and Transmission Characteristics of Customer Loop Plant." *Bell System Technical Journal* 57 (January 1978): 35–39.
3. Chin, Wu-Jhy, et al, "Loop Survey in the Taiwan Area and Feasibility Study for HDSL." IEEE Selected Areas in Communications 9, no. 6 (August 1991): 801–809.
4. Reeve, Whitham D., *Subscriber Loop Signalling and Transmission Handbook.* New York: IEEE Press, 1992.
5. Bellamy, John, *Digital Telephony.* New York: John Wiley and Sons, 1991.
6. Gang, Huang, and J.J. Werner, "Cable Characteristics." *ANSI Contribution T1E1.4/97-169*, Clearwater, FL, May 12, 1997.
7. Gilbert, Elliot and John Nellist, *Telecommunications Wiring for Commercial Buildings, A Practical Guide.* New York: IEEE Press, 1996.

8. Pickering, Ashley, "Primary Line Constants for US 24 and 26 AG Twisted Pair Cables." *ANSI Contribution T1E1.4/96-074.* Colorado Springs, CO, April 22, 1996.

9. Sadiku, M., *Elements of Electromagnetics.* Philadelphia: Saunders College Publishing, 1989.

10. Campbell, G., "Dr. Campbell's Memoranda of 1907 and 1912." *Bell System Technical Journal* 14, no. 4 (October 1935): 558–472.

11. AT&T Company, *Collected Papers of George A. Campbell.* New York: AT&T, 1937.

12. Cravis, H. and T.V. Carter, "Engineering of T1 Carrier System Repeated Lines." *Bell System Technical Journal* 42, no. 1 (March 1963): 431–486.

13. Miller, G., "The Effect of Longitudinal Imbalance on Crosstalk." *Bell System Technical Journal* 54, no. 7 (September 1975): 1227–1251.

14. Kerpez, K., "Near End Crosstalk Is Almost Gaussian." *IEEE Transactions on Communications* 41, no. 1 (January 1993): 670–672.

15. Werner, J.J., "The HDSL Environment." *IEEE Journal on Selected Areas in Communications* 9, no. 6 (August 1991): 785–800.

16. Gallager, Robert, *Information Theory and Reliable Communications.* New York: John Wiley and Sons, 1968.

CHAPTER 3

# Loop Analysis

In this chapter:

- A review of ABCD parameters
- Discussion of bridged taps
- Application of ABCD parameters to twisted pair characteristics
- Discussion of ADSL and VDSL test loops

This chapter presents methods to analyze transmission parameters of twisted pairs. This chapter provides a generic procedure using ADCD parameters that is applicable to almost every type of loop. Understanding the properties of twisted pairs is necessary for understanding many aspects of ADSL and VDSL modulation methods. Being able to characterize different types of loops is also necessary for benchmarking the limits of performance for a modulation. In this spirit, a discussion is given on the various test loops to be used when characterizing theoretical performance achievable on a twisted pair, as well as actual performance of different modulation schemes. Various standards bodies have defined these loops to benchmark proposed modulation schemes and define the performance goals of DSL technologies.[1,2] In addition, several sections deal with the definition and use of ABCD parameters, providing the tools to characterize virtually any loop configuration.

## ABCD Parameters

Before discussing various test loops, a review of *ABCD transmission parameters* will be helpful. ABCD parameters are useful in characterizing *two-port networks.* A two-port network is generically defined to have a single unique input port and a single unique output port with some transfer function between the two. These parameters mathematically relate the input voltage and current to the output voltage and current. Figure 3.1 shows a generic two-port network. In this figure, the output voltage and current $V_2$ and $I_2$ are related to the input voltage and current $V_1$ and $I_1$ in terms of ABCD parameters as in Eqtn. 3.1.

Figure 3.1 A generic two-port network.

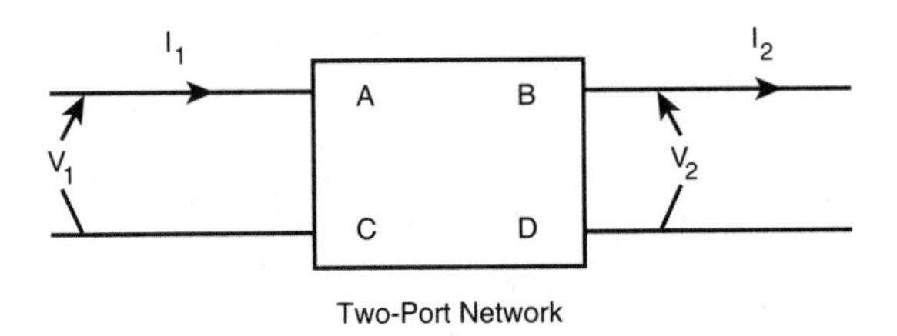


Eqtn. 3.1

$$V_1 = AV_2 + BI_2$$
$$I_1 = CV_2 + DI_2$$

Note that Eqtn. 3.1 characterizes the network without making any assumptions about the network's termination. The parameters A, B, C, and D are specific to the transfer function of the network.

Figure 3.2 shows two networks in tandem. Each network is independently characterized by its own ABCD parameters (denoted by subscript 1 and subscript 2).

Figure 3.2 Two two-port networks in tandem; each network has unique ABCD parameters.

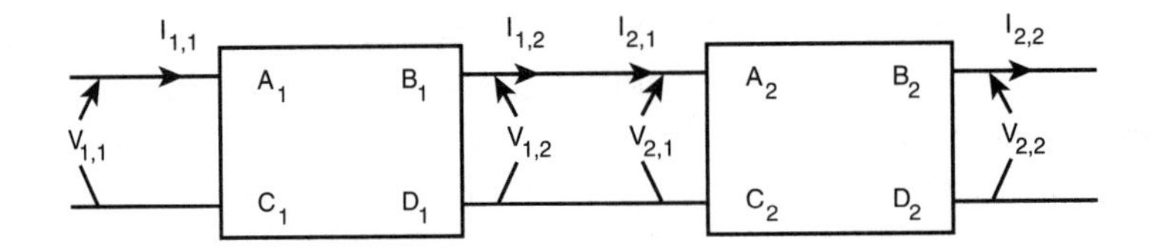


The effective ABCD parameters for the entire system can be given by a simple matrix multiplication of the individual parameters such that Eqtn. 3.2 is true.[3]

Eqtn. 3.2

$$\begin{bmatrix} A & B \\ C & D \end{bmatrix} = \begin{bmatrix} A_1 & B_1 \\ C_1 & D_1 \end{bmatrix} \begin{bmatrix} A_2 & B_2 \\ C_2 & D_2 \end{bmatrix}$$
$$= \begin{bmatrix} A_1 A_2 + B_1 C_2 & A_1 B_2 + B_1 D_2 \\ C_1 A_2 + D_1 C_2 & C_1 B_2 + D_1 D_2 \end{bmatrix}$$

If another two-port network was added to that in Figure 3.2 to the right of the existing networks, the new equivalent ABCD parameters of the entire system could be found by matrix multiplication of the ABCD matrix for the new network by the result of Eqtn. 3.2. Thus, if the parameters of each individual network are given in ABCD form, it is a trivial task to find the characteristics of the network as a whole.

ABCD parameters are useful for characterizing twisted-pair loops because not all loops consist of a uniform gauge and a perfect termination. For example, a loop might consist of 9 kft of #26 gauge cable followed by 2 kft of #24 gauge cable. The insertion loss analysis in Chapter 2 did not account for such a case.

## *ABCD Parameters of a Lumped Impedance*

One of the simplest cases of ABCD parameters is for a lumped impedance as shown in Figure 3.3.

Figure 3.3 A two-port network consisting of a lumped parallel impedance.

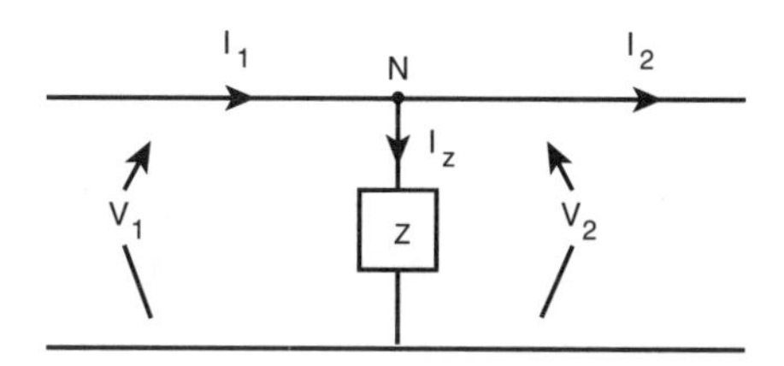


In this case, $V_2$=$V_1$ independent of $I_2$. Thus, the ABCD parameter equations can be written as in Eqtn. 3.3.

Eqtn. 3.3

$$V_1 = AV_2 + BI_2 = AV_1 + BI_2$$
$$I_1 = CV_2 + DI_2 = CV_1 + DI_2$$

Solving the first equation is trivial and yields Eqtn. 3.4.

Eqtn. 3.4

$$A = 1$$
$$B = 0$$

Eqtn. 3.3 is best solved using Kirkoff's current law at point N. This results in Eqtn. 3.5.

Eqtn. 3.5

$$\begin{aligned} I_1 &= I_Z + I_2 \\ &= \frac{V_1}{Z} + I_2 \end{aligned}$$

Comparing Eqtn. 3.5 and Eqtn. 3.3, it should be clear that Eqtn. 3.6 is true.

Eqtn. 3.6

$$C = \frac{1}{Z}$$
$$D = 1$$

Thus, the ABCD matrix for a two-port network consisting only of a shunt impedance is given by Eqtn. 3.7.

Eqtn. 3.7

$$\begin{bmatrix} 1 & 0 \\ \frac{1}{Z} & 1 \end{bmatrix}$$

Z in Eqtn. 3.7 represents the shunt impedance. Note that Z may be complex because no restriction is placed on it to be real.

## *ABCD Parameters of a Uniform Twisted Pair*

For a uniform twisted pair, finding the ABCD parameters is a bit more involved than for a simple lumped element. Chapter 2 showed that a wave traveling on a twisted pair is attenuated exponentially with distance and that the exponential depends on the pair's propagation constant. The voltage at a point on the line is the sum of a forward traveling wave at that point and the backward traveling wave. Figure 3.4 shows a twisted pair with the forward and backward traveling waves labeled at the end of the pair (x=L). Eqtn. 3.8 gives the voltage at the other end of the twisted pair (x=0).

Figure 3.4 A twisted pair transmission line; the forward (V+) and backward (V–) traveling waves are labeled.

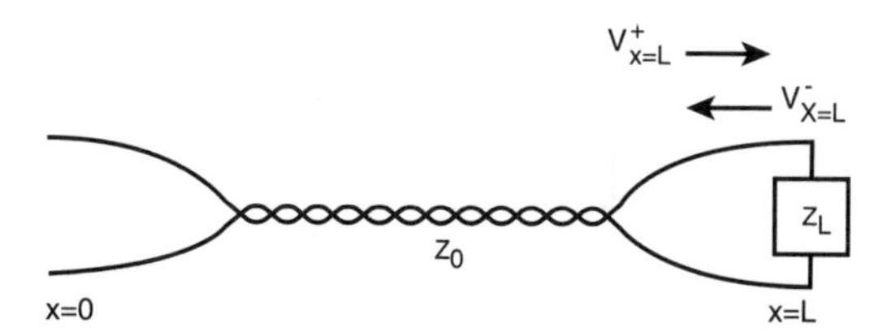


Eqtn. 3.8

$$V(f,0) = \left(V^{+}_{x=L}e^{\gamma(f)L} + V^{-}_{x=L}(f)e^{-\gamma(f)L}\right)$$
$$= V^{+}_{x=L}\left(e^{\gamma(f)L} + \Gamma_L e^{-\gamma(f)L}\right)$$

The reflection coefficient in Eqtn. 3.8 was given in Eqtn. 2.10 and is repeated in Eqtn. 3.9.

Eqtn. 3.9

$$\Gamma_L(f) = \frac{Z_L(f) - Z_o(f)}{Z_L(f) + Z_o(f)}$$

Note that in Eqtn. 3.8, the voltage at x=0 is the sum of a forward and backward traveling wave seen at x=L. The forward traveling wave will be larger at x=0 because it propagates from x=0 to x=L (indicated by the positive exponent in Eqtn. 3.8). The backward traveling wave will be larger at x=L because it propagates from x=L to x=0 (indicated by the negative exponential exponent in Eqtn. 3.8).

For ABCD parameters, it is helpful to write Eqtn. 3.8 in terms of the two-port output voltage and current rather than in terms of the positive traveling wave voltage, $V^{+}_{x=L}$. Figure 3.5 shows the equivalent two-port network representation of the uniform twisted pair.

Figure 3.5 An equivalent two-port network for a twisted pair.

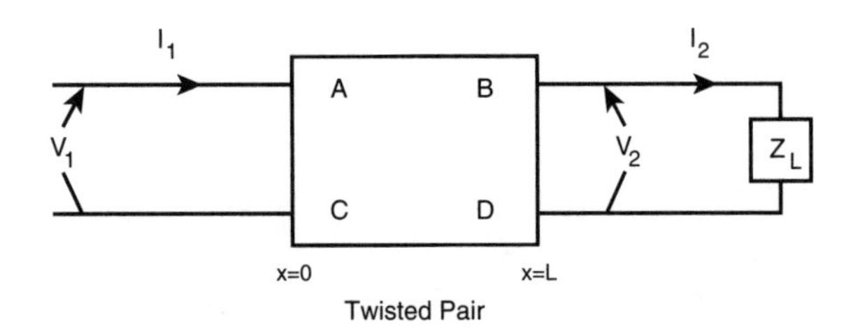


Dependence of all values on frequency has been dropped for notational simplicity. Because the output voltage is the sum of forward and backward wave voltages at x=L, you can immediately find the expression for $V_2$ given in Eqtn. 3.10.

Eqtn. 3.10

$$\begin{aligned} V_2 &= V^+_{x=L} + V^-_{x=L} \\ &= V^+_{x=L}\left(1+\Gamma_L\right) \end{aligned}$$

Solving for $V^+_{x=L}$ is then relatively straightforward as shown in Eqtn. 3.11.

Eqtn. 3.11

$$V^+_{x=L} = \frac{V_2}{1+\Gamma_L}$$

To find $V_1$, Eqtn. 3.11 can be substituted into Eqtn. 3.8 and upon simplifying, results in Eqtn. 3.12.

Eqtn. 3.12

$$V_1 = \frac{V_2}{1+\Gamma_L}\left(e^{\gamma(f)L} + \Gamma_L e^{-\gamma(f)L}\right)$$

To further simplify Eqtn. 3.12, Eqtn. 3.9, the definition of reflection coefficient, is used. After substituting and simplifying, Eqtn. 3.12 can be rewritten as Eqtn. 3.13.

Eqtn. 3.13

$$
\begin{aligned}
V_1 &= V_2\left(\left(\frac{1}{2}+\frac{Z_o}{2Z_L}\right)e^{\gamma(f)L}+\left(\frac{1}{2}-\frac{Z_o}{2Z_L}\right)e^{-\gamma(f)L}\right)\\
&= V_2\left(\left(\frac{1}{2}+\frac{Z_oI_2}{2V_2}\right)e^{\gamma(f)L}+\left(\frac{1}{2}-\frac{Z_oI_2}{2V_2}\right)e^{-\gamma(f)L}\right)\\
&= \left(\frac{e^{\gamma(f)L}+e^{-\gamma(f)L}}{2}\right)V_2+Z_o\left(\frac{e^{\gamma(f)L}-e^{-\gamma(f)L}}{2}\right)I_2
\end{aligned}
$$

Recall that $Z_o$ is the characteristic impedance of the transmission line at x=L, and $Z_L$ is the load impedance. The second line in Eqtn. 3.13 results from simply using Ohm's law and the third line from combining like terms. Note that the definitions of hyperbolic cosine (cosh) and hyperbolic sine (sinh) are present in the last line of Eqtn. 3.13. Substituting for these yields Eqtn. 3.14.

Eqtn. 3.14

$$
\begin{aligned}
V_1 &= (\cosh(\gamma L))V_2+Z_o(\sinh(\gamma L))I_2\\
&= AV_2+BI_2
\end{aligned}
$$

Similarly, the second two ABCD parameters can be found (the details are left to you in Exercise 2 at the end of the chapter) as in Eqtn. 3.15.

Eqtn. 3.15

$$
\begin{aligned}
I_1 &= \left(\frac{\sinh(\gamma L)}{Z_o}\right)V_2+(\cosh(\gamma L))I_2\\
&= CV_2+DI_2
\end{aligned}
$$

Thus, the ABCD parameters for a uniform twisted pair (or any generic uniform transmission line) are as shown in Eqtn. 3.16.

Eqtn. 3.16

$$
\begin{bmatrix} A & B \\ C & D \end{bmatrix} = \begin{bmatrix} \cosh(\gamma L) & Z_o\sinh(\gamma L) \\ \dfrac{\sinh(\gamma L)}{Z_o} & \cosh(\gamma L) \end{bmatrix}
$$

## *Bridged Taps*

Sometimes, a twisted pair running to a subscriber might have another, unused twisted pair section connected at some point along its length. The other end of the unused section is left open circuited (see Figure 3.6). The unused section of twisted pair is called a *bridged tap.*

Figure 3.6 A common bridged tap configuration on a twisted pair.

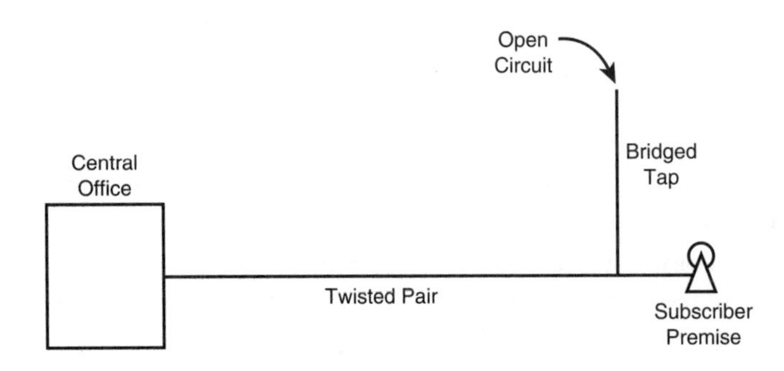


The purpose of a bridged tap is usually to leave flexibility in the location of a subscriber using the loop. This flexibility is necessary when twisted pairs are installed prior to being used because in many cases the actual location of the subscriber is not yet known (for example, when cables are laid in a neighborhood under development). A bridged tap causes reflections at the open circuit end producing dips in the transfer function of the loop to which it is attached. Note that the propagation constant and characteristic impedance of the bridged tap might or might not be the same as the loop to which the tap is connected. Also note that the main loop itself might have different line characteristics on each side of the bridged tap.

The ABCD parameters of a bridged tap are the same as any other uniform twisted pair as given in Eqtn. 3.16. In the bridged tap case, however, you do not need to be concerned with the transfer function of the tap (because the signal does not propagate along the bridged tap) but rather the effect that the tap's presence has on the main loop's transfer function. Using Eqtn. 3.16, a bridged tap's ABCD equations are shown in Eqtn. 3.17.

Eqtn. 3.17

$$V_1 = \cosh(\gamma L_{tap})V_2 + Z_{o,tap}\sinh(\gamma L_{tap})I_2 = \cosh(\gamma L_{tap})V_2$$

$$I_1 = \frac{\sinh(\gamma L_{tap})}{Z_{o,tap}}V_2 + \cosh(\gamma L_{tap})I_2 = \frac{\sinh(\gamma L_{tap})}{Z_{o,tap}}V_2$$

The simplification in both equations comes because the bridged tap, by definition, is open circuited at one end, and thus, $I_2=0$.

Dividing the two equations in Eqtn. 3.17 yields Eqtn. 3.18.

Eqtn. 3.18

$$Zin_{tap} = \frac{V_1}{I_1} = \frac{\cosh(\gamma L_{tap})V_2}{\frac{\sinh(\gamma L_{tap})}{Z_{o,tap}}V_2} = Z_{o,tap}\coth(\gamma L_{tap})$$

In Eqtn. 3.18, the hyperbolic cotangent (coth) is used. From the main loop's point of view, the effect of the bridged tap would be the same as the effect of a lumped impedance across the loop with a value of $Zin_{tap}$. The ABCD parameters of a lumped impedance were derived in Eqtn. 3.7. Substituting the effective shunt impedance of a bridged tap from Eqtn. 3.17 into Eqtn. 3.5 yields Eqtn. 3.19.

Eqtn. 3.19

$$ABCD_{tap} = \begin{bmatrix} 1 & 0 \\ \frac{1}{Zin_{tap}} & 1 \end{bmatrix} = \begin{bmatrix} 1 & 0 \\ \frac{1}{Z_{o,tap}\coth(\gamma L_{tap})} & 1 \end{bmatrix}$$

These ABCD parameters do not describe the parameters along the length of the bridged tap but rather the effect of the bridged tap on the main loop. The following section, "Loop Analysis," illustrates how this information can be used to model a loop with any structure.

## *Loop Analysis*

At this point, you can analyze a wide variety of loops by finding the ABCD parameters of each section of the loop and performing matrix multiplications. For example, consider the loop shown in Figure 3.7.

Figure 3.7 An example twisted pair to illustrate the usefulness of ABCD parameters.

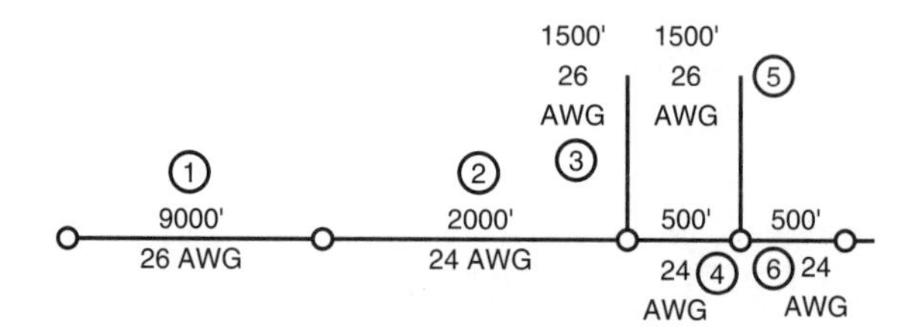


This loop consists of 9 kft of #26 gauge cable followed by a total of 3 kft of #24 gauge cable. Because two bridge taps exist on the #24 section (each 1.5 kft of #26 gauge cable), the #24 section is broken into one 2 kft section and two 500 foot sections. The ABCD parameters for the entire cable are found by multiplying the ABCD parameters for the individual sections along the loop. This is shown generically in Eqtn. 3.20.

Eqtn. 3.20

$$ABCD_{tot} = [9kft, \#26][2kft, \#24][BT1][500ft, \#24][BT2][500ft, \#24]$$

The values shown in each set of brackets represent the ABCD parameters of the listed section. Note that six sets of ABCD parameters are necessary to analyze this loop as labeled in the figure and shown in Eqtn. 3.20. This loop is T1.601 #9 and is analyzed in detail later in the chapter.

The ABCD parameters for each term in Eqtn. 3.20 were discussed in the preceding sections. The information necessary to find the ABCD parameters includes the gauge and length of each section (along with whether the section is part of the main loop or a bridged tap). A cable of this type, with varying gauge and multiple bridged taps, illustrates the usefulness of ABCD parameters.

## *Input Impedance, Transfer Functions, and Insertion Loss*

Given that you can characterize any twisted pair in terms of ABCD parameters, you can see how to find three important characteristics about a loop, mainly the *input impedance*, *transfer function*, and *insertion loss*. In each case, an assumption is made that the loop's ABCD parameters have been found.

For analysis of a loop, specific information beyond the ABCD parameters of the loop is necessary. This information includes the impedance of the source providing a signal to the loop and the impedance of the loop termination. These are sometimes referred to as *source impedance* (or generator impedance) and *termination impedance* (or load impedance). Figure 3.8 shows a generic end-to-end loop model.

Figure 3.8 A generic end-to-end model of a loop including the generator impedance at the modulator and the load or terminating impedance at the receiver.

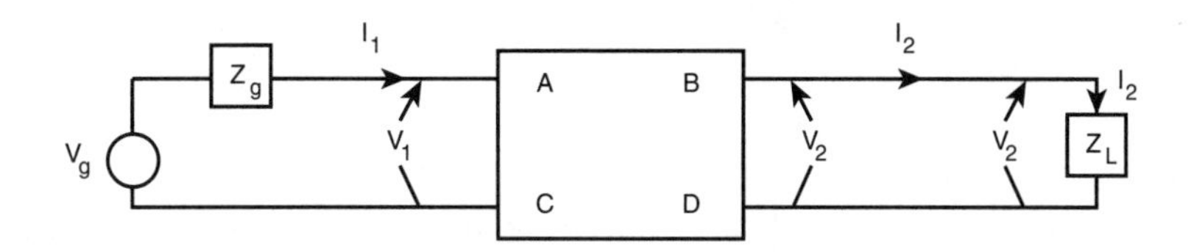


$Z_g$ is the impedance of the generator, and $Z_L$ is the load impedance (this analysis assumes that both are 100 ohms). Note that in ADSL and VDSL, $Z_g$ would normally be part of the transmitter design and $Z_L$ part of the receiver design. The input impedance will be a function of the impedance of the loop and of the load (but not $Z_g$). The transfer function will be from $V_1$ to $V_2$. The insertion loss will be a function of the loop, the load impedance, and the generator impedance.

The input impedance of the loop is simple to derive. Given the ABCD parameter equations as shown in Eqtn. 3.1, the input impedance is found by simply dividing the first equation by the second. Regarding Figure 3.8, by Ohm's law, it follows that $V_2=I_2Z_L$. Thus, the input impedance can be found as in Eqtn. 3.21.

Eqtn. 3.21

$$Z_{in} = \frac{V_1}{I_1} = \frac{AV_2 + BI_2}{CV_2 + DI_2}$$

$$= \frac{AI_2Z_L + BI_2}{CI_2Z_L + DI_2} = \frac{AZ_L + B}{CZ_L + D}$$

The transfer function of a loop can be found from the first ABCD equation. Once again, using Ohm's law, the equation can be written as in Eqtn. 3.22.

Eqtn. 3.22

$$V_1 = AV_2 + BI_2 = AV_2 + B\frac{V_2}{Z_L}$$

Ohm's law was used to replace $I_2$ by $V_2/Z_L$.

Eqtn. 3.22 can be solved algebraically for the transfer function, $V_2/V_1$ yielding the simple expression in Eqtn. 3.23.

Eqtn. 3.23

$$T = \frac{V_2}{V_1} = \frac{1}{A + \frac{B}{Z_L}} = \frac{Z_L}{AZ_L + B}$$

The insertion loss of the loop is the ratio of power delivered to the load to the power that would have been delivered to the load without the loop present. Essentially, this is the loss caused by inserting the loop between the source and termination. Insertion loss is normally expressed in dB.

If no loop were present ($Z_G$ was connected directly to $Z_L$), the voltage across the load could be found by a simple voltage divider as shown in Eqtn. 3.24.

Eqtn. 3.24

$$V_L = V_G \frac{Z_L}{Z_G + Z_L}$$

The power delivered to the load in this case would then follow simply as in Eqtn. 3.25.

Eqtn. 3.25

$$P_{\text{No Loop}} = \frac{V_L^2}{Z_L} = \frac{V_G^2 Z_L^2}{Z_L (Z_G + Z_L)^2}$$

When the loop is present, the voltage $V_1$ can be written as in Eqtn. 3.26.

Eqtn. 3.26

$$V_1 = AV_2 + BI_2 = V_G - I_1 Z_G$$

The last part of this equation follows from simple circuit theory (taking into account the voltage drop across $Z_G$). Substituting in the second generic ABCD equation for $I_1$ then yields Eqtn. 3.27.

Eqtn. 3.27

$$AV_2 + BI_2 = V_G - (CV_2 + DI_2) Z_G$$

The entire equation can be written in terms of voltages by substituting $V_2/Z_L$ in for the two $I_2$ terms. This is done in Eqtn. 3.28.

Eqtn. 3.28

$$AV_2 + B\frac{V_2}{Z_L} = V_G - \left(CV_2 + D\frac{V_2}{Z_L}\right)Z_G$$

Finally, algebraic manipulation can be used to solve Eqtn. 3.28 for $V_2$. The result is shown in Eqtn. 3.29.

Eqtn. 3.29

$$V_2 = \frac{V_G Z_L}{AZ_L + B + Z_G(CZ_L + D)}$$

Now that $V_2$ is known, the power delivered to the load with the loop present can be found as shown in Eqtn. 3.30.

Eqtn. 3.30

$$P_{\text{With Loop}} = \frac{V_2^2}{Z_L} = \frac{V_G^2 Z_L^2}{Z_L(AZ_L + B + Z_G(CZ_L + D))^2}$$

The insertion loss of the loop is then defined by Eqtn. 3.31.

Eqtn. 3.31

$$IL = 10\log\left(\left|\frac{P_{\text{No Loop}}}{P_{\text{With Loop}}}\right|\right)$$

$$= 10\log\left(\left|\frac{\dfrac{V_G^2 Z_G^2}{Z_L(Z_G + Z_L)^2}}{\dfrac{V_G^2 Z_L^2}{Z_L(AZ_L + B + Z_G(CZ_L + D))^2}}\right|\right)$$

The results of Eqtn. 3.25 and Eqtn. 3.30 have been used in Eqtn. 3.31 for the power expressions. The insertion loss expression in Eqtn. 3.31 can be algebraically simplified and results in Eqtn. 3.32.

Eqtn. 3.32

$$IL = 20\log\left(\left|\frac{AZ_L + B + Z_G(CZ_L + D)}{(Z_G + Z_L)}\right|\right)$$

This expression for insertion loss will be used extensively later in the book to predict the performance of ADSL and VDSL systems on various loops. For more discussion on ABCD parameters, Appendices C, D, and E of the *Subscriber Loop Signalling and Transmission Handbook* are recommended.[4]

# *Basic Loop Configurations*

To set requirements, compare different modulation proposals, and demonstrate compliance, standards bodies have defined fixed loops for ADSL and VDSL. (VDSL loops are still being defined. Discussed here are what have been agreed on at the present time.) The following sections present these loops and show the input impedance, transfer function, and insertion loss of each. A generator impedance and load impedance of 100 ohms is assumed throughout.

**Note**

A null loop, often called CSA #0, which is less than 10 feet of #26 gauge cable, is not discussed. It is sometimes used to show basic operational capability.

## *ADSL Loops*

Figure 3.9 shows the loops used for ADSL.[1] The variety of different loop topologies and gauges lends weight to the value of the previous discussions on ABCD parameters. For these loops, only #24 and #26 gauge cables are used.

The numerical models for sections of #24 and #26 gauge twisted pairs were presented in Chapter 2, "Twisted Pair Environment," and will be used when analyzing the loops. Specifically, the characteristic impedances and propagation constants over frequency of each cable will be necessary for the analysis. In addition, the basic concepts of ABCD parameters are needed along with the input impedance, transfer function, and insertion loss results derived in Eqtn. 3.21, Eqtn. 3.23, and Eqtn. 3.32, respectively.

Figures 3.10–3.17 show graphs of impedance over frequency as well as transfer function and insertion loss for all ADSL loops (except the null loop). The responses are shown up to around 1.2 MHz encompassing the entire ADSL band. For the input impedance and insertion loss graphs, a source and load impedance of 100 ohms is assumed. A few notes about the graphs follow Figure 3.9.

Figure 3.9 Test loops specified for ADSL.

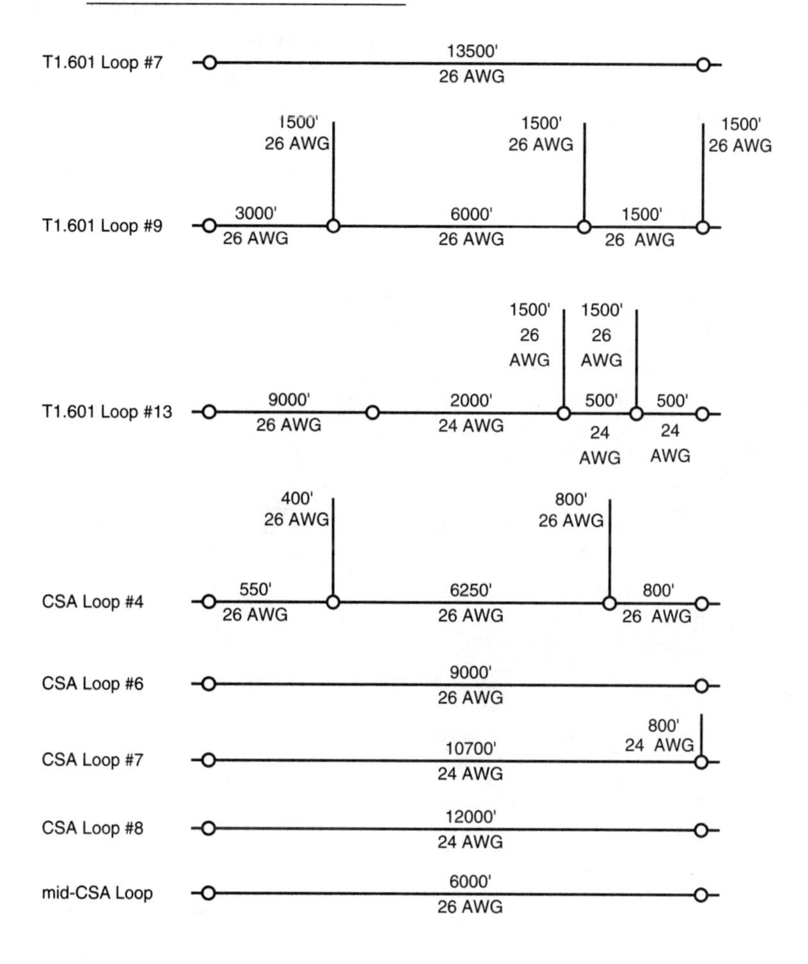


- The real part of the input impedance for all loops converges to around 100 ohms, and the imaginary part to zero as frequency increases.
- The insertion loss for the lines with no bridged taps is monotonically decreasing with frequency. The loops with bridged taps have an insertion loss exhibiting more fluctuation. T1.601 Loop #9 is a good example of the effect that bridged taps can have on insertion loss.

- The transfer function (graphed on a log scale) very much follows the shape of the insertion loss. The insertion loss is dependent on the choice of generator impedance (100 ohms in our case), whereas the transfer function is independent of this impedance. Insertion loss and transfer function curves are notably different when the input impedance strays from 100 ohms. Such behavior can be seen in the CSA Loop #4 graphs.

Figure 3.10 Input impedance, transfer function magnitude, and insertion loss for T1.601 Loop #7.

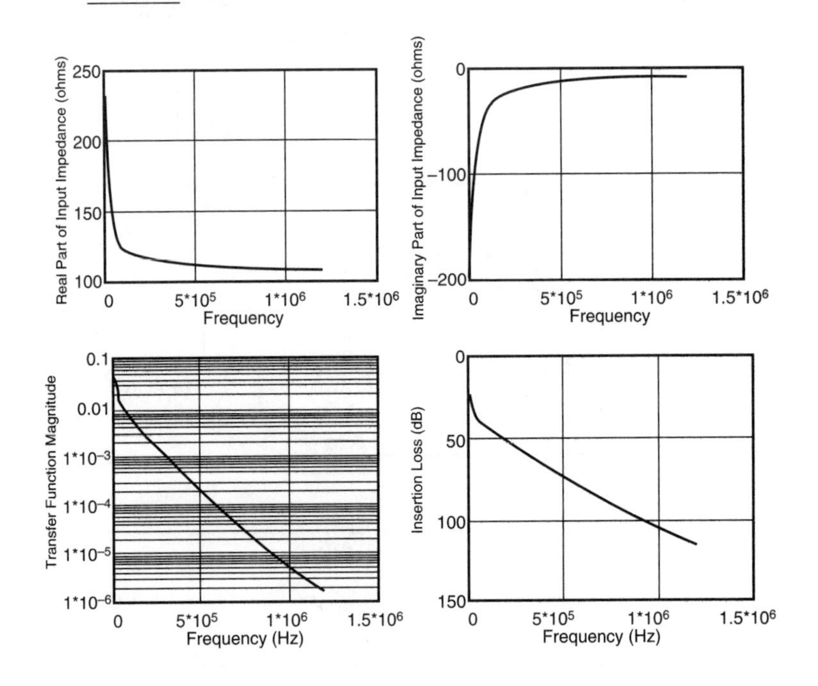

Figure 3.11 Input impedance, transfer function magnitude, and insertion loss for T1.601 Loop #9.

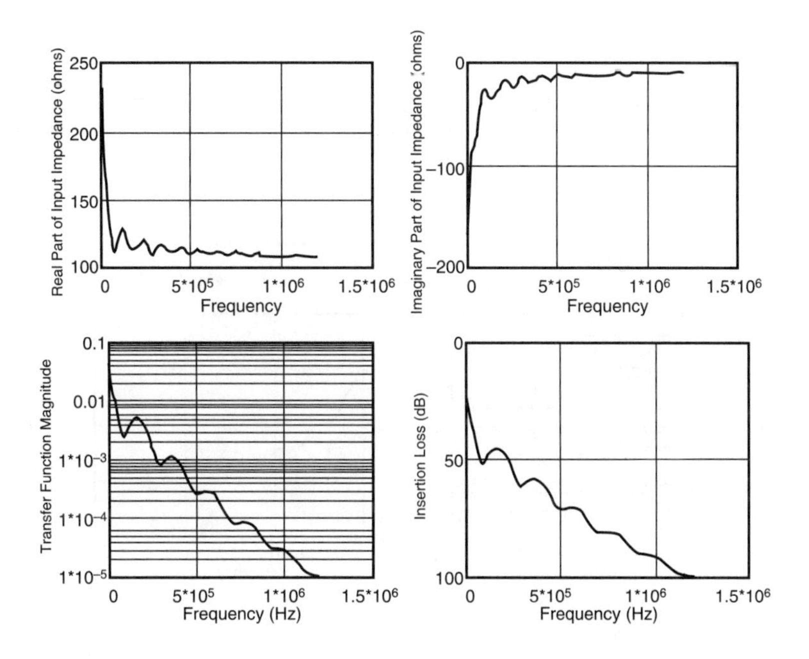


Figure 3.12 Input impedance, transfer function magnitude, and insertion loss for T1.601 Loop #13.

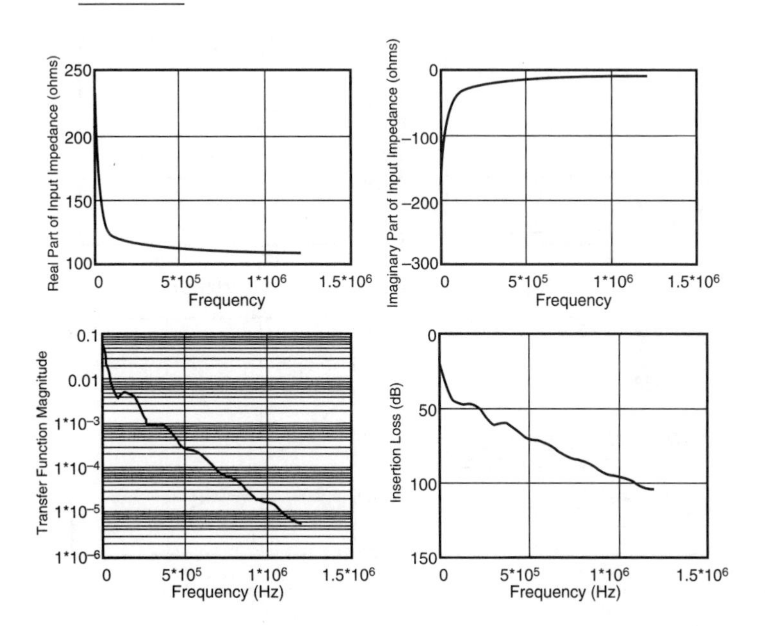

Figure 3.13 Input impedance, transfer function magnitude, and insertion loss for CSA Loop #4.

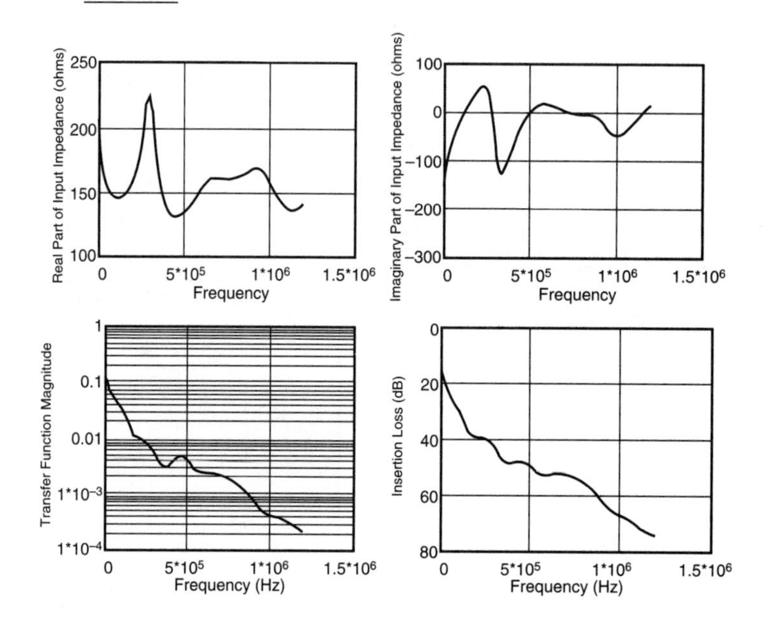


Figure 3.14 Input impedance, transfer function magnitude, and insertion loss for CSA Loop #6.

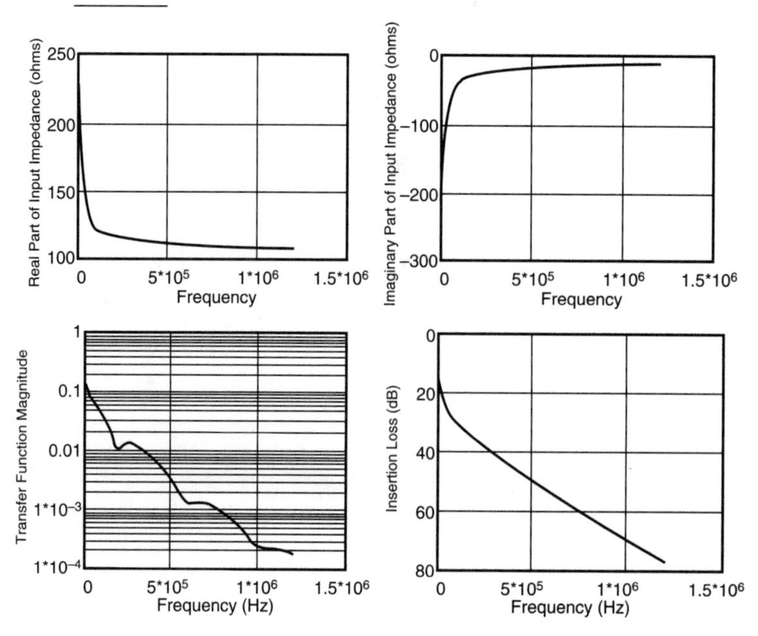

Figure 3.15 Input impedance, transfer function magnitude, and insertion loss for CSA Loop #7.

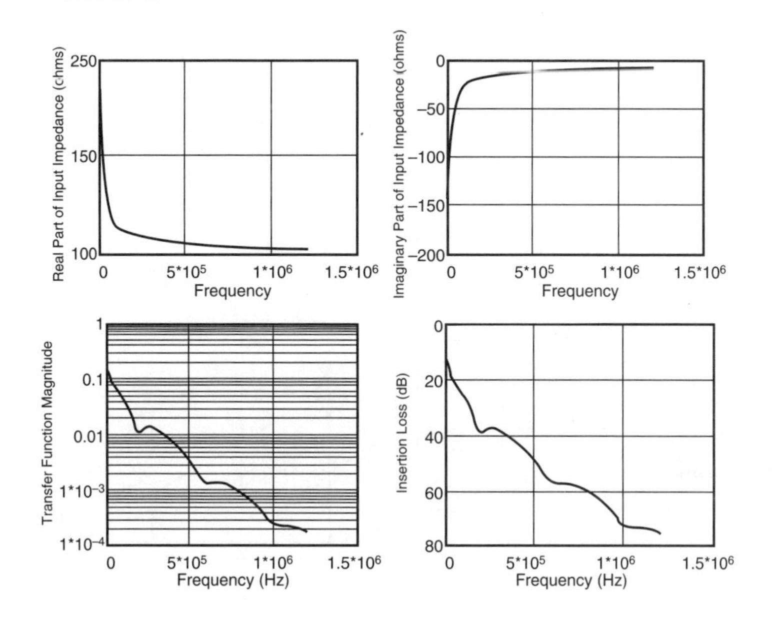


Figure 3.16 Input impedance, transfer function magnitude, and insertion loss for CSA Loop #8.

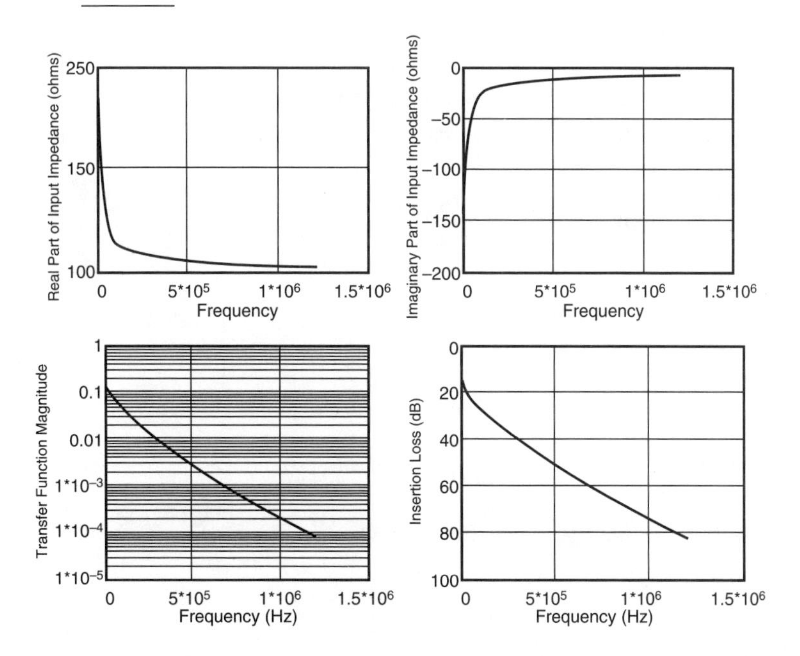

Figure 3.17 Input impedance, transfer function magnitude, and insertion loss for the Mid CSA Loop.

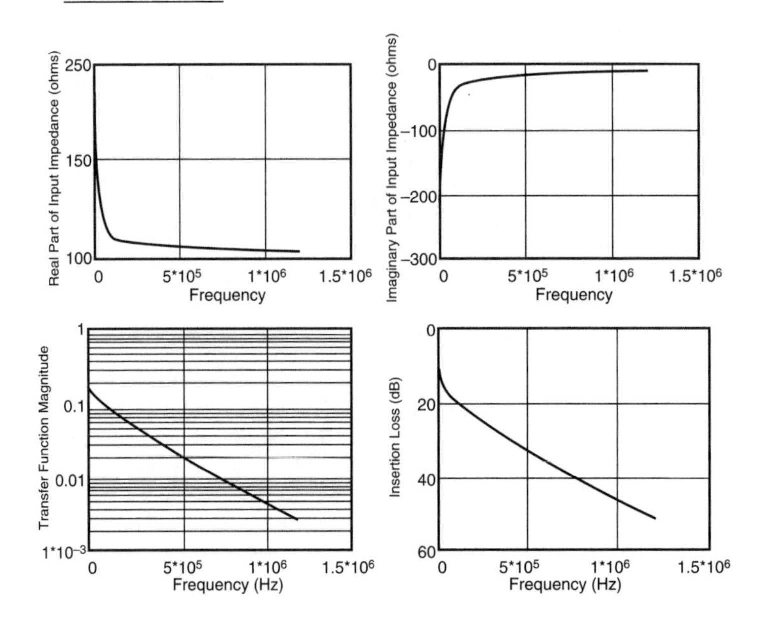


## *VDSL Loops*

Figure 3.18 shows the loops currently designated for VDSL. For these three loops, two new types of cables are shown, called DW10 and FP (DW10 stands for Drop Wire #10). DW10 consists of 0.5 mm untwisted conductors insulated with PVC. FP represents untwisted Flat-Pair wires. Short sections of these wires are often used to connect twisted pairs to homes and buildings. These wires are not included in the ADSL models because their effect is negligible.

Figure 3.18 Test loops specified for VDSL.

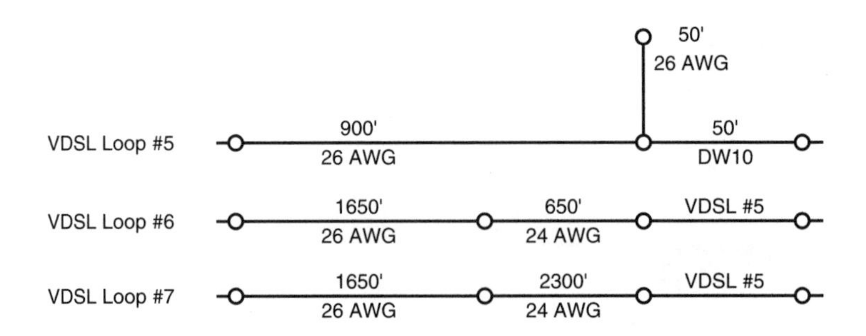

For VDSL, these wires are not negligible because VDSL loops are short, and the frequency range of VDSL signals is higher. Table 3.1 provides the numerical parameter models for DW10 and FP wires.

Table 3.1 Parameters for the Numeric Twisted Pair Model for DW10 and Flat-Pair Cable

| Parameter | DW10 | FP |
|---|---|---|
| $r_{oc}$ (ohms/km) | 180.93 | 41.6 |
| $r_{os}$ (ohms/km) | $\infty$ | $\infty$ |
| $a_c$ (ohms$^4$/km$^4$Hz$^2$) | 0.0497223 | 0.001218 |
| $a_s$ (ohms$^4$/km$^4$Hz$^2$) | 0 | 0 |
| $l_o$ (H/km) | $728.87*10^{-6}$ | $1000*10^{-6}$ |
| $l_\infty$ (H/km) | $543.43*10^{-6}$ | $911*10^{-6}$ |
| $f_m$ (Hz) | 718888 | |
| b | 0.75577086 | 1.195 |
| $g_o$ (Siemen/Hz*km) | 89*10[–9] | 53*10[–9] |
| $g_e$ | 0.856 | 0.88 |
| $c_\infty$ (nF/km) | 63.8 | 22.8 |
| $c_o$ (nF/km) | 51 | 31.78 |
| $c_e$ | –0.11584622 | 0.1109 |

VDSL Loops #5, #6, and #7 have more than one length defined for them as shown in Figure 3.18. Essentially, the different lengths represent a short, medium, and long length loop for VDSL testing. Note that VDSL Loop #5 and VDSL Loop #6 are similar except for the short section at the end, essentially exercising the different types of end drops. VDSL Loop #7 is meant to test operation in the presence of bridged taps. The other test loops are configurations representing common loop topologies.

Figures 3.19–3.22 show graphs for the selected VDSL loops. Again, for input impedance and insertion loss graphs, the source and load impedances are assumed to be 100 ohms. The graphs are shown up to around 15 MHz, which encompasses most of the usable VDSL band. In some cases (Figure 3.20 and Figure 3.22), more than one loop is shown on a graph to draw a comparison between loops. For loops with a selectable length, the longest length is shown. The following list provides a few notes about the VDSL loop graphs:

- VDSL Loop #2 and VDSL Loop #3 have very similar impedance characteristics, but VDSL Loop #3 is a bit more lossy. This suggests that DW10 is more lossy than FP cable.

- The input impedance of VDSL Loop #6 and VDSL Loop #7 is almost identical. VDSL Loop #7 has more loss because the middle section of #24 gauge twisted pair is almost four times as long as that for VDSL Loop #6.
- Properties similar to the ADSL loops are observed. The real and imaginary parts of the impedances converge similarly. Also, the insertion loss graphs generally decrease with frequency, although they do contain some ripple due to the presence of bridged taps.

Figure 3.19 Input impedance, transfer function magnitude, and insertion loss for the VDSL Loop #1.

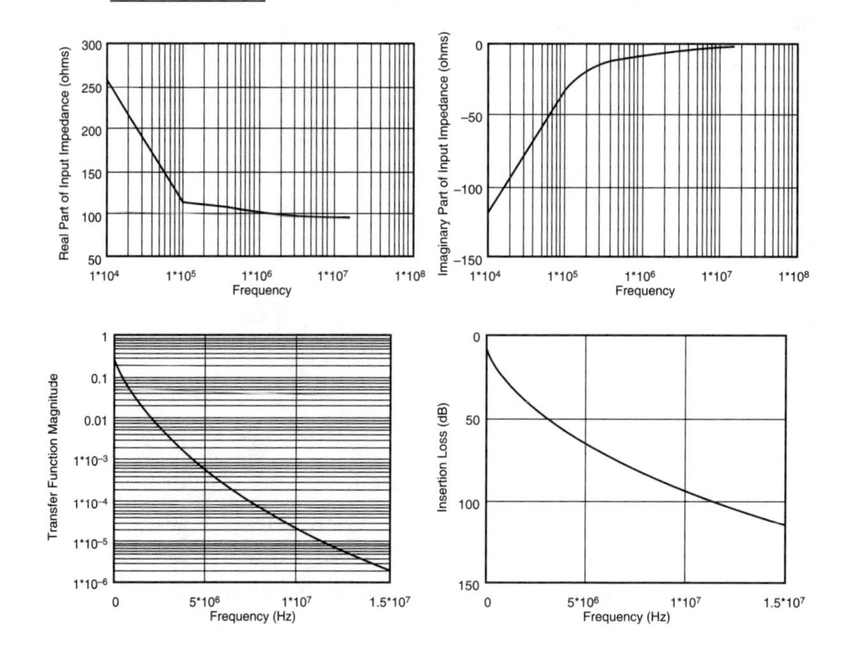

Figure 3.20 Input impedance, transfer function magnitude, and insertion loss for the VDSL Loop #2 and VDSL Loop #3.

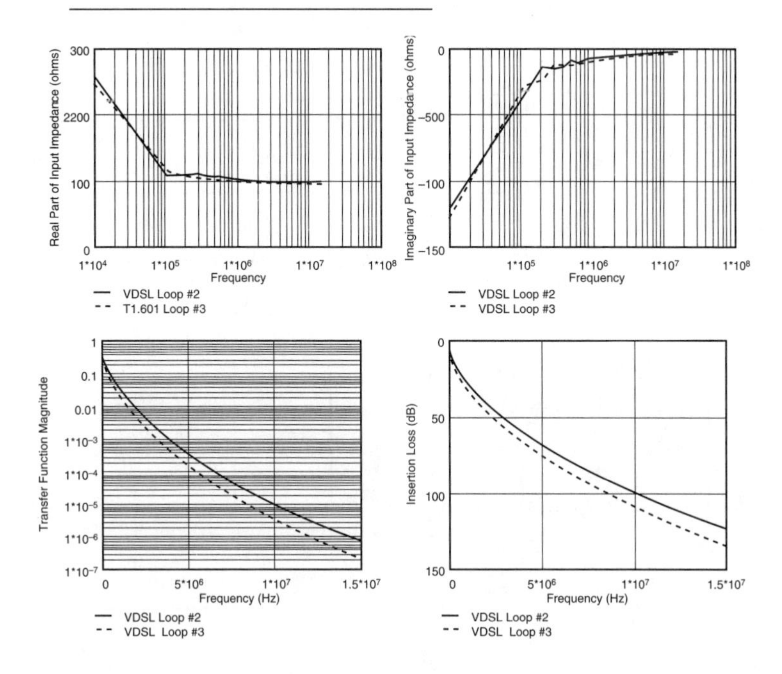


Figure 3.21 Input impedance, transfer function magnitude, and insertion loss for the VDSL Loop #4.

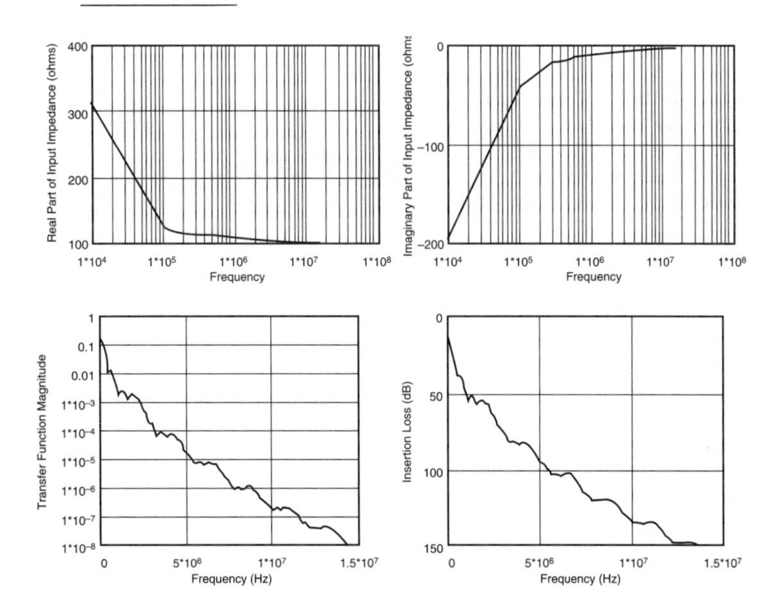

Figure 3.22 Input impedance, transfer function magnitude, and insertion loss for the VDSL Loop #6 and VDSL Loop #7.

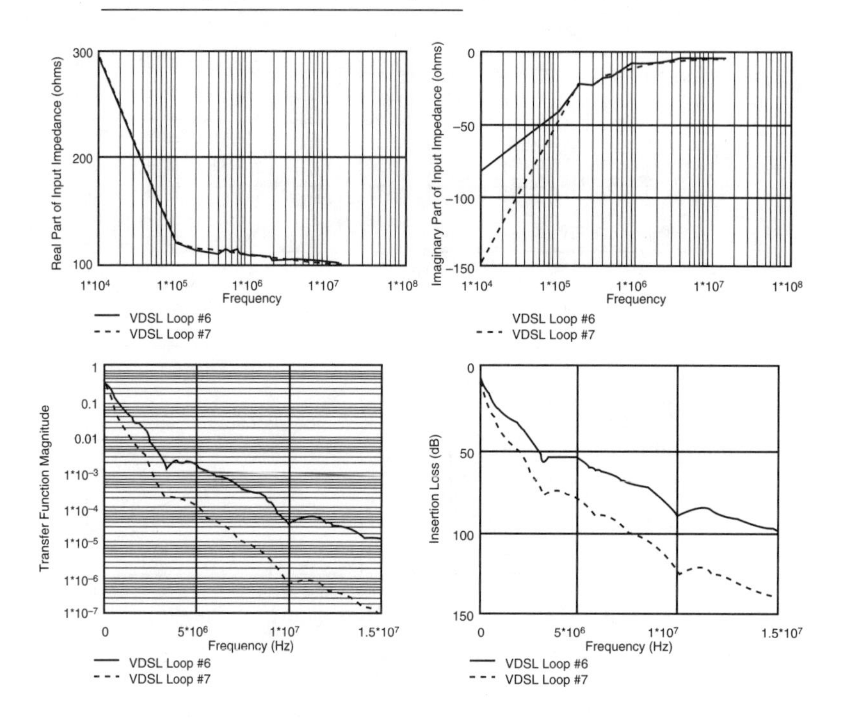


## Summary

This chapter described a method to analyze loops of many different types using ABCD parameters. Detailed derivations of ABCD parameters for transmission lines were presented, and expressions for input impedance, insertion loss, and transfer function were derived. Test loops used to benchmark ADSL systems and VDSL systems were presented, and ABCD parameters used to examine their characteristics were described in detail. These results are used in later chapters to derive the theoretical performance limits of ADSL and VDSL.

## Exercises

1. Prove the relationship given in Eqtn. 3.2 by writing out ABCD equations for each of the networks and solving them for the total network output.
2. Prove the relationship given in Eqtn. 3.15.

3. Simplify Eqtn. 3.14 for the special case where $Z_o=Z_L$.

4. Find the input impedance, transfer function magnitude, and insertion loss of a 9 kft, #26 gauge cable and a 9 kft, #26 gauge cable with a 2 kft, #26 gauge bridged tap located at 8 kft.

5. Find the transfer function phase response of CSA Loop #6. Why does the slope of the phase increase with increasing frequency?

6. Find the input impedance of CSA Loop #7 in the reverse direction (looking in from the subscriber's location). Assume a termination at the other end of 100 ohms. How does the input impedance behave when the bridged tap's length is varied? How does this compare with the input impedance for this loop looking in from the central office?

7. Consider a source generating a signal occupying the band from 0 Hz to 1.0 MHz with a flat power spectral density of ñ40 dBm/Hz. What is the total output power of the source? Compare the total power of the received signal if such a source were the transmitter on each of the ADSL loops. Compare the total power received in only the 0 Hz to 500 kHz range.

8. Consider a source generating a signal occupying the band from 0 Hz to 10 MHz with a flat power spectral density of –60 dBm/Hz. What is the total output power of the source? Compare the total power of the received signal if such a source were the transmitter on each of the VDSL loops. Compare the total power received in only the 0 Hz–500 MHz range.

## *Endnotes*


1. American National Standards Institute, *Asymmetric Digital Subscriber Line Metallic Interface Standard T1.431*. August 1995.

2. John Cioffi, *VDSL System Requirements Document.* T1E1.4 Contribution 96–153, November 1996.

3. J. David Irwin, *Basic Engineering Circuit Analysis.* New York: Macmillan, 1987.

4. Whitham D. Reeve, *Subscriber Loop Signalling and Transmission Handbook.* New York: IEEE Press, 1992.

CHAPTER 4

# *Power Spectral Densities and Crosstalk Models*

In this chapter:

- Explanation of a complete crosstalk model for DSLs
- Method of determining power spectral density (PSD)
- PSD derivations for many DSLs
- NEXT and FEXT models for common DSL disturbers

This chapter discusses crosstalk noises present on digital subscriber lines (DSLs) and applies the mathematical models for near-end crosstalk (NEXT) and far-end crosstalk (FEXT) for a general disturber (see Chapter 2, "Twisted Pair Environment") to specific types of common ADSL and VDSL disturbers. Chapter 5, "DSL Theoretical Capacity in Crosstalk Environments," combines these concepts with the analysis tools presented in Chapter 3, "Loop Analysis," to analyze the theoretical capabilities of any type of twisted pair under any type of environment.

This chapter also discusses power spectral density (PSD). The PSD of a line code defines the distribution of the line code's power in the frequency domain. A *PSD mask* is a template that specifies the maximum PSD allowable for a line code. PSD masks are used as both guidelines for the design and implementation of a DSL technology as well as for crosstalk modeling (as described in the next section) to simulate and benchmark performance.

The crosstalk derivations in this chapter are only statistical models and thus do not imply that every DSL will be corrupted by the predicted amounts of crosstalk. Instead, the models and the different types of crosstalk tests for ADSL and VDSL are meant to be reasonable as they are based on actual empirical data, yet somewhat pessimistic in that they assume near worst-case conditions. Designers hope that a system's positive performance in the presence of the crosstalk specified in the models will ensure adequate performance in the highly variable and diverse environment of the outside loop plant.

# Introduction

Chapter 2 gave the crosstalk loss function for both NEXT and FEXT. Eqtn. 4.1 repeats this function for NEXT.

Eqtn. 4.1

$$H_{Next}(f) = K_{NEXT} f^{3/2}$$

Here $K_{NEXT}$ is a constant that depends on the quality of the twisted pair type. Eqtn. 4.2 repeats the loss equation for FEXT.

Eqtn. 4.2

$$\begin{aligned} H_{FEXT}(f) &= K_{FEXT} f^2 e^{-2\alpha(f)L} L \\ &= K_{FEXT} f^2 \left| H_{channel}(f) \right|^2 L \end{aligned}$$

Here $K_{FEXT}$ is a constant, and L is the length of the line.

For analysis, Eqtn. 4.1 and Eqtn. 4.2 allow crosstalk to be modeled as a signal on a disturbing line, coupling onto the disturbed line through either a NEXT or FEXT transfer function (depending on whether the crosstalk is NEXT or FEXT). Figure 4.1 shows a simple model for NEXT, and Figure 4.2 shows a simple model for FEXT. In each case, the power of the interference depends on both the spectrum of the disturbing signal and the crosstalk loss function.

**Note**
Common disturbers include ISDN, HDSL, T1, ADSL, and VDSL lines that may be nearby.

Figure 4.1 A simple model of a NEXT disturber.

Figure 4.2 A simple model of a FEXT disturber.

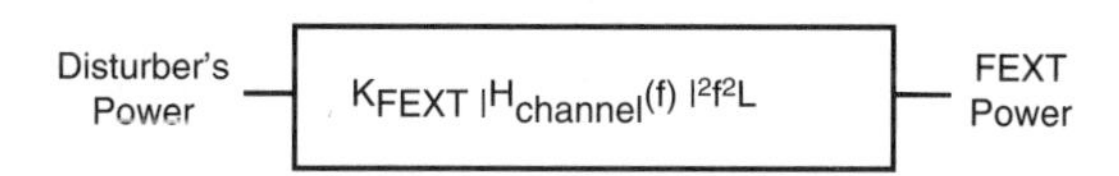


In general, you need be concerned only with the expected crosstalk power and not with the crosstalk voltage. The reason is simply that a statistical model of crosstalk power allows the prediction of signal-to-noise ratio (SNR) on the disturbed twisted pair. Chapter 5 shows that knowledge of the SNR enables engineers to predict the performance limit on the loop.

## Crosstalk Equations

The equations that describe crosstalk consist of three basic terms. The first term describes the PSD of the disturber. The PSD of the disturber is the expected power on the disturbing line and thus includes all transmitter filtering associated with the disturber. PSD is a function of frequency.

The second term of the crosstalk equations is the coupling term. This term is simply the NEXT or FEXT equation (Eqtn. 4.1 and Eqtn. 4.2) and depends on whether the crosstalk being described is NEXT or FEXT.

The final term of the crosstalk equations is derived from and reflects the number of crosstalkers being considered; this term is described in the next section.

Eqtn. 4.3 gives the general form of a crosstalk equation for some disturber.

Eqtn. 4.3

$$PSD_{xtalk} = [\text{Disturber PSD}][\text{NEXT/FEXT Equation}][\text{Total Disturbers Term}]$$

A model to illustrate the block diagram is shown in Figure 4.3. Note that Eqtn. 4.3 is valid for crosstalkers of the same type (for example, a number of ISDN crosstalkers). If more than one type of crosstalker exists, Eqtn. 4.3 would also have to be calculated for the second type, using the second crosstalker's PSD as well as the number of disturbers of that type.

Figure 4.3 A simple model for multiple NEXT or FEXT disturbers.

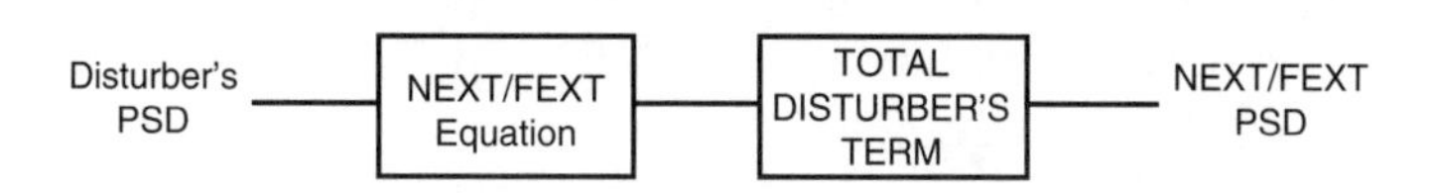

## *Number of Crosstalkers*

At first glance, adjusting crosstalk for different numbers of disturbers by simply multiplying the total crosstalk for one disturber by the total number of disturbers may seem trivial. Such a method is incorrect, however, and will lead to an overly pessimistic crosstalk model in that the predicted crosstalk power will be too high. The reason is that crosstalk between twisted pairs has been statistically modeled. The value used for $K_{NEXT}$ (or $K_{FEXT}$) for a cable type is empirically determined so that for a single crosstalker, sometimes called "pair-to-pair crosstalk," the crosstalk power predicted is worse than 99% of any measured pair-to-pair crosstalk power. Put another way, for pair-to-pair crosstalk, the actual crosstalk is worse than that predicted by the mode only 1% of the time.

As more crosstalkers are included in the model, however, to assume that they are also worse than 99% of all lines is overly pessimistic. They are certainly more likely to be better (have less effect) than the first crosstalker. Thus as the number of disturbers increases, the cumulative effect of each additional disturber should be less than that of the previous disturber. For NEXT, the model for incorporating the number of disturbers is normally combined with the NEXT coupling equation (dependence on $f^{3/2}$) and is given by Eqtn. 4.4

Eqtn. 4.4

$$x_N = [\text{NEXT Equation}][\text{Number of Disturbers Model}]$$
$$= 0.882 \times 10^{-14} \times N^{0.6} \times f^{3/2}$$

In Eqtn. 4.4, N is the total number of disturbers. Thus two of the three terms needed for finding the PSD of crosstalk are known.

## *Types of Disturbers*

Two methods are available for finding the PSDs of common disturbers for ADSL and VDSL. The first method, used on the more "classical" baseband disturbers including ISDN, HDSL, and T1, entails using the autocorrelation function. In this method, the PSD is given by Eqtn. 4.5.

Eqtn. 4.5

$$PSD = \frac{2\Im[R_{xx}(\tau)]}{ZT}$$

Here T is the length of the baseband symbol, Z is the reference impedance across which the signal is measured, $R_{xx}$ is the autocorrelation function of the line code, and the fourier transform function is shown in Eqtn. 4.6.

Eqtn. 4.6

$$\Im[R_{xx}(t)] = \int_{-\infty}^{\infty} R_{xx}(\tau)e^{-j2\pi ft}dt$$

The factor of two in Eqtn. 4.5 reflects the use of the single-sided PSD, as opposed to the double-sided PSD (which is defined over both positive and negative frequencies). Thus the PSDs obtained from Eqtn. 4.5 are valid only on (0, ∞). The proper unit for PSD is Watts/Hz, and the power in any frequency region is found by integrating the PSD between the edges of the region. The autocorrelation of a wide-sense stationary waveform x is given by Eqtn. 4.7.

**Note**

You can also talk about a PSD having units of dBm/Hz. This terminology tends to be convenient in graphically showing PSD as well as when comparing crosstalk PSD with received-signal PSDs.

Eqtn. 4.7

$$R_{xx}(\tau) = E[x(t)x(t-\tau)]$$

Here x(t–τ) is simply x delayed by time τ, and E[*] is the expected value operator.

The second method for defining the PSD of a disturber is to simply specify a PSD mask for the disturber type. A PSD mask is a template defining the maximum value of the technology's PSD at each frequency. Sometimes a mask is used because the PSD of certain technologies is not mathematically convenient. Other times the mask allows for flexibility in the implementation of DSL designs while ensuring that all designs adhere to the same PSD restrictions. ADSL and VDSL PSD specifications for both CAP- and DMT-based line codes use this method. In this case, the PSD is again the single-sided PSD, the unit is still Watts/Hz, and the power in any frequency band is found by the same method described previously.

Note that it is possible to use both methods to bound a PSD. Often the in-band PSD (the energy in the main part of the PSD) is bounded by the first method , and the out-of-band energy (sometimes called the side lobes) is bounded by the second method. In practical DSL systems, it is customary to use analog filtering to comply with out-of-band PSD requirements.

## *ISDN and HDSL*

ISDN and HDSL use the *2B1Q line code* in which two bits are encoded into a pulse that can take on one of four levels. The pulse, also called a *quat,* is then sent over the twisted pair. The method in which two bits are coded into a quat is shown in Table 4.1. An example of a 2B1Q line code for a string of bits appears in Figure 4.4. Each quat is independent of the preceding and succeeding quats, which simplifies the calculation of $R_{xx}$ for 2B1Q.

Table 4.1 Method of Encoding Two Bits into One Quat for 2B1Q Line Codes

| First Bit | Second Bit | Normalized Level |
|---|---|---|
| 0 | 0 | -3 |
| 0 | 1 | -1 |
| 1 | 0 | +3 |
| 1 | 1 | +1 |

Figure 4.4 An example of a 2B1Q line code; two bits make up one quat, and each quat takes one of four discrete levels.

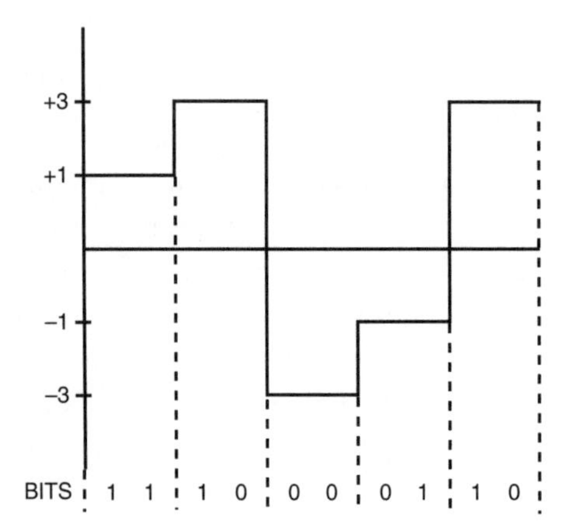


A 2B1Q data stream is best viewed as a sequence of impulses, $v_k$ sent at the system's baud rate and passed through a filter with impulse response p. The impulse response of p forms the pulses of the line code. Such an implementation appears in Figure 4.5.

Figure 4.5 A simple 2B1Q modulator.

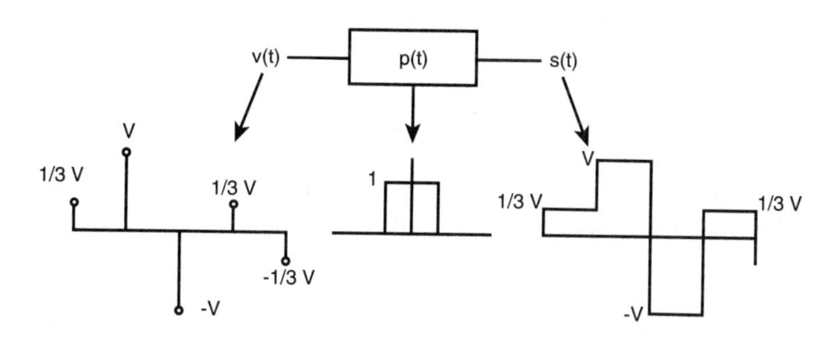


The height of each impulse can be modeled as a random variable that can take on one of four values: +V, –V, +1/3V, and –1/3V.

The second moment of an impulse is given by Eqtn. 4.8.

Eqtn. 4.8

$$E\left[v^2\right] = \frac{1}{4}\left(V^2 + \left(\frac{1}{3}V\right)^2 + (-V)^2 + \left(-\frac{1}{3}V\right)^2\right)$$
$$= \frac{1}{4}\left(\frac{20}{9}V^2\right) = \frac{5}{9}V^2$$

Note that with the assumption of equal probabilities over all four possible values of v, the mean of v is zero and thus the variance of v is the same as the second moment given in Eqtn. 4.8.

A train of impulses is then given by v as in Eqtn. 4.9.

Eqtn. 4.9

$$v(t) = \sum v_k \partial(t - kT)$$

Here T is the period of the line code quats. Eqtn. 4.10 gives the autocorrelation function of such a train of impulses, assuming the random variable $v_k$, as being independent of one another.

Eqtn. 4.10

$$R_{vv}(\tau) = E[v(t)v(t-\tau)]$$
$$= \frac{5}{9}V^2\partial(\tau)$$

Now if the string of impulses v is passed though a filter with impulse response p, the output s is given by the convolution of the two quantities as in Eqtn. 4.11.

Eqtn. 4.11

$$s(t) = \int_{-\infty}^{\infty} v(t-\alpha)p(\alpha)d\alpha$$

The autocorrelation function of s is given by Eqtn. 4.12.

Eqtn. 4.12

$$R_{ss}(\tau) = E[s(t)s(t-\tau)]$$
$$= E\left[\int_{-\infty}^{\infty} v(t-\alpha)p(\alpha)d\alpha \int_{-\infty}^{\infty} v(t+\tau-\alpha)p(\alpha)d\alpha\right]$$
$$= E\left[\int_{-\infty}^{\infty}\int_{-\infty}^{\infty} v(t-\alpha)v(t+\tau-\beta)p(\alpha)p(\beta)d\alpha d\beta\right]$$
$$= \int_{-\infty}^{\infty}\int_{-\infty}^{\infty} E[v(t-\alpha)v(t+\tau-\beta)]p(\alpha)p(\beta)d\alpha d\beta$$
$$= \int_{-\infty}^{\infty}\int_{-\infty}^{\infty} R_{vv}(\tau+\alpha-\beta)p(\alpha)p(\beta)d\alpha d\beta$$

The fourier transform of Eqtn. 4.12 is given by Eqtn. 4.13.

Eqtn. 4.13

$$S(f) = \int_{-\infty}^{\infty} R_{ss}(\tau)e^{-j2\pi f\tau}d\tau$$
$$= \int_{-\infty}^{\infty}\left(\int_{-\infty}^{\infty}\int_{-\infty}^{\infty} R_{vv}(\tau+\alpha-\beta)p(\alpha)p(\beta)d\alpha d\beta\right)e^{-j2\pi f\tau}d\tau$$
$$= \int_{-\infty}^{\infty}\int_{-\infty}^{\infty}\int_{-\infty}^{\infty} R_{vv}(\tau+\alpha-\beta)p(\alpha)p(\beta)e^{-j2\pi f\tau}d\alpha d\beta d\tau$$

If you change the variables, let $t=\tau+\alpha-\beta$, and recognize that $dt=d\tau$, then Eqtn. 4.13 can be simplified to Eqtn. 4.14.

Eqtn. 4.14

$$
\begin{aligned}
S(f) &= \int_{-\infty}^{\infty} R_{ss}(\tau)e^{-j2\pi f\tau}d\tau \\
&= \int_{-\infty}^{\infty}\left(\int_{-\infty}^{\infty}\int_{-\infty}^{\infty} R_{vv}(t)p(\alpha)p(\beta)d\alpha d\beta\right)e^{-j2\pi f(t+\alpha-\beta)}d\tau \\
&= \int_{-\infty}^{\infty} R_{vv}(t)^2\,\partial(t)e^{-j2\pi ft}dt\int_{-\infty}^{\infty} p(\alpha)e^{-j2\pi f\alpha}d\alpha\int_{-\infty}^{\infty} p(\beta)e^{j2\pi f\beta}d\beta \\
&= \int_{-\infty}^{\infty}\frac{5}{9}V^2\partial(t)e^{-j2\pi ft}dt\int_{-\infty}^{\infty} p(\alpha)e^{-j2\pi f\alpha}d\alpha\int_{-\infty}^{\infty} p(\beta)e^{j2\pi f\beta}d\beta \\
&= \frac{5}{9}V^2\int_{-\infty}^{\infty} p(\alpha)e^{-j2\pi f\alpha}d\alpha\int_{-\infty}^{\infty} p(\beta)e^{j2\pi f\beta}d\beta
\end{aligned}
$$

Note that last line in Eqtn. 4.14 resulted from the sifting property. The first integration in the last line is the fourier transform of the impulse response of the pulse filter. The second is the complex conjugate of the pulse filter impulse response fourier transform (the exponent is positive instead of negative). Thus Eqtn. 4.14 can be rewritten as in Eqtn. 4.15.

Eqtn. 4.15

$$
\begin{aligned}
S(f) &= V(f)P(f)P^*(f) \\
&= \frac{5}{9}V^2|P(f)|^2
\end{aligned}
$$

Assuming a square pulse impulse response for p, Eqtn. 4.15 can be written as shown in Eqtn. 4.16.

Eqtn. 4.16

$$
S(f) = \frac{5}{9}V^2\left|\frac{\sin(\pi fT)}{\pi f}\right|^2
$$

Using Eqtn. 4.5, the PSD of the 2B1Q signal is given as Eqtn. 4.17.

Eqtn. 4.17

$$S_{PSD}(f) = \frac{2S(f)}{ZT} = \frac{10}{9}\frac{V^2}{ZT}\frac{\sin^2(\pi f T)}{\pi^2 f^2}$$

For an ISDN system, Z is assumed to be 135 ohms, and T, the pulse period, is $1/80*10^3$ s. For an HDSL system, Z is 100 ohms, and T is $1/392*10^3$ s. The resulting PSDs appear in Figure 4.6.

Figure 4.6 ISDN and HDSL PSDs.

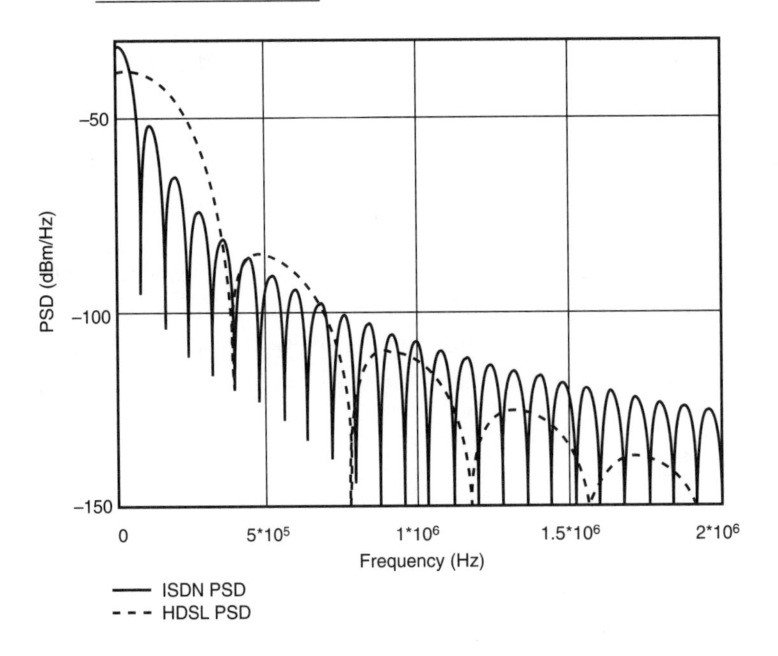


Along the way to deriving Eqtn. 4.17, an important result has been illustrated. If a signal x with autocorrelation function $R_{xx}$ is passed through a linear, time-invariant filter with impulse response h, the fourier transform of the autocorrelation at the output of the filter will provide the results given by Eqtn. 4.18.

Eqtn. 4.18

$$\Im[R_{yy}(\tau)] = \Im[R_{xx}(\tau)]|H(f)|^2$$

Figure 4.7 illustrates this result, which will be helpful later in the chapter.

**Figure 4.7** The effect of passing a signal through a linear, time-invariant filter on the signal's PSD; the input and output PSD expressions are shown.

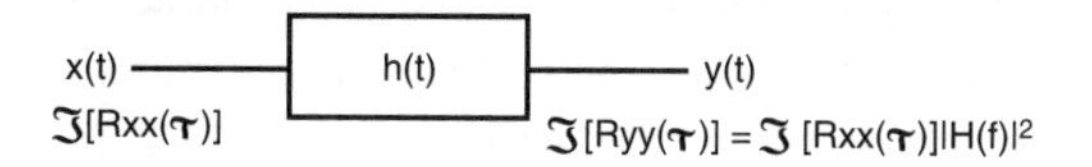


## *T1 Lines*

T1 lines use a binary return to zero (RZ) *alternate mark inversion* (AMI) line code. In this scheme, a binary 1 is encoded as a pulse with either a positive or negative level, and a binary 0 is encoded as a 0 V symbol (in other words, no pulse at all). An example of an AMI line code is shown in Figure 4.8. Note that the polarity of each pulse representing a binary 1 is opposite of the previous binary 1. This condition gives rise to the name "alternate mark inversion."

**Figure 4.8** An example of an AMI line codes such as that used by a T1 line.

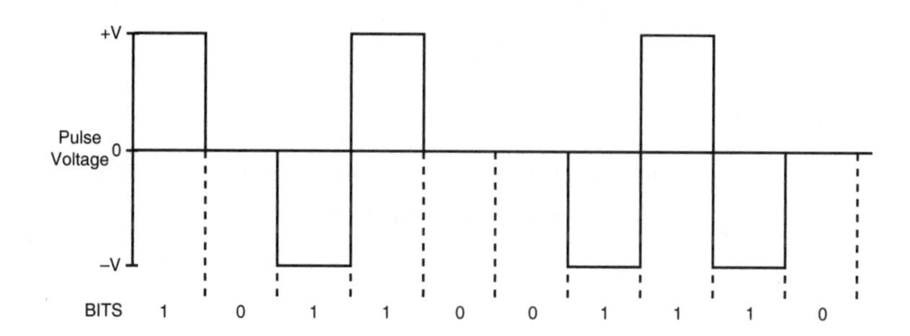


Similar to the 2B1Q line code, the T1 line code can be viewed as a sequence of impulses fed through a pulse filter with impulse response p. This is shown in Figure 4.9.

**Figure 4.9** A simple T1 line modulator using a bipolar AMI line code.

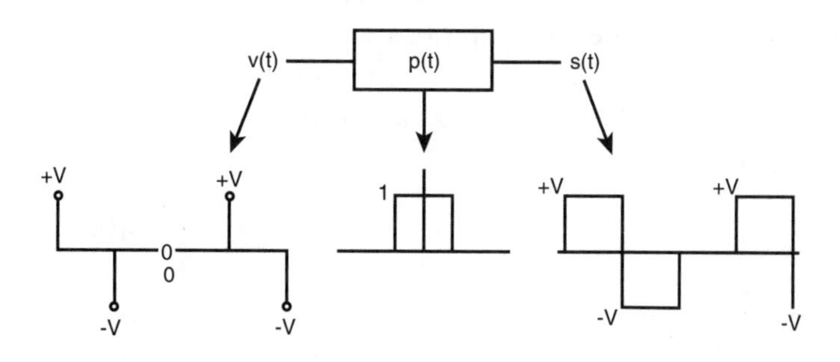

In this case, the impulses can have a height of +V, –V, or 0. The first two impulse heights represent binary 1s, and the third represents a binary 0. (It is assumed that the network element producing the impulses encodes the proper polarity pulse for a binary 1.) The final two impulses represent binary 1s. In contrast to ISDN and HDSL, T1 lines typically do not employ scrambling at the transmitter and descrambling at the receiver. As a result, the relative density of ones and zeros being encoded is not necessarily uniform. The interference models for both ADSL and VDSL assume that the density is uniform; this discussion proceeds on that assumption. If $v_k$ is the sequence of impulse, the probability density function (pdf) of each impulse is given by Eqtn. 4.19.

Eqtn. 4.19

$$f_v(v) = \begin{cases} \frac{1}{4} & v = -V \\ \frac{1}{2} & v = 0 \\ \frac{1}{4} & v = +V \\ 0 & \text{otherwise} \end{cases}$$

A continuous time representation of the impulse train $v_k$ can be represented by Eqtn. 4.20.

Eqtn. 4.20

$$v(t) = \sum v_k \partial(t - kT)$$

Here T is the bit period of the line code. The waveform v is then passed through the pulse filter to form the T1 waveform. Using the convolution method similar to what was done when analyzing 2B1Q, the actual line code waveform is given by Eqtn. 4.21.

Eqtn. 4.21

$$s(t) = \int_{-\infty}^{\infty} v(t - \alpha) p(\alpha) d\alpha$$

Eqtn. 4.21 is identical to Eqtn. 4.11 though the process v is certainly different. The line code PSD is then given by the fourier transform of the autocorrelation function of s. The analysis is identical to that done in the previous section for 2B1Q, and the two main quantities necessary for the result include the fourier transform of the autocorrelation function of v and the fourier transform of the impulse response p. The resulting power spectrum of s is then given by Eqtn. 4.22.

Eqtn. 4.22

$$
\begin{aligned}
S(f) &= \Im(R_{vv}(\tau))P(f)P^*(f) \\
&= V(f)P(f)P^*(f) \\
&= V(f)|P(f)|^2
\end{aligned}
$$

Finding V first requires finding the autocorrelation of v, which is given by Eqtn. 4.23.

Eqtn. 4.23

$$
R_{vv}(\tau) = \begin{cases} \dfrac{-V^2}{4} & \tau = \text{-T} \\ \dfrac{V^2}{2} & \tau = 0 \\ \dfrac{-V^2}{4} & \tau = \text{T} \\ 0 & \text{otherwise} \end{cases}
$$

The value at τ=0 is simply the second moment of discrete random variable v, and the values at τ=–T and τ=T result in shifting the waveform by one symbol. The nonzero values of the autocorrelation function at shifts of one symbol are a result of the memory in the encoding of AMI. (The derivation of Eqtn. 4.23 is left as an exercise for the end of this chapter.) The fourier transform of $R_{vv}$ is given by Eqtn. 4.24.

Eqtn. 4.24

$$
\begin{aligned}
V(f) &= \int_{-\infty}^{\infty} R_{vv}(\tau)e^{-j2\pi f\tau}d\tau \\
&= \int_{-\infty}^{\infty}\left(\frac{-V^2}{4}\partial(\tau+T) + \frac{V^2}{2}\partial(\tau) + \frac{-V^2}{4}\partial(\tau-T)\right)e^{-j2\pi f\tau}d\tau \\
&= \frac{-V^2}{4}e^{j2\pi fT} + \frac{V^2}{2} + \frac{-V^2}{4}e^{-j2\pi fT} \\
&= \frac{V^2}{2}(1-\cos(2\pi fT))
\end{aligned}
$$

The last line in Eqtn. 4.24 results simply from Euler's formula and some factorization.

Now that the fourier transform of $R_{vv}$ is known, the fourier transform of p is necessary. Because p is again a pulse of width T (as was the case for 2B1Q), its spectrum is given by Eqtn. 4.25.

Eqtn. 4.25

$$P(f) = \frac{\sin(\pi f T)}{\pi f}$$

The spectrum of s is given by Eqtn. 4.26.

Eqtn. 4.26

$$\begin{aligned} S(f) &= V(f)P(f)P^*(f) \\ &= V(f)|P(f)|^2 \\ &= \frac{V^2}{2}(1-\cos(2\pi f T))\left|\frac{\sin(\pi f T)}{\pi f}\right|^2 \end{aligned}$$

The PSD of s is then given by Eqtn. 4.27.

Eqtn. 4.27

$$\begin{aligned} S_{PSD}(f) &= \frac{2S(f)}{ZT} \\ &= \frac{2V^2}{2ZT}(1-\cos(2\pi f T))\left(\frac{\sin(\pi f T)}{\pi f}\right)^2 \\ &= \frac{V^2}{ZT}\frac{\sin^2(\pi f T)}{\pi^2 f^2}(1-\cos(2\pi f T)) \end{aligned}$$

For T1 lines, the termination reference impedance Z is 100 ohms, and the bit period is $1/1.544*10^6$ s. The voltage magnitude V of an AMI pulse is typically 3.6V for a T1 line. ADSL and VDSL interference models make two extra assumptions about a T1 disturber. First, they assume that a T1 transmitter has a low-pass shaping filter to limit higher-frequency components on the line. The shaping filter is modeled as a third-order Butterworth filter with a 3.0 MHz, 3 dB point. The magnitude of the transfer function of this filter is given by Eqtn. 4.28.

Eqtn. 4.28

$$\left|H_{shaping}(f)\right|^2 = \frac{1}{1+\left(\frac{f}{3\times10^6}\right)^6}$$

Second, the T1 transmitter is coupled through a transformer that can be modeled as a high-pass filter with a 40 kHz, 3 dB point. Eqtn. 29 gives the magnitude of the transfer function of the transformer model.

Eqtn. 4.29

$$\left|H_{transformer}(f)\right|^2 = \frac{f^2}{f^2+\left(40\times10^3\right)^2}$$

By passing the T1 signal through the shaping filter and transformer, the resultant PSD becomes (using Eqtn. 4.28 and Eqtn. 4.29) Eqtn. 4.30.

Eqtn. 4.30

$$\begin{aligned} PSD_{T1}(f) &= S_{PSD}(f)\left|H_{shaping}(f)\right|^2\left|H_{transformer}(f)\right|^2 \\ &= \frac{V^2}{ZT}\frac{\sin^2(\pi fT)}{\pi^2 f^2}\left(1-\cos(2\pi fT)\right)\left(\frac{1}{1+\left(\frac{f}{3\times10^6}\right)^6}\right)\left(\frac{f^2}{f^2+\left(40\times10^3\right)^2}\right) \end{aligned}$$

Here T and Z are the same as those defined in Eqtn. 4.27. The PSD of a T1 signal with a binary 1's density of 50% is shown in Figure 4.10.

Figure 4.10 PSD of an AMI T1 line with a binary 1's density of 50%.

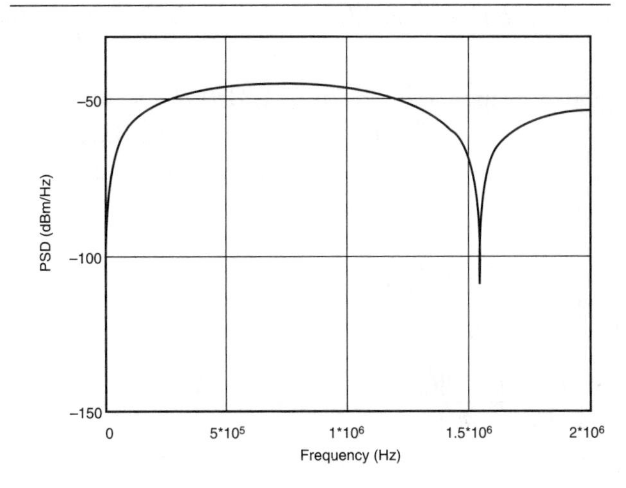

## *DMT ADSL (Downstream)*

For ADSL separate models are necessary for the upstream and downstream directions as a result of different data rates and modulation parameters for each direction. Several methods can be used to find the downstream PSD of a DMT-based ADSL system. One reasonable method is to simply find the theoretical PSD for DMT modulation. A second method is to make an assumption about how the transmitter is implemented and create a mathematical model similar to those for 2B1Q and T1. Both of these methods have some shortcomings. The former is not always applicable to practical implementations. The latter assumes a type of implementation that can vary depending on the actual design. Neither method adequately addresses out-of-band energy. The value, however, of these models is that they form a good basis on which to specify a generic PSD mask. An implementation can then be engineered to keep the actual output power spectrum within the PSD mask. This section looks at and compares these two methods and then compares the standard DMT PSD mask for the downstream direction to the models.

Later chapters show that an ADSL DMT symbol is merely a number of sign and cosine waves at different frequencies being sent at a baud rate of 4.3125 kHz. The lowest frequency usable is 34.5 kHz, and the highest is 1.104 MHz. All frequency multiples of 4.3125 kHz between these limits are also used. Mathematically, it is correct to say that a sine and cosine wave are sent having a frequency of nx4.3125 kHz with n=8,9...256. A theoretical way to create a DMT symbol is to create the cosine and sine waves at all the necessary frequencies and then sum them. Eqtn. 4.31 gives the summation for $f_o$=4.3125 kHz.

Eqtn. 4.31

$$s(t) = \sum_{n=8}^{256} \left( a_n \cos(2\pi n f_o t) + b_n \sin(2\pi n f_o t) \right) \quad \text{for } 0 \le t < \frac{1}{f_o} \qquad (4.31)$$

For a single cosine and sine summation in Eqtn. 4.31 (say, for n=m), the expression for the power spectrum is derived later in the chapter in detail. Eqtn. 4.32 gives a generalized version of this PSD, representing the PSD due to term m of Eqtn. 4.31.

Eqtn. 4.32

$$\begin{aligned} PSD_m &= \frac{4V^2 \sin^2(\pi T(f - mf_o))}{\pi^2 ZT(f - mf_o)^2} \\ &= K_m \frac{\sin^2(\pi T(f - mf_o))}{\pi^2 (f - mf_o)^2} \end{aligned}$$

If the sine and cosine waves are independent for a DMT symbol, the total PSD is equal to the sum of the PSDs due to each term in Eqtn. 4.31. Thus the theoretical PSD for a downstream DMT symbol is given in Eqtn. 4.33.

Eqtn. 4.33

$$PSD_{DS} = \sum_{i=8}^{256} K_i \frac{\sin^2(\pi T(f - if_o))}{\pi^2 (f - if_o)^2}$$
$$= K_{DS} \sum_{i=8}^{256} \frac{\sin^2(\pi T(f - if_o))}{\pi^2 (f - if_o)^2}$$

In Eqtn. 4.33, $K_{DS}$ represents the expected value of all scalar terms in a typical PSD equation (termination impedance, expected voltage level, and so on). For a typical DSL system, this value is approximately $10^{-7}$.

A design that produces a PSD in accordance with the theoretical PSD given in Eqtn. 4.33 may not be in the best interest of DSL technology. In some areas, the mask may be too difficult to achieve (for example, near the edges of the pass band). In the out-of-band regions, the mask may not be stringent enough. These regions could be attenuated more with reasonable filters, reducing the amount of crosstalk coupled into other DSLs. Under the circumstances, basing the PSD model of a downstream DMT ADSL symbol solely on this model is not very wise. As a second point of reference, a basic implementation can be assumed for the front end of a DMT modulator. A DMT symbol can be formed by passing an impulse train $v_k$ though a pulse filter p, just as in the 2B1Q and T1 case. The resulting waveform is then high-pass and low-pass filtered and sent out of the transmitter. A block diagram of this process appears in Figure 4.11. The impulses are assumed to be independent and Gaussian distributed with a mean of zero and a variance of $V^2$. Common notation for each $v_k$ would be $N(0,V^2)$.

**Note**

The interpreted impulse heights for this model are not the same as those for T1 and 2B1Q line codes. That is, the height of a single impulse does not correspond to a single bit or set of bits, and further processing is necessary to decode bits from sequences of samples. This topic is discussed again in Chapter 6, "DSL Modulation Basics."

Figure 4.11 A common method of producing an ADSL signal.

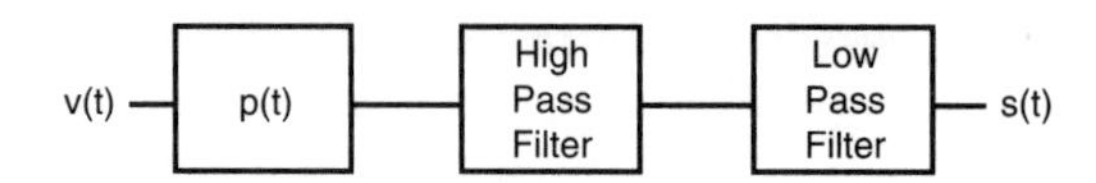


As in the previous cases, the continuous time function v, describing the train of impulses, can be written as in Eqtn. 4.34.

Eqtn. 4.34

$$v(t) = \sum_{\infty} v_k \partial(t - kT)$$

Assuming an independent $v_k$, Eqtn. 4.35 gives the autocorrelation function, $R_{vv}$, of v.

Eqtn. 4.35

$$R_{vv}(\tau) = V^2 \partial(\tau)$$

As previously shown, the PSD at the output of the transmitter is simply the fourier transform of Eqtn. 4.35 multiplied by the squared magnitude spectrums of each of the filters between the v and the output. Given that the filters are the pulse filter, the high-pass filter (HPF), and the low-pass filter (LPF), the PSD of a downstream ADSL signal is given by Eqtn. 4.36.

Eqtn. 4.36

$$PSD_{ADSL}(f) = \frac{2\Im[R_{vv}(\tau)]P(f)|^2 |HPF(f)|^2 |LPF(f)|^2}{ZT}$$

A reasonable LPF used to filter the ADSL downstream PSD is a fourth-order Butterworth filter with a cutoff frequency of 1.104 MHz. The square of its magnitude spectrum is given by Eqtn. 4.37.

Eqtn. 4.37

$$|LPF(f)|^2 = \frac{1}{1 + \left(\frac{f}{1.104 \times 10^6}\right)^8}$$

A reasonable HPF used to filter the ADSL downstream PSD is a fourth-order filter with a 3 dB point at 20 kHz. This filter would serve the purpose of limiting the low end of the ADSL spectrum to separate ADSL from normal Plain Old Telephone Service (POTS) traffic. The squared magnitude of its spectrum is provided by Eqtn. 4.38.

**Note**

Do not interpret the specific transfer functions for the HPF and LPF necessities to use for any ADSL implementation. These functions are chosen for PSD-modeling purposes only.

Eqtn. 4.38

$$|HPF(f)|^2 = \frac{f^8}{f^8 + \left(20 \times 10^3\right)^8}$$

All together, the PSD of a downstream ADSL signal then becomes Eqtn. 4.39.

Eqtn. 4.39

$$PSD_{ADSL}(f) = \frac{2\Im[R_w(\tau)]|P(f)|^2 |HPF(f)|^2 |LPF(f)|^2}{ZT}$$
$$= \frac{2V^2}{ZT}\left(\frac{\sin(\pi f T)}{\pi f}\right)^2 \left(\frac{1}{1+\left(\frac{f}{1.104\times10^6}\right)^8}\right)\left(\frac{f^8}{f^8+\left(20\times10^3\right)^8}\right)$$

For ADSL, Z—the reference impedance—is assumed to be 100 ohms, and T is given by 1/(2.208 MHz). The theoretical PSD of a downstream DMT ADSL line code as well as the PSD calculated in this model appear in Figure 4.12.

Figure 4.12 Theoretical and implementable ADSL downstream PSDs with a linear frequency scale and a logarithmic frequency scale.

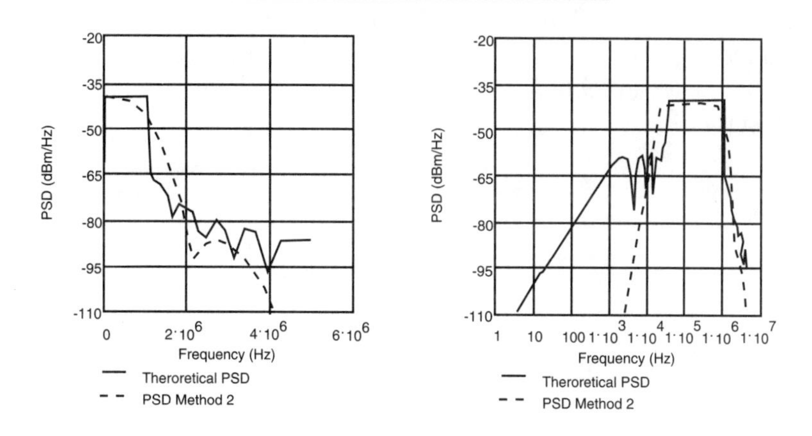


Note that in some areas the theoretical PSD is the higher (primarily outside of the passband) of the two; at other times it is lower (primarily near the edge of the passband). A tighter specification in the out-of-band regions is desired to reduce crosstalk into other DSLs.

These two masks provide insight to the theoretical downstream PSD and to an achievable PSD due to a feasible implementation. These two masks lead to the PSD mask specified by ANSI for an ADSL downstream implementation. This mask is shown in Figure 4.13.

Figure 4.13 PSD mask of a downstream DMT ADSL signal in an echo-canceled system with a linear frequency scale and a logarithmic frequency scale.

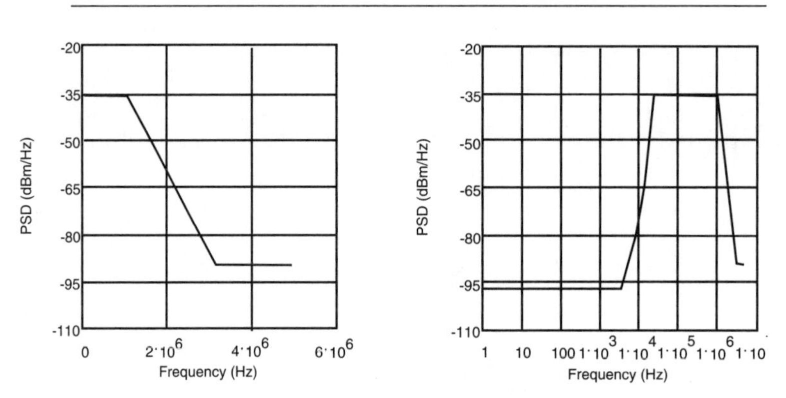


Basically, the mask is a piecewise continuous function. Conforming to this mask should not cause unreasonably complex designs for ADSL implementation, but should minimize interference in the out-of-band region.

The PSD mask in Figure 4.13 is for downstream ADSL employing echo cancellation. A second downstream mask, called the "reduced NEXT mask" is defined for systems not using echo cancellation. This mask reduces the power allowed in the lower passband region. The mask is illustrated in Figure 4.14. An ADSL system operating in frequency division multiplexing (FDM) mode would use this PSD mask. See Chapter 7, "ADSL Modulation Specifics," for additional information of the two types of downstream ADSL modes for DMT.

Figure 4.14 PSD mask of a downstream DMT ADSL signal in an FDM system with a linear frequency scale and a logarithmic frequency scale.

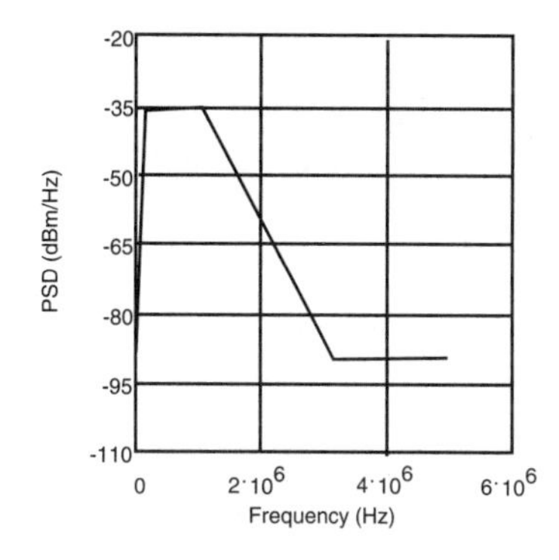


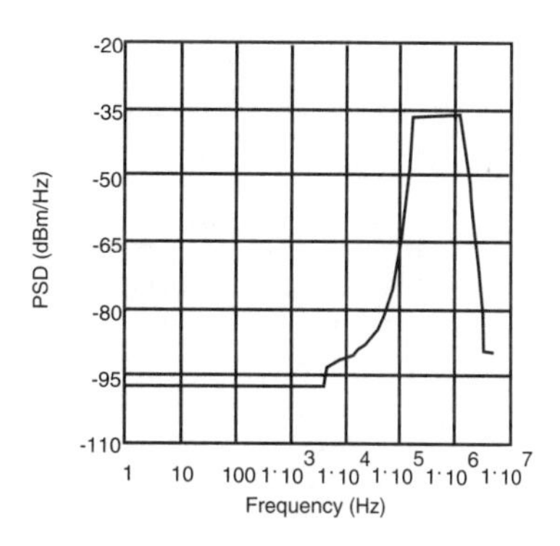


## *DMT ADSL (Upstream)*

Normally, the ADSL upstream signal occupies a frequency range from 25 kHz to 138 kHz. For an echo-canceled system, the upstream PSD overlaps the downstream PSD in this region. If you use the same type of analysis to find the theoretical PSD as you used for downstream ADSL, the expression results in Eqtn. 4.40.

**Note**

In some FDM systems, the upper frequency may be less than 138 kHz if those frequencies are to be used by the downstream ADSL signal.

**Note**

If POTS does not reside on the twisted pair, some engineers have suggested using some of the bandwidth between DC and 25 kHz for upstream ADSL (and downstream ADSL in the case of an echo-canceled system), which would then extend the frequency range down.

Eqtn. 4.40

$$PSD_{US} = \sum_{i=8}^{32} K_i \frac{\sin^2(\pi T(f - if_o))}{\pi^2 (f - if_o)^2}$$

$$= K_{US} \sum_{i=8}^{32} \frac{\sin^2(\pi T(f - if_o))}{\pi^2 (f - if_o)^2}$$

The expression in Eqtn. 4.40 has the same basic components as the expression in Eqtn. 4.32, and $K_{US}$ is given by approximately $10^{-7}$.

Because of the deficiencies associated with using the theoretical PSD expression as the PSD mask specification, the PSD from an experimental implementation is again helpful. An upstream DMT symbol can be created by passing an impulse train through a pulse filter at the rate of 276 kHz. The output of the pulse filter is then sent through a shaping filter. Figure 4.15 shows such a system.

Figure 4.15 A method of producing an ADSL upstream signal.

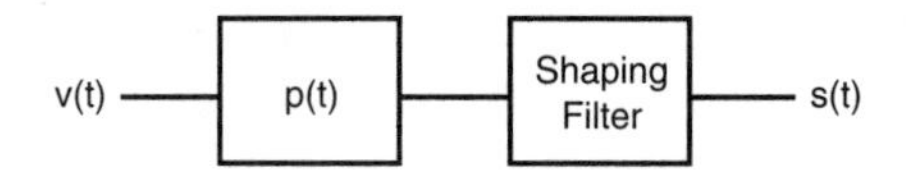


If the impulses are Gaussian with statistics given by $N(0,V^2)$, then Eqtn. 4.34 and Eqtn. 4.35 apply as the train of impulses and autocorrelation function, respectively. Following the treatment of a signal passed through a filter used in the previous sections, the output PSD of Figure 4.15 can be written as Eqtn. 4.41.

Eqtn. 4.41

$$PSD_{ADSL,US}(f) = \frac{2\Im[R_{vv}(\tau)]|P(f)|^2|H_{US}(f)|^2}{ZT} = \frac{2V^2}{ZT}\left(\frac{\sin(\pi f T)}{\pi f}\right)^2 |H_{US}(f)|^2$$

In Eqtn. 4.41, Z is given by 100 ohms, and T is $1/(276*10^3)$ s. For a downstream disturber, the shaping filter was explicitly defined so that calculation of the downstream PSD was straightforward. In the original ADSL upstream PSD specification, a generic mask was used as the upstream shaping filter magnitude. Mathematically, this mask is given in Eqtn. 4.42.

Eqtn. 4.42

$$|H_{US}(f)|^2 = \begin{cases} 10^{\frac{-38-30}{10}} & f < 138\,\text{kHz} \\ 10^{\frac{-38-24\left(\frac{f-13800}{43125}\right)-30}{10}} & f > 138\,\text{kHz} \end{cases}$$

The resulting PSD due to the implementation along with the mask defined in Eqtn. 4.42 is shown in Figure 4.16. Overlaid on this figure is the theoretical expression from Eqtn. 4.40.

Figure 4.16 Theoretical and implementation ADSL upstream PSDs with a linear frequency scale and a log frequency scale.

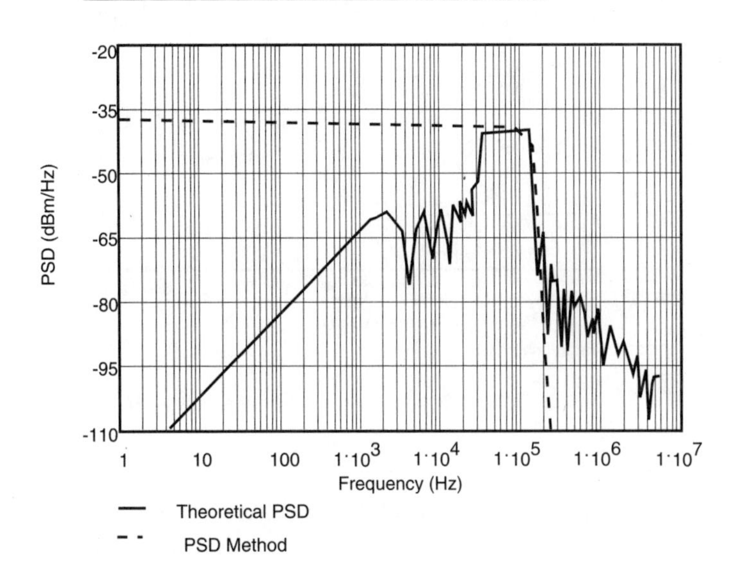


To more fully specify the upstream ADSL PSD, a piecewise continuous mask was created. This mask is based on the masks shown in Figure 4.16 and adds more complete out-of-band specifications. The piecewise continuous mask is shown in Figure 4.17.

One caution about the ADSL upstream and downstream masks is that they continue to evolve. As new services are studied, requests made for the out-of-band PSD regions may be tightened. In some cases, because of implementation problems, loosening of the requirements is proposed. These activities may continue for many years to come.

Figure 4.17 Piecewise continuous PSD for an upstream DMT ADSL signal.

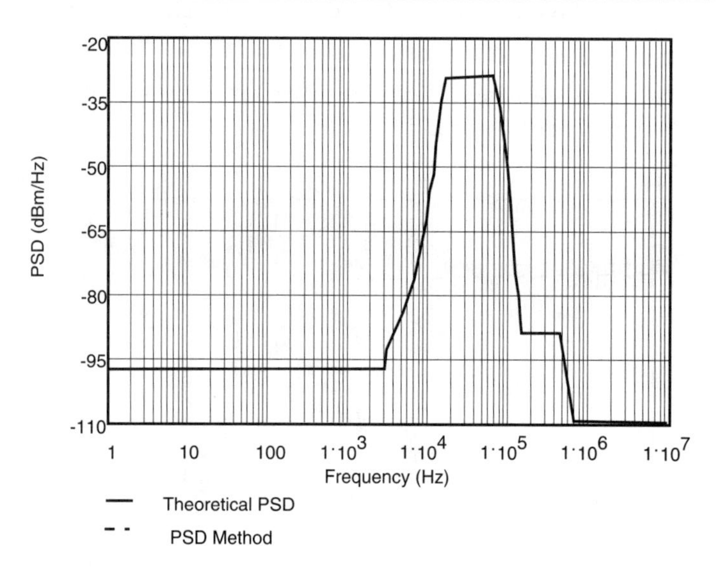


## *CAP/QAM ADSL*

Carrierless Amplitude Modulation/Phase Modulation (CAP) and Quadrature Amplitude Modulation (QAM) technologies have very similar (and in some cases identical) time-domain and frequency-domain representations. For that reason, CAP and QAM are treated together from a PSD and crosstalk point of view. All further discussion in this section is applicable to both CAP and QAM. Both an unshaped (unfiltered) CAP signal and a shaped (filtered) CAP signal are discussed. Normal implementations use a shaped signal; however, discussing unshaped CAP first is valuable for in-depth understanding.

An unshaped CAP signal can be produced by combining the outputs of two filters, g and h, each excited by an input impulse at the baud rate of the CAP system. See Figure 4.18.

The input impulse trains $a_k$ and $b_k$ have independent and identically distributed heights. The impulse height probability density function has zero mean and variance $V^2$ for both $a_k$ and $b_k$. If the continuous time waveforms for the two pulse trains are given by a and b, then the autocorrelation functions are as shown in Eqtn. 4.43 and Eqtn. 4.44.

Figure 4.18 Block diagram for a CAP/QAM modulator.

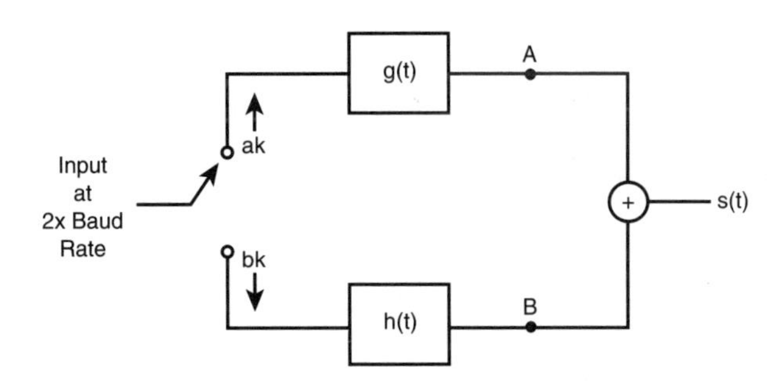


Eqtn. 4.43

$$R_{aa}(\tau) = V^2\delta(\tau)$$

Eqtn. 4.44

$$R_{bb}(\tau) = V^2\delta(\tau)$$

Because all $a_k$ and $b_k$ are independent, the cross-correlation function between a and b is given by Eqtn. 4.45.

Eqtn. 4.45

$$R_{ab}(\tau) = 0$$

For an unfiltered system, the impulse responses of the two filters consist of either a cosine wave or a sine wave for a length of time T. Eqtn. 4.46 and Eqtn. 4.47 show these impulse responses.

Eqtn. 4.46

$$g(t) = \begin{cases} \cos(2\pi f_c t) & |t| \le \dfrac{T}{2} \\ 0 & |t| > \dfrac{T}{2} \end{cases}$$

Eqtn. 4.47

$$h(t) = \begin{cases} \sin(2\pi f_c t) & |t| \le \dfrac{T}{2} \\ 0 & |t| > \dfrac{T}{2} \end{cases}$$

In Eqtn. 4.47, T is the period of a CAP symbol. The fourier transform is used to find the frequency-domain response of a time-domain impulse response. The fourier transform of g is given by Eqtn. 4.48.

Eqtn. 4.48

$$
\begin{aligned}
G(f) &= \int_{-\frac{T}{2}}^{\frac{T}{2}} \cos(2\pi f_c t) e^{-j2\pi f t} dt \\
&= \int_{-\frac{T}{2}}^{\frac{T}{2}} \left( \frac{e^{j2\pi f_c t} + e^{-j2\pi f_c t}}{2} \right) e^{-j2\pi f t} dt \\
&= \frac{1}{2} \int_{-\frac{T}{2}}^{\frac{T}{2}} e^{j2\pi t(f_c - f)} + e^{-j2\pi t(f_c + f)} dt \\
&= \frac{1}{2} \left( \frac{e^{j2\pi t(f_c - f)}}{j2\pi(f_c - f)} + \frac{e^{-j2\pi t(f_c + f)}}{-j2\pi(f_c + f)} \right)_{-\frac{T}{2}}^{\frac{T}{2}} \\
&= \frac{1}{2} \left( \frac{e^{j\pi T(f_c - f)} - e^{-j\pi T(f_c - f)}}{2j\pi(f_c - f)} + \frac{e^{-j\pi T(f_c + f)} - e^{j\pi T(f_c + f)}}{-2j\pi(f_c + f)} \right)
\end{aligned}
$$

Here Euler's formula was used to get to the second line. Note that in the last line of Eqtn. 4.48, each term inside the parenthesis has an embedded sine term. Thus the result of Eqtn. 4.48 can be simplified as in Eqtn. 4.49.

Eqtn. 4.49

$$
\begin{aligned}
G(f) &= \frac{1}{2} \left( \frac{e^{j\pi T(f_c - f)} - e^{-j\pi T(f_c - f)}}{2j\pi(f_c - f)} + \frac{e^{-j\pi T(f_c + f)} - e^{j\pi T(f_c + f)}}{-2j\pi(f_c + f)} \right) \\
&= \frac{1}{2} \left( \frac{\sin(\pi T(f - f_c))}{\pi(f - f_c)} + \frac{\sin(\pi T(f + f_c))}{\pi(f + f_c)} \right)
\end{aligned}
$$

Euler's formula was again used to get the second line in Eqtn. 4.49. Note that the terms in the subtraction and sine of the first term were interchanged. The resulting sign change in both the numerator and denominator cancel out each other. Note also that the result is the addition of two sinc functions, one centered at $f_c$ and one centered at $-f_c$. Because G will be used to help find the PSD of a CAP signal, the single-sided magnitude squared of the transfer function G is of concern. Assuming that the frequency $f_c$, sometimes called the *center frequency*, is much larger than 1/T, then the tails of each sinc function are negligible near the peak of the other sinc function. With this assumption, the single-sided response squared magnitude response of G is given by Eqtn. 4.50.

Eqtn. 4.50

$$|G(f)|^2 \approx \left|\frac{\sin(\pi T(f - f_c))}{\pi(f - f_c)}\right|^2$$

With the knowledge of the autocorrelation of a, as well as the transfer function G, the PSD at point A in Figure 4.18 is given by Eqtn. 4.51.

Eqtn. 4.51

$$\begin{aligned} PSD_A(f) &= \frac{2\Im(R_{aa}(\tau))|G(f)|^2}{ZT} \\ &= \frac{2V^2 \sin^2(\pi T(f - f_c))}{\pi^2 ZT(f - f_c)^2} \end{aligned}$$

The PSD at point B in Figure 4.18 is identical to that given in Eqtn. 4.51. This result should not be very surprising given that the autocorrelations of the inputs are identical and the filters, g and h, are just phase-shifted versions of one another (and phase does not contribute the squared magnitude response). Because the inputs a and b are uncorrelated, the outputs at points A and B of Figure 4.18 are also uncorrelated. Thus the PSD at the output of the modulator, point C, is simply the sum of the PSDs at points A and B. Therefore, the PSD of an unfiltered CAP signal is given by Eqtn. 4.52.

Eqtn. 4.52

$$\begin{aligned} PSD_{CAP,un} &= PSD_A(f) + PSD_B(f) \\ &= \frac{4V^2 \sin^2(\pi T(f - f_c))}{\pi^2 ZT(f - f_c)^2} \end{aligned}$$

The normal reference impedance for CAP and QAM on a twisted pair is 100 ohms. The symbol period T and the center frequency $f_c$ have purposely been left unspecified until now. Different combinations of T and $f_c$ can be implemented to achieve different bit rates on the twisted pair. Also, different values will change the frequency band being used. For illustration purposes, an unfiltered CAP signal centered at 435.5 kHz with a baud rate of 340 ksymbols/s ($T=1/340*10^3$ s) is shown in Figure 4.19.

**Note**

One normally does not talk about center frequency with a CAP signal. Nevertheless, it is intuitive to the current discussion and is therefore used for illustrative purposes only.

Figure 4.19 A normalized PSD for an unfiltered 340 kbaud CAP/QAM signal centered at 435.5 kHz.

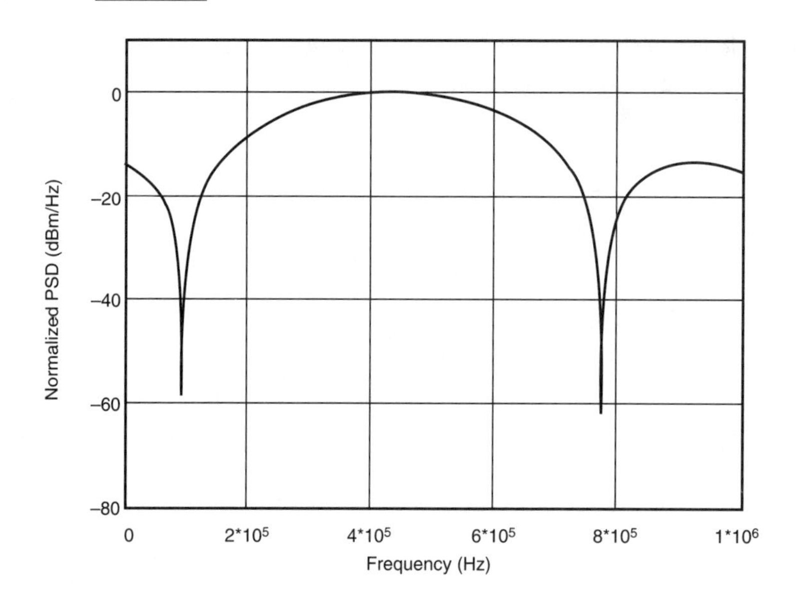


In Figure 4.19, the width of the center lobe is approximately twice the symbol baud rate. Shaping can reduce this wide bandwidth by a factor of two. Note also the level of the center lobe in Figure 4.19 compared to the side lobes. The side lobes, representing the out-of-band energy of the line code, are less than 20 dB below the center lobe, or in band energy. It is more desirable to have the sidebands attenuated more so that they do not cause severe crosstalk into other lines. Sideband reduction can also be obtained by using shaping.

A shaped CAP signal can be achieved by adding shaping, or windowing, to the impulse response g and h in the CAP modulator. The new filters are labeled $r_I$ and $r_Q$ and are shown in Figure 4.20.

Figure 4.20 Block diagram for a CAP/QAM modulator with shaping.

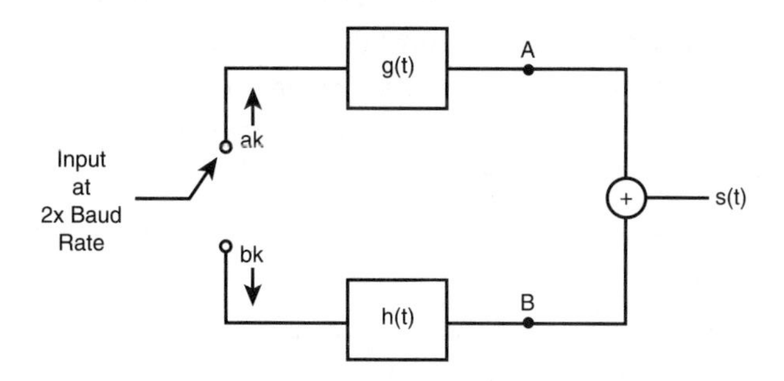


The main purpose of shaping is to reduce bandwidth of the signal's main lobe to approximately the symbol baud rate and to reduce the out-of-band energy in the CAP signal. *Out-of-band energy* refers to the frequency components outside the main lobe of the PSD that ripple away on either side.

Note that the impulse responses g and h in Figure 4.18 can be considered to have rectangular shaping given by Eqtn. 4.53.

Eqtn. 4.53

$$m(t) = u\left(t + \frac{T}{2}\right) - u\left(t - \frac{T}{2}\right)$$

More simply put, in this rectangular window the symbol baud rate is the width. Such a window tends to have poor properties in the frequency domain (which results in the wide bandwidth and high sidebands in Figure 4.19). The most commonly used shaping filter for DSL-based CAP and QAM is the square root raised cosine filter. A baseband normalized square root raised cosine filter has an impulse response given by Eqtn. 4.54.

Eqtn. 4.54

$$r_{srrc}(t) = \frac{\sin\left(\pi(1-\alpha)\frac{t}{T}\right) + 4\alpha\frac{t}{T}\sin\left(\pi(1+\alpha)\frac{t}{T}\right)}{\pi\frac{t}{T}\left(1 - \left(4\alpha\frac{t}{T}\right)^2\right)}$$

In Eqtn. 4.54, the parameter $\alpha$ represents the excess bandwidth that the signal will occupy in the frequency domain due to shaping. A larger $\alpha$ implies more bandwidth. For $\alpha=0$, the signal bandwidth is equal to exactly half the baud rate. This type of filter is difficult to implement, and so some bandwidth expansion is normally used. Typical values of $\alpha$ are 0.15 and 0.2, implying that 15% and 20% extra bandwidth, respectively, above the baud rate of the system is needed. The extra bandwidth is normally considered a fair trade-off for the reduced side-band energy. The impulse response for a square root raised cosine filter for various values of $\alpha$ along with a rectangular window used in an unshaped CAP modulator is shown in Figure 4.21.

Figure 4.21 Examples of square root raised cosine impulse response for various values of a.

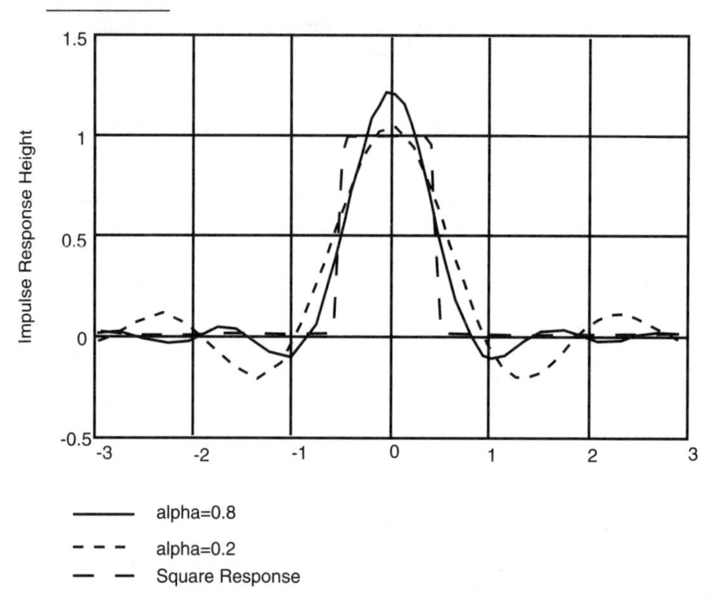


Note that impulses passed through the square root raised cosine filter at baud rate T produce symbols that overlap in time. The receiver can handle this overlap with equalization. When the square root raised cosine response is combined with the CAP modulator filters, the resulting filters become Eqtn. 4.55 and Eqtn. 4.56.

Eqtn. 4.55

$$r_I(t) = \frac{\sin\left(\pi(1-\alpha)\frac{t}{T}\right) + 4\alpha\frac{t}{T}\sin\left(\pi(1+\alpha)\frac{t}{T}\right)}{\pi\frac{t}{T}\left(1-\left(4\alpha\frac{t}{T}\right)^2\right)}\cos(2\pi f_c t)$$

Eqtn. 4.56

$$r_Q(t) = \frac{\sin\left(\pi(1-\alpha)\frac{t}{T}\right) + 4\alpha\frac{t}{T}\sin\left(\pi(1+\alpha)\frac{t}{T}\right)}{\pi\frac{t}{T}\left(1-\left(4\alpha\frac{t}{T}\right)^2\right)}\sin(2\pi f_c t)$$

Normally, the impulse response is several symbols long. Each filter will have the same squared magnitude frequency response. For an infinitely long impulse response, the response of $r_I$ can be approximated by Eqtn. 4.57.

Eqtn. 4.57

$$R_I(f) = \begin{cases} T & |f-f_c| < \frac{1}{2T}(1-\alpha) \\ \frac{T}{\sqrt{2}}\sqrt{1-\sin\left(\frac{\pi T}{\alpha}\left(|f-f_c|-\frac{1}{2T}\right)\right)} & \frac{1}{2T}(1-\alpha) \le |f-f_c| \le \frac{1}{2T}(1+\alpha) \\ 0 & |f-f_c| > \frac{1}{2T}(1+\alpha) \end{cases}$$

The expression in Eqtn. 4.57 has three parts. The first part is the in-band signal with an amplitude of T (normally, this would be scaled to the proper signal level desired). The second part of Eqtn. 4.57 is the bandwidth expansion region. The final part of the equation represents the out-of-band region and is approximated as zero.

For a truncated window, a closed-form solution to the frequency-domain template of the shaping is not possible. However, the numeric approximation in Figure 4.22 uses the same parameters as the unshaped example (Figure 4.19) uses.

Note the level of the sidebands in this figure as compared to Figure 4.19. Also note that the out-of-band energy is not zero (negative infinity on a dB scale) as predicted in Eqtn. 4.57. Given the response of $r_I$ and the autocorrelation function of a, the PSD at point A in Figure 4.20 is given by Eqtn. 4.58.

Figure 4.22 Approximation of the normalized spectrum of a QAM signal with square root raised cosine shaping.

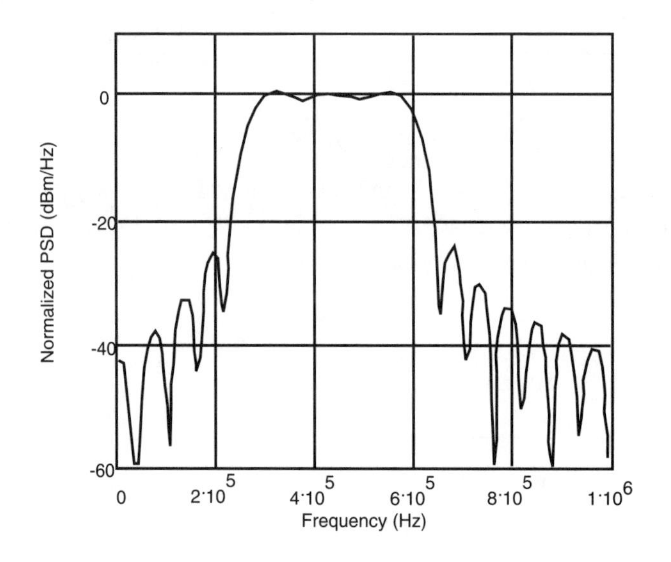


Eqtn. 4.58

$$PSD_A(f) = \frac{2\Im(R_{aa}(\tau))|R_I(f)|^2}{ZT} = \frac{2V^2|R_I(f)|^2}{ZT}$$

Once again, because of the independence of a and b, the output PSD of the shaped CAP modulator is simply the sum of the PSDs of each modulator leg. Thus the output PSD is given by Eqtn. 4.59.

Eqtn. 4.59

$$PSD_{CAP}(f) = \frac{2V^2|R_I(f)|^2}{ZT} + \frac{2V^2|R_Q(f)|^2}{ZT} = \frac{4V^2|R_I(f)|^2}{ZT}$$

In Eqtn. 4.59, the terminating impedance Z is 100 ohms, and T is purposely left unspecified. The center frequency $f_c$ is also not specified. The CAP/QAM modem developers have not yet finalized the exact parameters for ADSL lines.

Some candidate values are shown in Table 4.2.

Table 4.2 Candidate Symbol Rates and Bandwidths for CAP/QAM ADSL Signals

| Symbol Rate 1/T (ksymbols/s) | Excess Bandwidth (α) | Center Frequency (kHz) | Low-End 3 dB Point (kHz) | High-End dB Point (kHz) |
|---|---|---|---|---|
| 84 | 0.20 | 84.0 | 42.0 | 126.0 |
| 136 | 0.15 | 113.2 | 45.2 | 181.2 |
| 340 | 0.15 | 435.5 | 265.5 | 605.5 |
| 680 | 0.15 | 631.0 | 291.0 | 971.0 |
| 952 | 0.15 | 787.4 | 311.4 | 1263.4 |

The first two entries in Table 4.2 are for upstream ADSL, and the rest are for downstream. The PSDs for two of the CAP/QAM implementations shown are given in Figure 4.23.

Figure 4.23 PSDs for various CAP/QAM signals.

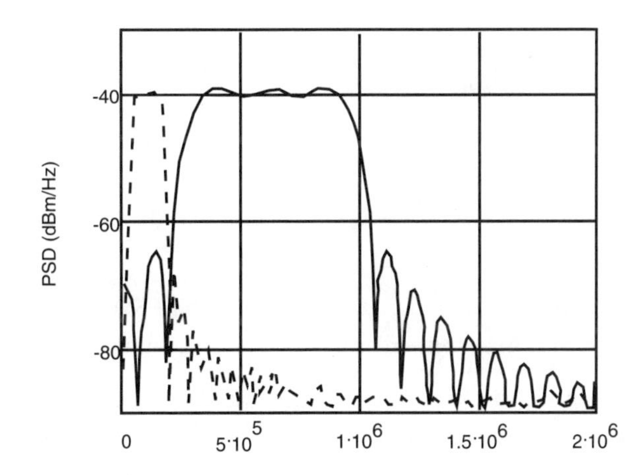


## *VDSL PSDs*

Standards for VDSL are still under development; although a PSD requirement has been agreed upon, it is likely to evolve as times goes on. The mask may change drastically depending on the modulation type chosen for VDSL (several modulation candidates exist, including a variation of DMT called "synchronous DMT(SDMT)," a variation of DMT called "zipper," and also CAP- and QAM-based techniques).

One major difference between ADSL and VDSL is that VDSL signals overlap the amateur radio bands as well as several other allocated frequency regions. Even though a VDSL signal is confined to a twisted pair and not radiated from an antenna, at high frequencies the emissions from the twisted pair can be large enough to interfere with receivers of these other services.

This type of emission is commonly referred to as radio frequency interference (RFI) egress. Depending on the region and type of VDSL installation, the RFI egress may or may not be a problem. When it is a problem, to reduce RFI egress, the VDSL PSD will have notches at identified frequency bands. A list of the frequency bands at which the VDSL PSD will need reduction appears in Table 4.3.

Table 4.3 Internationally Recognized Bands That Require PSD Reduction for VDSL

| Start Frequency (MHz) | Stop Frequency (MHz) |
|---|---|
| 1.8 | 2.0 |
| 3.4 | 4.0 |
| 7.0 | 7.3 |
| 10.1 | 10.15 |
| 14.0 | 14.35 |
| 18.068 | 18.168 |
| 21 | 21.45 |
| 24.890 | 24.990 |
| 28.0 | 29.7 |

Other possible bands needing PSD reduction are under study and may be added to those listed in Table 4.3.

Any type of VDSL modulation must be able to reduce the PSD to acceptable levels in these bands. The PSD mask requirement for VDSL is shown in Figure 4.24.

Similar to the ADSL masks, this mask is piecewise continuous. A requirement to reduce the PSD in the RFI regions given in Table 4.3 is also imposed (but not shown in Figure 4.24). The PSD in these regions must be reduced to –80 dBm/Hz (or lower if desired). Note that the VDSL band is much wider than the bands of any other DSL technology and that the starting point of the passband is above the highest frequency used in ADSL. This arrangement would allow ADSL and VDSL to coexist in a binder group without major interference. An option to allow the VDSL band's lower limit to extend down into the ADSL band is being discussed for situations in which VDSL is deployed, but ADSL is not.

Figure 4.24 PSD mask for VDSL.

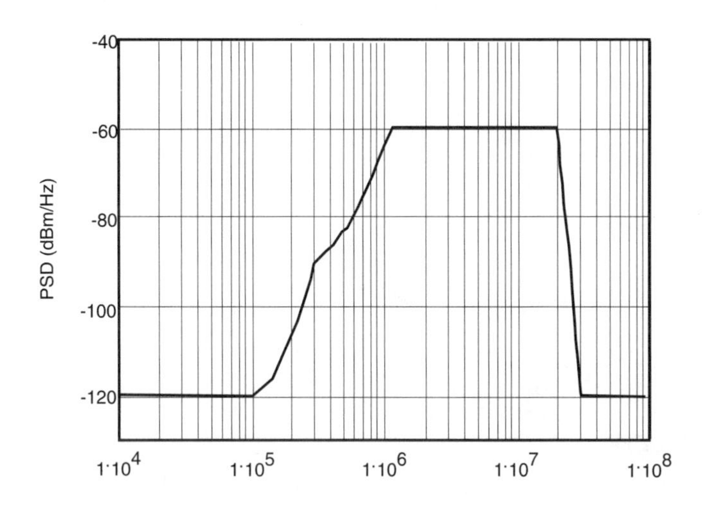


## *Summary of PSDs*

In the previous sections, PSDs of most of the common types of signals in the DSL environment were derived. These PSDs, along with channel transfer functions, are useful for finding the received signal power of a signal that propagates along a twisted pair. Also, the PSDs are the basis for finding NEXT and FEXT expressions for different types of disturbers. The former quantity defines the signal power at a receiver; the latter defines the noise power at a receiver. Together they define the SNR at a receiver. This quantity predicts system performance, including the maximum data rate achievable or the SNR margin for a fixed data rate.

PSDs are also useful for finding the total power in a given frequency band. Integration of the PSD over the band of interest yields the total power in that band due to the signal. Integrating over all frequencies gives the total power of the signal.

# *NEXT and FEXT Expressions*

In previous sections, the PSDs of various types of DSL line codes were derived. The PSDs of the line codes are different from the PSD of a disturber coupled onto a disturbed DSL. The latter PSD is often called the NEXT PSD or the FEXT PSD, depending on the type of crosstalk. NEXT and FEXT PSDs are found by simply passing the disturbing line's PSD (as derived previously) through the NEXT or FEXT transfer functions derived in Chapter 2. Figure 4.25 shows the combination of this crosstalk model with the channel transfer function of the disturbed line. The NEXT and FEXT PSDs must also be adjusted for the number of disturbers present, as discussed earlier in this chapter.

Figure 4.25 Additive model for NEXT and FEXT into the disturbed line's signal.

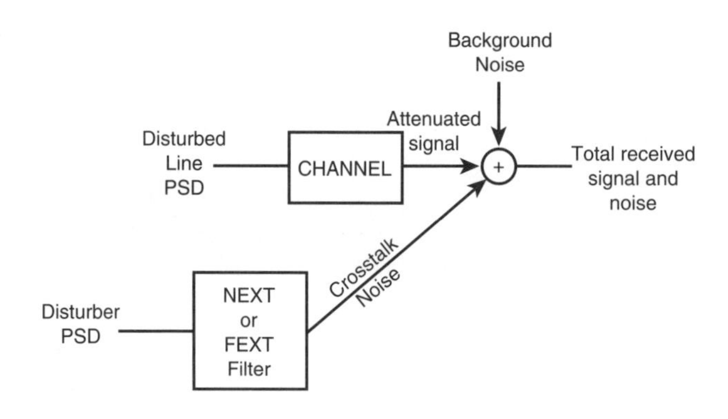


## *NEXT Disturbers*

The general expression for a NEXT disturber, based on the disturbing systems PSD, is given by Eqtn. 4.60.

Eqtn. 4.60

$$PSD_{NEXT}(f) = PSD_{DIST}(f) \times X_N f^{3/2}$$

Here $X_N$ is given by Eqtn. 4.61.

Eqtn. 4.61

$$X_N = 0.882 \times 10^{-14} \times N^{0.6}$$

Here N is the number of disturbers having a PSD given by $PSD_{DIST}$. Note that as the number of disturbers increases, the effect of each additional disturber decreases. This outcome takes into account worst-case modeling and the fact that all disturbers will not be on the worst-case disturbing line.

It is generally assumed that all crosstalk comes from inside the binder group of the disturbed twisted pair; the coupling values in Eqtn. 4.61 are based on this assumption. For some DSL technologies, common practice prohibits T1 disturbers from being in the same binder group as the DSL line. However, T1 lines do tend to be in the vicinity of many DSL technologies and so some adjustment to Eqtn. 4.60 is necessary to account for T1 disturbances from outside the DSL binder group.

The adjustment is a reduction by 15.5 dB of the T1 NEXT calculated using Eqtn. 4.60. This is equivalent to multiplying Eqtn. 4.60 by 0.0282. Graphs of the NEXT PSD for ISDN, HDSL, and T1 disturbers are shown in Figure 4.26 and Figure 4.27.

Figure 4.26 10 ISDN, HDSL, and T1 NEXT disturbers.

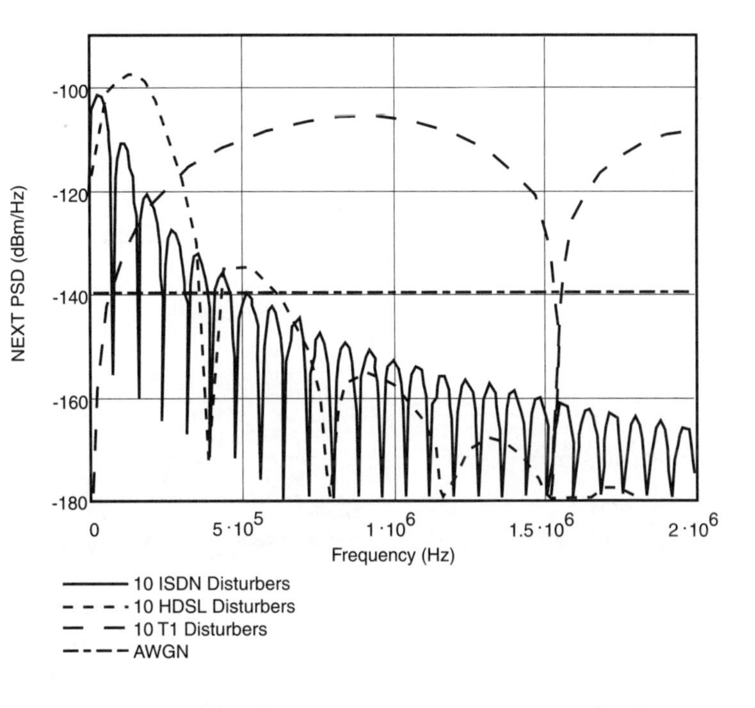


Figure 4.27 10, 24, and 49 HDSL NEXT disturbers.

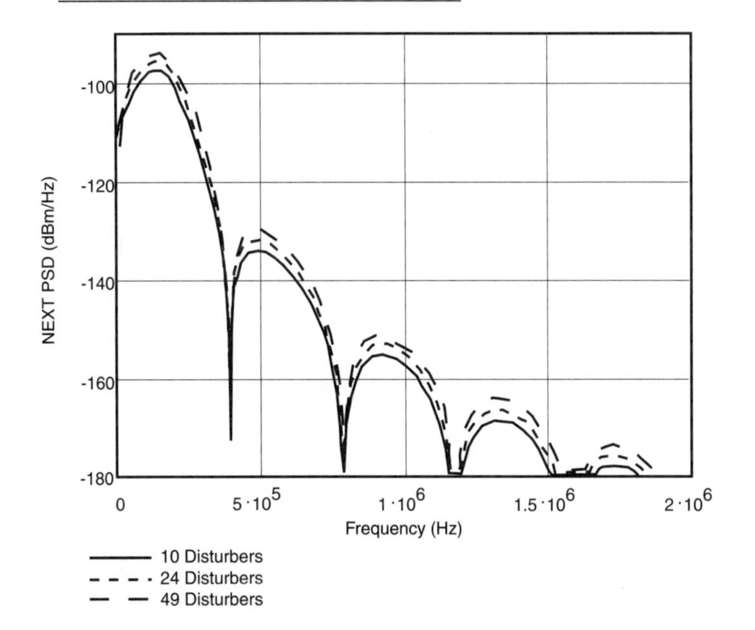

In Figure 4.26, 10 NEXT disturbers exist for each type of disturber. Figure 4.27 compares 10, 24, and 49 disturbers for HDSL. In addition, the normally assumed *additive white Gaussian noise* (AWGN) floor for a DSL line of –140 dBm/Hz is shown.

Figure 4.28 shows NEXT curves for upstream and downstream ADSL signals (both DMT based and CAP/QAM based).

Figure 4.28 10 DMT and CAP/QAM NEXT disturbers.

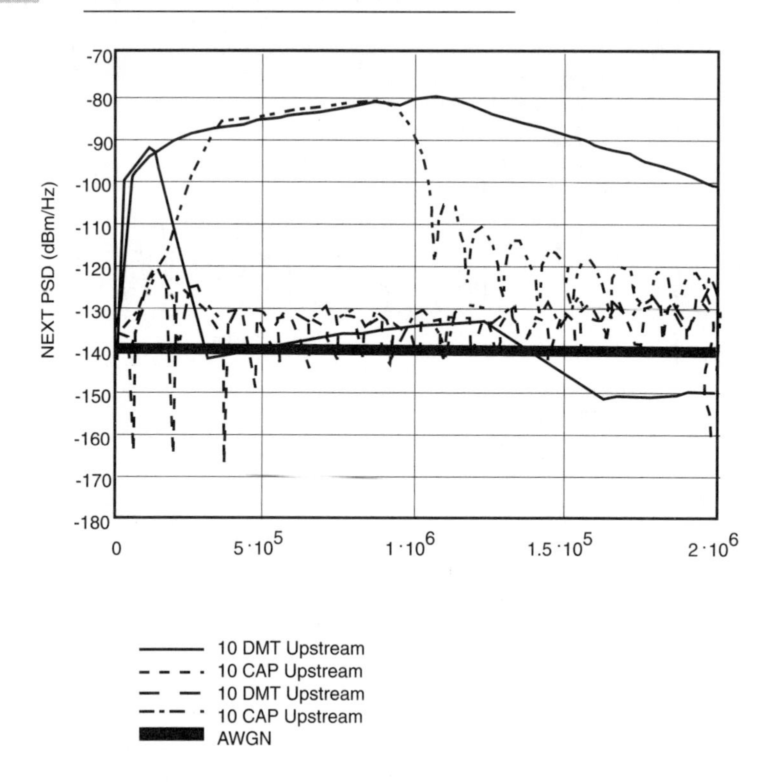


Figure 4.29 compares the NEXT from DMT-based ADSL to HDSL crosstalk. The next chapter looks at the effect of NEXT disturbers on the data rate achievable over the DSL.

Figure 4.29 Comparison of DMT downstream NEXT to HDSL NEXT.

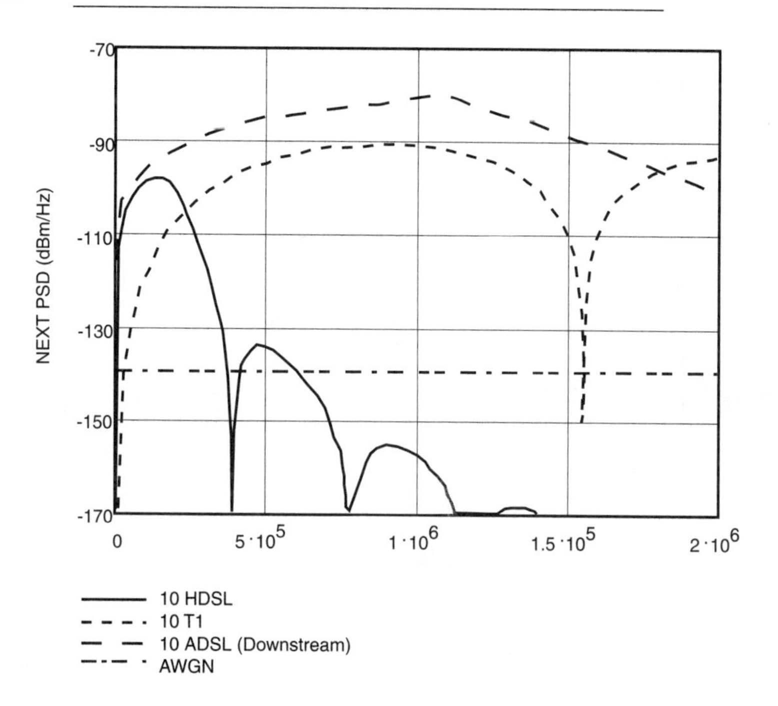


## *FEXT Disturbers*

The general expression for a FEXT disturber, based on the PSD of the disturbing signal, is given by Eqtn. 4.62.

Eqtn. 4.62

$$PSD_{FEXT}(f) = PSD_{DIST}(f) \times |H_{channel}(f)|^2 \times K_{FEXT} \times l \times f^2$$

Here $H_{channel}$ is the response of the channel, $K_{FEXT}$ is a coupling constant, and l is the coupling path length. Normally, NEXT is much greater than FEXT, and so FEXT is often ignored.

However, FEXT from other ADSL signals (ADSL self-FEXT) is used in some performance benchmarking. Specifically, FEXT from downstream ADSL is often used. FEXT for 10 downstream ADSL disturbers assuming a CSA #6 Loop is shown in Figure 4.30. The coupling constant $K_{FEXT}$ for 10 disturbers is given by $3.083*10^{-20}$. Also shown is the –140 dBm/Hz AWGN noise floor.

Figure 4.30 10 DMT ADSL FEXT disturbers on CSA #6 Loop.

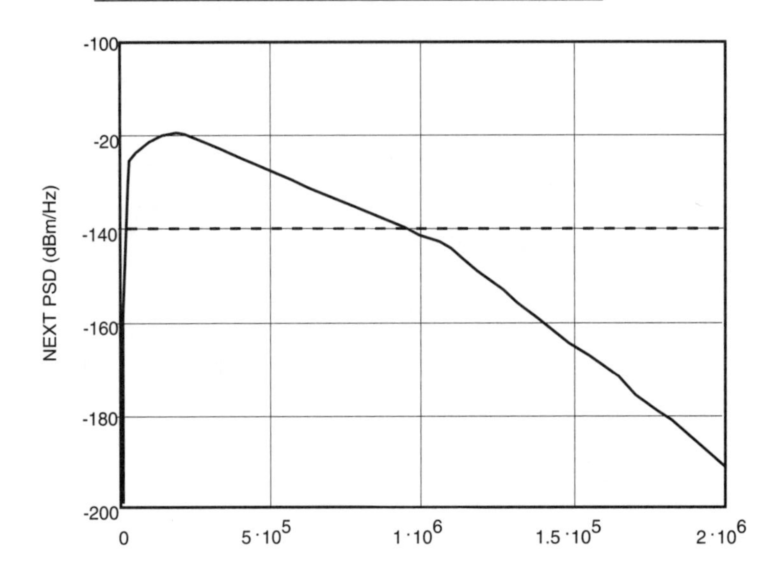


## *Summary*

This chapter discussed complete crosstalk models, both NEXT and FEXT, for DSLs, and the PSDs of ISDN, HDSL, T1, and both CAP/QAM and DMT-based ADSL and VDSL were derived. These PSDs can be "dropped" into a generic NEXT or FEXT equation that, along with the coupling coefficient and the number of disturbers, gives the PSD of the crosstalk. This PSD is noise to an ADSL or VDSL receiver. The treatment of NEXT is a bit more in-depth than that of FEXT because NEXT is usually a more dominant source of crosstalk noise.

Chapter 3 derived channel transfer functions for various types of twisted pairs. Passing the PSDs derived in this chapter through the twisted pair transfer functions derived in previous chapters results in the signal PSD at the DSL receiver. Given that signal PSD and the noise PSD due to crosstalk, the performance of the DSL can be predicted.

# Exercises

1. Derive the general result given in Eqtn. 4.17 for the autocorrelation of the output of a linear, time-invariant filter. Assume the input autocorrelation to be $R_{xx}$.
2. Derive the autocorrelation result given in Eqtn. 4.19. Use the probability density function given in Eqtn. 4.18.
3. Approximate the total power for an ISDN signal by integrating the PSD. Compare this value to the average power of the time-domain ISDN waveform.
4. Assume that the binary 1's density on a T1 line is 100%. What will the time-domain waveform look like? Calculate the PSD for this scenario.
5. Verify that the PSD at point B in Figure 4.18 is identical to the PSD given in Eqtn. 4.51.
6. Another common shaping pulse is the raised cosine pulse. Its baseband time-domain waveform is given by Eqtn. 4.63.

Eqtn. 4.63

$$g_{rc}(t) = \frac{\sin\left(\frac{\pi t}{T}\right)}{\frac{\pi t}{T}} \frac{\cos\left(\alpha \frac{\pi t}{T}\right)}{1-\left(2\alpha \frac{\pi t}{T}\right)^2}$$

   Plot this pulse for various values of $\alpha$. For $\alpha=0$, compare this pulse to a sinc pulse. Plot passband versions of the raised cosine pulse.

7. Plot the passband versions of the square root raised cosine pulse. For various values of $\alpha$, compare these the passband plots of the raised cosine pulse.
8. Plot all of the upstream and downstream candidate PSDs for CAP/QAM based VDSL. Which upstream/downstream combinations would effectively eliminate NEXT (assuming that all lines in the binder group were using the same parameters)?
9. Find the total power for 10 NEXT HDSL disturbers and compare to the total power of 24 HDSL disturbers.
10. Find the NEXT PSD of a line having 10 T1 NEXT disturbers and 10 ISDN NEXT disturbers. Why is this assumption overly pessimistic?
11. Compare the PSD of NEXT due to 10 ADSL downstream signals with the FEXT due to 10 ADSL downstream signals on a CSA #6 Loop.

12. A downstream ADSL DMT signal is passed through a CSA #6 Loop. Find the PSD at the ADSL receiver. Find the NEXT PSD of 20 HDSL disturbers. Using the received PSD and the NEXT PSD, find the SNR over frequency at the receiver.

13. Repeat the previous exercise for a mid-CSA loop with 10 T1 NEXT disturbers. Repeat with a T1.601 Loop #13 and 24 ISDN disturbers.

CHAPTER 5

# *DSL Theoretical Capacity in Crosstalk Environments*

In this chapter:

- Review and discussion of channel capacity
- Evaluation of channel capacity for ADSL loops
- Evaluation of channel capacity for VDSL loops
- Case study highlighting channel capacity over frequency

This chapter examines the potential of a twisted pair channel to carry information—better known as the channel capacity of a twisted pair channel. The discussion here applies much of the information presented in Chapters 2, "Twisted Pair Environment," and 3, "Loop Analysis," regarding twisted pair loss, and in Chapter 4, "Power Spectral Densities and Crosstalk Models," regarding crosstalk between twisted pairs. This chapter also reviews the basic concepts of capacity and the derivation of capacity under different channel conditions. In addition, this chapter presents a method for finding capacity in channels with intersymbol interference and colored noise and applies that method to many ADSL and VDSL scenarios. For further reading, refer to "Achievable Information Rates on Digital Subscriber Loops: Limiting Information Rates with Crosstalk"[1] and "Asynchronous Multiple-Access Channel Capacity"[2] for well-written discussions on capacity, as well as for discussions on the methods of calculating capacity used in this chapter.

Chapters 2 and 3 discussed the loss on a twisted pair channel as well as the crosstalk coupling between two twisted pairs that are physically located near one another. Both of these conditions are factors in determining the amount of information that a twisted pair channel can reliably carry. Simply put, these factors determine the data rate capability of the twisted pair. It makes sense that twisted pair channels with more loss support lower data rates than do twisted pairs with less loss. Similarly, a twisted pair with more crosstalk disturbance supports a lower data rate than a twisted pair with little or no crosstalk. Mathematically bounding channel capacity provides a benchmark for the performance of various types of coding and modulation techniques. This chapter sets the benchmarks; later chapters compare the performance of modulation techniques, namely, DMT- and CAP/QAM-based systems, to the benchmarks.

# *A Review of Capacity*

For a continuous time channel, capacity is a function of signal power and noise power at the receiver of the channel. If a specific application uses only a certain frequency band on the channel, the frequency band used (the bandwidth) also affects capacity.

Truly understanding channel capacity begins with an understanding of entropy and mutual information, two topics often discussed by information theory texts.[3,4] The following sections review the mathematical specifics of capacity.

## *Entropy and Information*

Information theory defines the capacity of a channel as the maximum mutual information between a channel's output and a channel's input, or I(X;Y), where X is the input and Y the output. Written in terms of entropy for discrete random variables X and Y, this definition is equivalent to Eqtn. 5.1.

Eqtn. 5.1

$$\max\left(I(X;Y)\right) = \max\left(H(X) - H(X/Y)\right)$$

In Eqtn. 5.1, H is the entropy, or uncertainty, of input X, and H(X/Y) is the entropy or uncertainty of input X given output Y. A trivial example is best used to demonstrate the relationship between information and uncertainty.

Consider the following situation: A person flips a coin (a fair one) and tells the result to another person. In this example, the input can be either a head or a tail; each is equally likely. The uncertainty of the input H is one bit. The output of the channel Y is what the listener hears. At one extreme, the listener clearly hears what the "flipper" says is the result of the coin toss. In this case, H(X/Y)=0 because no uncertainty about the outcome of the toss remains. Thus the speaker conveyed one bit of information over the channel.

> **Note**
>
> A *bit* is one basic unit of information. You will see later that it involves using base 2 logarithms to define entropy. Another common unit of information is *nat*, which uses base e logarithms (or natural logs). All entropy and information expressions in this book use bits as units.

At the other extreme, the listener cannot hear what the speaker is saying. In this case, H(X/Y)=1 bit. In other words, the listener received no indication of whether a head or tail resulted and thus still has one bit of uncertainty. The information conveyed over the channel by the speaker is zero bits because uncertainty about the coin toss outcome was not reduced (there was one bit of uncertainty before the speaker spoke and one bit after). Between the extremes, the listener may have faintly understood or heard the speaker, and somewhere between zero and one bit of uncertainty was removed. The more clearly the speaker was heard, the more uncertainty was removed.

Returning to a more analytical treatment H, the entropy of the input in Eqtn. 5.1 is given by Eqtn. 5.2.

Eqtn. 5.2

$$H(X) = \sum_{x' \in \Gamma} \Pr(x') \log_2 \left( \frac{1}{\Pr(x')} \right)$$

Here Γ represents a set of all values that X can take on, and Pr(x') represents the probability that X takes on the value of x'. Likewise, H(X/Y), the uncertainty of the input given the output, can be written as Eqtn. 5.3.

Eqtn. 5.3

$$H(X/Y) = \sum_{x' \in \Gamma} \sum_{y' \in \Psi} \Pr[(x', y')] \log_2 \left( \frac{1}{\Pr[(x'/y')]} \right)$$

Here Γ represents a set of all possible values of X, and ψ represents a set of all possible values of Y. Note that if X and Y are independent, then the uncertainty of X given Y is the same as the uncertainty of X as illustrated in Eqtn. 5.4.

Eqtn. 5.4

$$H(X/Y) = H(X)$$

This equation is true because knowledge of Y clears up no uncertainty of X.

The expression for capacity given in Eqtn.5.1 can be equivalently written as Eqtn. 5.5.[3,4]

Eqtn. 5.5

$$\max(I(X;Y)) = \max(H(Y) - H(Y/X))$$

Whereas Eqtn. 5.1 allows for an intuitive understanding of capacity, Eqtn. 5.5 derives an analytical expression for the capacity of a channel with noise.

For a continuous random variable, the concepts of entropy and information can be extended, and the entropy of some continuous random variable Z defined as in Eqtn. 5.6.

**Note**

The extension of entropy to the continuous case is not so straightforward. In this context, it is not truly uncertainty, as uncertainty would be infinite for a continuous random variable, but rather as a quantity analogous to uncertainty. This condition is sometimes referred to as *differential entropy.*

Eqtn. 5.6

$$h(Z) = \int_{-\infty}^{\infty} p(z)\log_2\left(\frac{1}{p(z)}\right)dz$$

Here p is the probability density function of random variable Z. If Z is Gaussian and characterized by a zero mean and a variance $\sigma_z^2$ of $N_o/2$ (denoted by $(0,N_o/2)$), then h can be written as shown in Eqtn. 5.7.

Eqtn. 5.7

$$\begin{aligned} h(Z) &= \int_{-\infty}^{\infty} p(z)\log_2\left(\frac{1}{(p(z))}\right)dz \\ &= \int_{-\infty}^{\infty} p(z)\left(\log_2(\sqrt{2\pi\sigma_z^2}) + \frac{z^2}{2\sigma_z^2}\right)dz \\ &= \frac{1}{2}\log_2\left(2\pi\sigma_z^2\right) + \frac{1}{2} \\ &= \frac{1}{2}\log_2\left(2\pi N_o\right) \end{aligned}$$

The final expression in Eqtn. 5.7 will be useful in calculating the capacity of an additive white Gaussian noise (AWGN) channel.

## *Gaussian Channels*

Consider an AWGN channel now with input X, output Y, and additive noise Z, where Z is independent of X. Eqtn. 5.8 gives the output of the channel.

Eqtn. 5.8

$$Y = X + Z$$

The capacity of the channel can then be written as Eqtn. 5.9.

Eqtn. 5.9

$$C = \max\left(I(X;Y)\right) = \max\left(h(Y) - h(Y / X)\right)$$
$$= \max\left(h(Y) - h\left((X + Z) / X\right)\right)$$

Because all uncertainty about X can be cleared up if X is known, the h(X+Z/X) term in Eqtn. 5.9 can be simplified and the equation can be rewritten as Eqtn. 5.10.

Eqtn. 5.10

$$C = \max\left(h(Y) - h(Z / X\right)$$

Remember that the AWGN term Z and the input signal X are independent by assumption. As discussed previously, no uncertainty can be cleared with knowledge of an independent entity. Thus knowledge of X resolves no uncertainty about Z, and Eqtn. 5.10 becomes Eqtn. 5.11.

Eqtn. 5.11

$$C = \max\left(h(Y) - h(Z)\right)$$

If Z is AWGN with statistics $(0, \sigma_z^2)$, Eqtn. 5.7 can be substituted for the second entropy term in Eqtn. 5.10, yielding Eqtn. 5.12.

Eqtn. 5.12

$$C = \max\left(h(Y)\right) - \frac{1}{2}\log_2\left(4\pi\sigma_z^2\right)$$

Maximizing Eqtn. 5.11 is equivalent to maximizing h, which is a function of the probability distribution of X, namely, p. For a given variance, it can be shown that the Gaussian distribution with zero mean will maximize entropy. [3] Because Z is Gaussian and Y=X+Z, if X is Gaussian, then Y is also Gaussian (the sum of two Gaussian random variables is also a

Gaussian random variable). Assuming a Gaussian distributed X with statistics $(0,\sigma_x^2)$, Y then has statistics $(0,\sigma_x^2+\sigma_z^2)$ and h can be written as Eqtn. 5.13.

Eqtn. 5.13

$$h(Y) = \frac{1}{2}\log_2\left(4\pi\left(\sigma_x^2 + \sigma_z^2\right)\right)$$

And Eqtn. 5.13 can be simplified to Eqtn. 5.14.

Eqtn. 5.14

$$\begin{aligned} C &= \frac{1}{2}\log_2\left(4\pi\left(\sigma_x^2 + \sigma_z^2\right)\right) - \frac{1}{2}\log_2\left(4\pi\sigma_z^2\right) \\ &= \frac{1}{2}\log_2\left(\frac{4\pi\left(\sigma_x^2 + \sigma_z^2\right)}{4\pi\sigma_z^2}\right) \\ &= \frac{1}{2}\log_2\left(1 + \frac{\sigma_x^2}{\sigma_z^2}\right) \\ &= \frac{1}{2}\log_2\left(1 + \frac{E\left[x^2\right]}{E\left[z^2\right]}\right) \\ &= \frac{1}{2}\log_2\left(1 + SNR\right) \end{aligned}$$

Here SNR is the average signal-to-noise ratio at the receiver. Thus the channel capacity of an AWGN channel increases logarithmically with the average SNR at the receiver of the channel.

## Channels with Memory and Interference

When a channel is not "memoryless," that is, the channel has some impulse response that is nonzero at more than one point in time, the analysis of channel capacity is a bit more involved. Furthermore, if colored noise rather than white noise is added to the channel, the AWGN channel results are not easily applied. A channel model including memory and colored noise is representative of a twisted pair channel with crosstalk noise. This transfer function over frequency of the twisted pair gives rise to the memory of the channel. The presence of crosstalk that varies with frequency gives rise to the colored-noise model. Note that a channel with memory is also called a channel with *intersymbol interference* or ISI.

This discussion considers a channel with input x, output y, a colored-noise component representing crosstalk resulting from AWGN source q passed through a linear time invariant system (LTI) with impulse response h, and an AWGN noise component z. It is also assumed that x, q, and z are mutually independent and that the channel itself has an impulse response given by g. The channel model appears in Figure 5.1.

At first glance, the channel in Figure 5.1 may seem very different from the AWGN channel. Through transformations, however, it is possible to create an equivalent series of independent AWGN channels. Then you can use Eqtn. 5.14 to find the capacity of each AWGN channel. Because the channel is independent, the sum of the capacities of the individual channels is equivalent to the capacity of the original, colored, ISI channel.

Figure 5.1 Block diagram for system with ISI, AWGN, and colored-noise corruption.

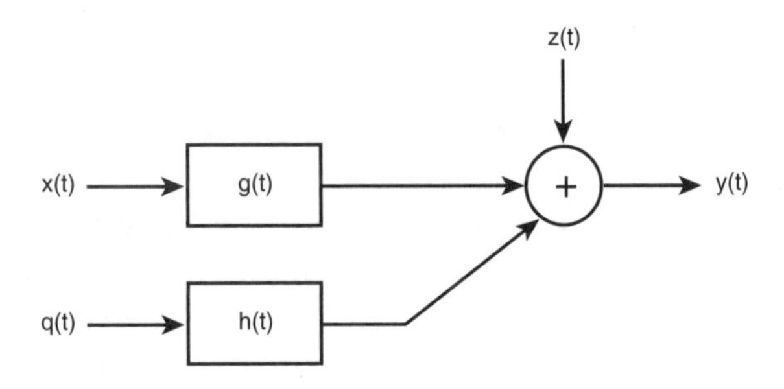


Assuming that the channel shown in Figure 5.1 is sampled, we can write the output at time n as Eqtn. 5.15.

Eqtn. 5.15

$$y_n = \sum_{j=0}^{kg-1} g_j x_{n-j} + \sum_{j=0}^{kh-1} h_j q_{n-j} + z_n$$

Eqtn. 5.15 also assumes that the impulse response $g_j$ of the twisted pair channel is kg samples long and that of the noise coupling function $h_j$ is kh samples long.

Given Eqtn. 5.15, for a block of data beginning at n=0, some of the channel outputs may require knowledge of inputs x or n at indexes less than zero (past values of the inputs). Strictly for analysis, Verdu showed that the memory of the system could be assigned arbitrarily for analysis.[5] By assigning memory properly, the summations in Eqtn. 5.15 can be written as circular convolutions, and the N block output can be written as Eqtn. 5.16.

**Note**

Memory assigned in this manner is sometimes called a *cyclic prefix*. For very large blocks, the length of the assigned prefix will be negligible compared to the block length.

Eqtn. 5.16

$$y^N = g^N \otimes x^N + h^N \otimes q^N + z^N$$

Here ⊗ represents the circular convolution operator superscript, and superscript N represents N length sample vectors of each quantity. It is assumed that N is larger than the channel transfer function lengths kg and kh and that zero padding is used to extend the vectors $g_j$ and $h_j$ to length N.

A discrete fourier transform (DFT) of Eqtn. 5.16 can be taken and results in 5.17.

Eqtn. 5.17

$$Y^N = G^N X^N + H^N Q^N + Z^N$$

In Eqtn. 5.17, each capital letter represents the DFT of its respective lowercase time-domain counterpart. In general, each discrete point in Eqtn. 5.17 contains both a real and an imaginary component. Also, if all values in Eqtn. 5.16 are real (real values in the time domain), a DFT has the property that the second half of the transformed sequence will have complex conjugate symmetry with the first half of the sequence.[6] Thus the second half of the quantities after the transform can be dropped, and enough information remains to reconstruct the original untransformed sequence. The truncated version of Eqtn. 5.17 is shown in Eqtn. 5.18.

**Note**

Depending on whether N is odd or even, the actual symmetry may be around the center point (odd N) or around a point between the two most-center points (even N).

Eqtn. 5.18

$$Y^L = G^L X^L + H^L Q^L + Z^L$$

Here L represents half the length of the original transformed sequence. Hereafter, for notation convenience only, the superscript L will be dropped when referring to each quantity in Eqtn. 5.18.

Where we originally had a sequence of correlated samples, we now have a sequence of independent samples. (Each discrete frequency point is one channel; thus L independent channels exist.)[7] In a sense, each point in the sequence represents the output of one channel. The transformed set of channels appears in Figure 5.2.

Figure 5.2 Original channel with ISI (top) and a transformed set of independent channels with no ISI (bottom).

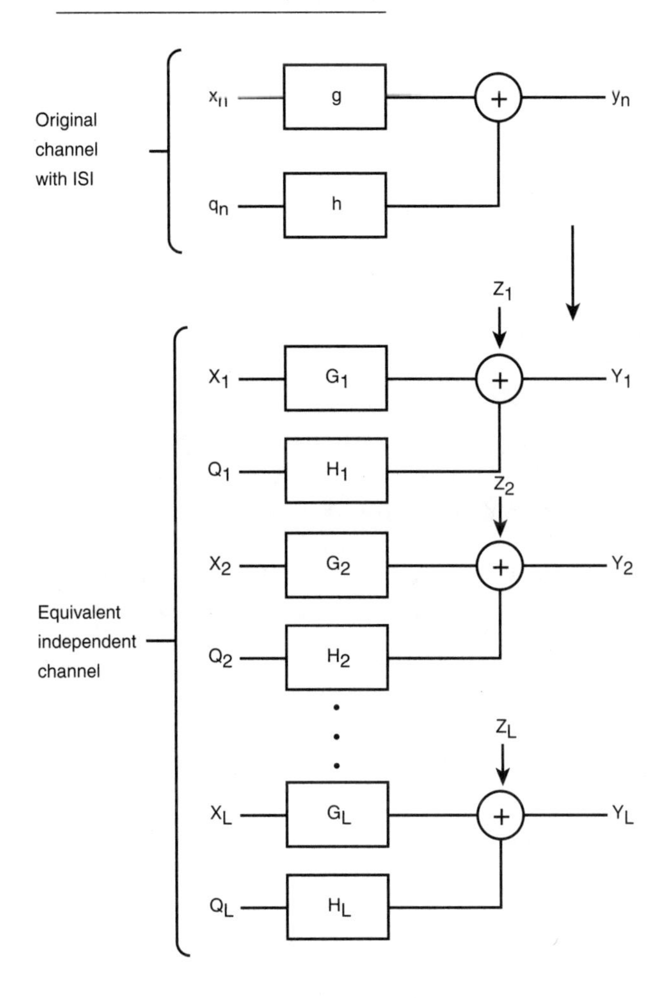


For channel m, the received signal is denoted by $Y_m$ as shown in Eqtn. 5.19.

Eqtn. 5.19

$$\begin{aligned} Y_m &= G_m X_m + (H_m Q_m + Z_m) \\ &= \text{Signal} + (\text{Noise}) \end{aligned}$$

In Eqtn. 5.19, $G_m X_m$ represents the received signal on channel m, and $H_m Q_m + Z_m$ represents the received noise on channel m. Thus the SNR for channel m at the receiver is given by Eqtn. 5.20.

Eqtn. 5.20

$$SNR_m = \frac{|G_m X_m|^2}{|H_m Q_m + Z_m|^2}$$

If the power allotted to channel m is $S_m$, then from Eqtn. 5.14 the capacity on channel m is given by Eqtn. 5.21.

Eqtn. 5.21

$$\begin{aligned} C_m &= \frac{1}{2}\log_2(1 + SNR_m) \\ &= \frac{1}{2}\log_2\left(1 + \frac{|G_m X_m|^2}{|H_m Q_m + Z_m|^2}\right) \\ &= \frac{1}{2}\log_2\left(1 + \frac{|G_m|^2 S_m}{|H_m Q_m + Z_m|^2}\right) \end{aligned}$$

Because the subchannels are independent of one another, the total capacity of the system is the sum of the capacities of the individual subchannels. Thus the total capacity is given by Eqtn. 5.22.

Eqtn. 5.22

$$\begin{aligned} C &= \sum_{i=1}^{L} C_i = \sum_{i=1}^{L} \frac{1}{2}\log_2(1 + SNR_i) \\ &= \sum_{i=1}^{L} \frac{1}{2}\log_2\left(1 + \frac{|G_i|^2 S_i}{|H_i Q_i + Z_i|^2}\right) \end{aligned}$$

In this equation, $S_i$ is the power allotted to channel i. Note that the capacity expression given in Eqtn. 5.22 makes no assumption about bandwidth. The units of Eqtn. 5.22 are bits per channel usage. If the original channel has a bandwidth of B Hz, then each subchannel will have a bandwidth of B/L Hz. For such a channel, it is possible to send a symbol over the channel 2B/L times per second. By sending symbols at this rate, the maximum rate of the channel $R_{max}$ is given by Eqtn. 5.23.

Eqtn. 5.23

$$R_{max} = \sum_{i=1}^{L} \frac{B}{L}\log_2\left(1 + \frac{|G_i|^2 S_i}{|H_i Q_i + Z_i|^2}\right)$$

At this point, if $S_i$ is known for all i subchannels, the total capacity of the channel is given by Eqtn. 5.23. This condition is true for most DSL technologies and is known as the power spectral density (PSD) limited channel. For the PSD limited channel, the sum of power on all the subchannels, as well as the power on each individual subchannel, is constrained. Both ADSL and VDSL fall under this category. The analysis in this chapter is based on Eqtn. 5.23.

For completeness, a derivation is given (see Appendix 5A) to find the capacity for the total power limited case also. In the total power limited case, the sum of power on all of the subchannels is constrained, but how the power is distributed among the channels is not constrained. This approach allows for one additional level of optimization.

**Note**

Optimization of this type in an environment prone to crosstalk would make performance prediction quite difficult. This situation is investigated later in the chapter.

## *Framework for Capacity Calculations for ADSL and VDSL*

The result of Eqtn. 5.23, along with the results from Chapter 3 and Chapter 4, allow capacity calculations for any loop with any combination DSL crosstalk disturbers. Table 5.1 relates the quantities in Eqtn. 5.23 to the entities they represent.

Table 5.1 Description of Each Term in Eqtn. 5.23 for Calculating the Capacity of a Twisted Pair Channel

| Quantity | Description |
|---|---|
| B | The entire bandwidth of the modulation scheme—1.1 MHz for ADSL and 15 MHz for VDSL. If L is the number of subchannels, then B/L is the bandwidth of each subchannel. |
| $S_i$ | The signal power for each subchannel at the transmitter. |
| $G_i$ | The loss between the transmitter and receiver for each subchannel. (ABCD parameters are used to find this quantity.) |
| $Q_i$ | The signal power at a disturbers transmitter. |
| $H_i$ | The transfer function for a disturber given by the NEXT or FEXT equations in Chapter 4. |
| $Z_i$ | The background AWGN on a subchannel. |

For ideal calculations, the number of subchannels would approach infinity, and the summation in Eqtn. 5.23 would become an integration. For practical performance calculations, however, the summation is adequate. For ADSL, capacity is found with the following variations:

- Different loops with the same noise and interference
- A single loop with different types of crosstalk
- Loop and noise called out in the ADSL standard

For VDSL, a slightly less vigorous examination will be carried out, focusing on the capacity of different VDSL loops under some limited crosstalk noises.

Also, a case study on a loop shows the SNR as a function of frequency and the information-carrying ability at different points in the frequency band of the channel. Note that capacity is independent of the type of modulation scheme used on the twisted pair. How close each modulation method comes to operating at the maximum theoretical rate (capacity) is investigated in later chapters.

## *ADSL Capacities*

For ADSL, the signal power at the input to the channel is simply given by the PSD derived for ADSL. The capacity will be found for both the upstream and the downstream directions of data transmission. For the downstream direction, the capacity summation uses 1,000 points evenly spaced between 30 kHz and 1.106 MHz. These frequencies correspond to the actual usable band of an ADSL signal. Frequencies below 30 kHz are typically not used because of the Plain Old Telephone Service (POTS) underlay, and above 1.106 MHz the PSD for a downstream ADSL signal drops off quickly.

For the upstream direction, the capacity summation uses 1,000 points between 30 kHz and 138 kHz. The PSD of the upstream signal is specified to fall off above 138 kHz, and most usable signal energy is within this frequency range.

### *Loop Capacities with Identical Crosstalk*

To compare the different types of loops, a single crosstalk will be applied to a variety of different loops, and the capacities calculated. In general, the crosstalk will consist of 24 HDSL disturbers and a background AWGN of -140 dBm/Hz. The capacity for each case will be found as well as the theoretical maximum rate of each line with a 6 *SNR margin*. The SNR margin, or simply margin, is similar to a cushion given for performance. With a 6 dB margin specified, the maximum achievable rate is equivalent to the capacity of a line having noise 6 dB greater than the actual noise (or equivalently, multiplying the SNR by

0.251). When calculating the actual operating rate of an ADSL modem given some crosstalk configuration, a 6 dB margin is prudent to use because small changes in the channel transfer function or crosstalk noise will not affect the quality of the ADSL link. The loop performances are shown in Table 5.2.

**Note**

The changes in the channel transfer function can be the result of weather (heat, cold, rain, and so on) or of other types of DSL lines being added or removed from the binder group.

Table 5.2 Loop Performance with Fixed Crosstalk and Noise

| | **Channel Capacity** | | **Achievable Rate with 6 dB Margin** | |
|---|---|---|---|---|
| **Loop** | **Upstream (kbit/s)** | **Downstream (Mbit/s)** | **Upstream (kbit/s)** | **Downstream (Mbit/s)** |
| T1.601 #7 | 591 | 3.91 | 396 | 2.45 |
| T1.601 #13 | 554 | 5.05 | 370 | 3.35 |
| CSA #4 | 1113 | 12.5 | 899 | 10.4 |
| CSA #6 | 1086 | 12.2 | 872 | 10.1 |
| CSA #7 | 1196 | 12.2 | 979 | 10.5 |
| CSA #8 | 1156 | 11.4 | 941 | 9.22 |
| Mid CSA | 1425 | 18.4 | 1208 | 16.2 |

Note several things about Table 5.2. First, the T1.601 loops do not support data rates as high as the CSA loops in either the upstream or downstream direction. Second, the data rate reduction varies when a 6 dB margin is imposed. The "costs" in the downstream direction are roughly 1.5 Mbps–2.5 Mbps, and in the upstream direction are in the neighborhood of 200 Kbps. Finally, note the high data rates supported by the mid-CSA loop. Because this loop is shorter than the others, you would expect it to support higher data rates in both directions.

## *Comparison of Different Types of Crosstalk*

This section uses a CSA #6 Loop (9 kft of #26) to compare different types of crosstalk. This loop is good for comparing crosstalk because the loop contains no bridged taps and thus has a monotonically increasing insertion loss. It is often used for benchmarking in xDSL reports and publications.

The capacity of the loop is found with NEXT from ISDN, HDSL, and T1 lines, as well as NEXT and FEXT from other ADSL lines. For the downstream direction, FEXT will be from other downstream ADSL signals and ADSL NEXT will be from upstream ADSL signals. The opposite will be true for the upstream direction. In all cases, 24 disturbers will be present, and again, an AWGN of -140 dBm/Hz will be assumed. As a baseline simulation, the capacity of the loop with only the AWGN source will be given. Table 5.3 gives the resulting capacities, as well as the achievable rates for each scenario with a 6 dB margin.

Table 5.3 Loop Performance with Different Types of Crosstalk

| | Channel Capacity | | Achievable Rate with 6 dB Margin | |
|---|---|---|---|---|
| **Disturbers** | **Upstream (kbit/s)** | **Downstream (Mbit/s)** | **Upstream (kbit/s)** | **Downstream (Mbit/s)** |
| AWGN Only | 2601 | 16.8 | 2385 | 14.7 |
| 24 ISDN NEXT | 1485 | 14.4 | 1271 | 12.2 |
| 24 HDSL NEXT | 1089 | 12.2 | 872 | 10.1 |
| 24 T1 NEXT | 2338 | 7.29 | 2124 | 5.88 |
| 24 ADSL NEXT | 1126 | 14.8 | 910 | 11.9 |
| 24 ADSL FEXT | 2072 | 14.1 | 1156 | 12.6 |
| Downstream ADSL NEXT | 1109 | 2.45 | 894 | 1.67 |

Several things are worth highlighting in Table 5.3. First, not including the last line in the table, T1 NEXT is by far the worst disturber for the downstream direction. This result includes the 15.5 dB reduction in the T1 interfering model. This result is due to the fact that the T1 NEXT PSD is significant over the entire downstream ADSL bandwidth.

Note, however, that the T1 NEXT is not as detrimental to the upstream channel as are the other disturbers. HDSL and ADSL are the next most damaging to the upstream ADSL. Both of these disturbers span the entire upstream bandwidth.

Finally, note that ADSL FEXT is not as damaging to the downstream direction as downstream ADSL NEXT. ADSL systems have been designed so that this disturbance does not occur. In fact, this is one of the reasons that the upstream ADSL channel uses a far smaller bandwidth than the downstream channel uses. Put another way, there is no self-NEXT for a downstream ADSL channel. Some proposed "reverse ADSL" schemes in which the symmetry of the channels is reversed would cause self-NEXT to be present and severely impair the "regular" ADSL lines. Because different types of DSLs having varying effects on one another, spectral planning is essential for any major deployment of ADSL.

## *Industry Standard Loop and Crosstalk Configurations*

Some of the standard test loops and test noise configurations appear in Table 5.4. Upstream and downstream theoretical maximum rates with a 6 dB margin are also shown. For the HDSL, ISDN, and T1 NEXT, the crosstalk was injected at the near end for both the upstream and downstream measurements.

Table 5.4 Performance over Different Standard Specified Loops

| | Type and Number of Disturbers | | | | Achievable Rates with 6 dB Margin (except where noted) | |
|---|---|---|---|---|---|---|
| Loop | ADSL NEXT And FEXT | HDSL NEXT | ISDN NEXT | T1 NEXT | Upstream (kbit/s) | Downstream (Mbit/s) |
| T1.601 #7 | 0 | 0 | 24 | 0 | 767 | 4.46 |
| T1.601 #13 | 0 | 0 | 24 | 0 | 730 | 5.39 |
| CSA #4 | 24 | 0 | 24 | 0 | 887 | 10.6 |
| CSA #6 | 0 | 20 | 0 | 0 | 889 | 10.2 |
| Mid CSA (3 dB margin only) | 0 | 0 | 0 | 10 | 2325 | 12.2 |
| T1.601 #13 | 0 | 0 | 24 | 0 | 629 | 5.80 |
| CSA #4 | 10 | 10 | 24 | 0 | 852 | 9.59 |
| CSA #6 | 10 | 10 | 24 | 0 | 824 | 9.45 |
| CSA #8 | 10 | 10 | 24 | 0 | 892 | 8.85 |
| CSA #6[1] | 0 | 0 | 0 | 4 | 1821 | 6.88 |
| Mid CSA | 0 | 0 | 10 | 24 | 1981 | 9.43 |

[1] 0 dB margin. Originally, if T1 disturbers were detected, a 6 dB power boost in the downstream PSD was recommended to boost the rate of this configuration. The power boost was dropped because of the added complexity it gave to the analog portion of the ADSL modem. The result shown assumes no power boost.

## *VDSL Capacities*

For VDSL, the capacity of different lines will be found with no crosstalk and then with several different types of disturbers. The loops investigated are the VDSL loops discussed in Chapter 3, and the reduction of capacity due to RFI is disregarded for the time being. The PSD used for the VDSL transmitter is that given in Chapter 4. From 1.1 MHz to 20 MHz, 1,000 equally spaced points approximate Eqtn. 5.23. In contrast to the ADSL performance calculations, a single result, rather than an upstream and a downstream capacity, is shown. If VDSL systems employed echo cancellation, one might argue that the calculated rates would be achievable bidirectionally. This discussion assumes, however, that the systems are not echo canceled and that the resultant bit rate must be split between an upstream and a downstream rate (which could be done by frequency-division techniques or time-division techniques).

**Note**

The computational complexity of echo-canceled VDSL is much larger than that for ADSL, so VDSL systems are normally not echo canceled. In addition, this complexity may present some serious spectral compatibility problems with other DSLs.

Table 5.5 shows the capacities for VDSL loops given a 6 dB margin. For loops VDSL #1 through VDSL #4, which have a variable length and gauge segments, a 4.5 kft, #24 loop was used. Note the large rate results for VDSL Loop #5 as compared to the others. This result makes sense because the VDSL loop is much shorter than the other two.

Table 5.5 VDSL Loop Performance with Different Types of Crosstalk

| | Achievable Rate with 6 dB Margin (Mbit/s) | | | | | | |
|---|---|---|---|---|---|---|---|
| **Disturber** | **Loop #1** | **Loop #2** | **Loop #3** | **Loop #4** | **VDSL Loop #5** | **VDSL Loop #6** | **VDSL Loop #7** |
| **AWGN Only** | 36.1 | 30.6 | 22.4 | 9.98 | 306 | 48.1 | 15.8 |
| **HDSL NEXT** | 36.1 | 30.6 | 22.4 | 9.98 | 306 | 48.1 | 15.8 |
| **ADSL NEXT** | 16.5 | 11.6 | 5.51 | 1.20 | 285 | 28.2 | 26.6 |
| **T1 NEXT** | 10.3 | 7.68 | 4.48 | 0.69 | 270 | 17.2 | 2.33 |
| **VDSL NEXT** | 0.91 | 0.69 | 0.31 | 0.02 | 26.6 | 1.62 | 0.19 |
| **VDSL FEXT** | 30.0 | 26.1 | 20.1 | 9.74 | 130 | 40.0 | 14.4 |

Several other points worth mentioning about Table 5.5 are the following:

- HDSL NEXT has a negligible effect on VDSL because the main energy for HDSL is well below the lower edge of the VDSL band. HDSL sidebands that overlap the VSDL band will be sufficiently attenuated so that no performance degradation is realized.
- T1 NEXT severely limits the rate of VDSL. The T1 NEXT includes the 15 dB reduction in coupling discussed in Chapter 4, reflecting binder group and distance separation. In general, T1 and VDSL deployment will have to be carefully coordinated to reduce crosstalk problems.

- ADSL has a large effect on VDSL rates. VDSL Loop #7 is especially limited by ADSL crosstalk.
- VDSL NEXT severely limits the achievable rate. In some cases, the achievable rate is below achievable ADSL rates. This condition drives the requirement for VDSL system designs to have no self-NEXT.

A comparison of achievable rates for the different VDSL Loop #1 possibilities for three different noise configurations appears in Table 5.6. As expected, shorter loops can support higher data rates. Also, loops consisting of #24 gauge twisted pair have higher achievable rates than the same length pair utilizing #26 gauge. Note that for 4.5 kft lengths, the total achievable rates get very low depending on the type of disturber (remember that the achievable rate shown in this table is the aggregate upstream and downstream rates).

Table 5.6 Performance of VDSL Loop #1 Variations

| | Achievable Rate with 6 dB Margin (Mbit/s) | | | | | |
|---|---|---|---|---|---|---|
| **Disturber** | **1.5 kft #24** | **3.0 kft #24** | **4.5 kft #24** | **1.0 kft #26** | **3.0 kft #26** | **4.5 kft #26** |
| **AWGN Only** | 274 | 101 | 36.1 | 304 | 57.7 | 18.3 |
| **ADSL NEXT** | 253 | 79.6 | 16.4 | 283 | 37.0 | 3.17 |
| **VDSL FEXT** | 117 | 69.4 | 29.9 | 128 | 45.4 | 16.7 |

## *Usable Frequencies on Crosstalk Channels*

This section focuses on a case study of a single loop with a specified crosstalk disturbance. The SNR at the receiver of the loop as a function of frequency is investigated to show which parts of the band are most capable of carrying information. The loop appears in Figure 5.3.

Figure 5.3 Loop for SNR and channel-capacity case study.

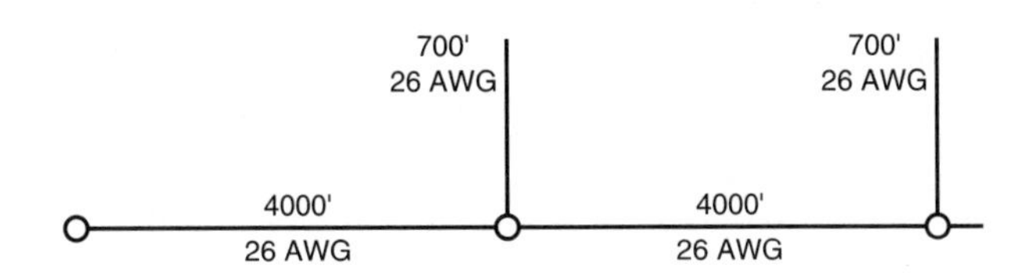

This study assumes that 20 T1 NEXT disturbers and 20 ISDN NEXT disturbers exist. In addition, a background AWGN of -140 dBm/Hz will be added to the crosstalk noise. Assuming that this loop is carrying downstream DMT ADSL, the signal power at the loop's receiver is shown in Figure 5.4.

Figure 5.4 Received signal and noise on case study loop.

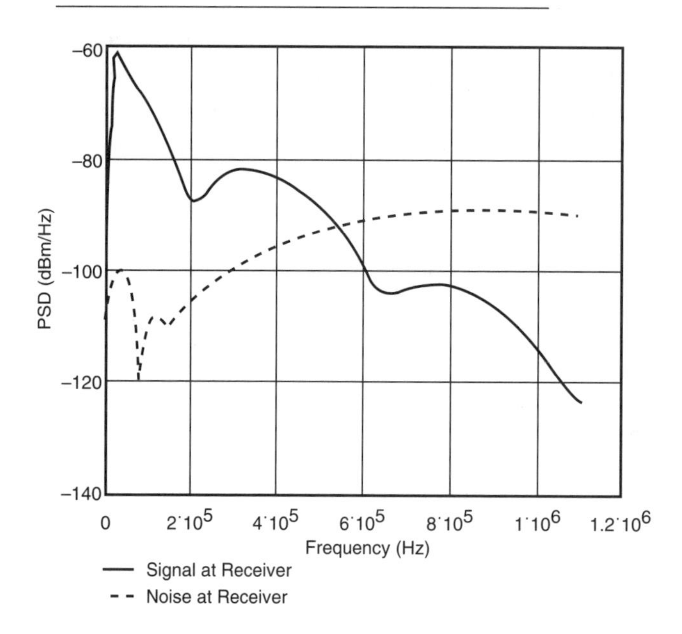


Figure 5.4 also shows the noise due to the crosstalk as well as the AWGN. You might expect areas in the frequency band at which the received signal power is greater than the noise power to be able to carry more information than those areas where the noise is greater.

Figure 5.5 shows the SNR as a function of frequency on the loop. You should be able to verify that the SNR curve is simply the difference of the two curves shown in Figure 5.4.

## Note

Recall that when working in mW, you can find the SNR by doing a division, as opposed to a subtraction used when the powers are given in dBm.

Figure 5.5 SNR at the receiver on case study loop.

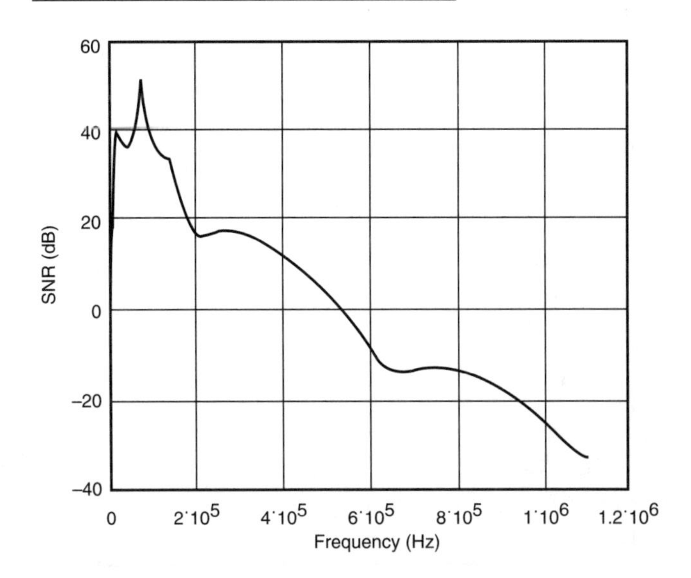


Figure 5.6 shows the information-carrying ability of the channel as a function of frequency. A 6 dB margin is assumed for this graph. The units of the y-axis of this graph are bits/Hz. Notice that the channel is able to carry more bits/Hz at frequencies where the SNR is high. At frequencies where the SNR is very low (including when it is negative), the bits/Hz that the channel can carry approaches zero.

Some systems may not be able to support areas with high information-carrying capability to the limit. For example, in one area the plot in Figure 5.6 can carry about 15 bits/Hz. An implementation may be designed to only send up to 10 bits/Hz (it would still send information at this frequency, but at 10 bits/Hz instead of 15 bits/Hz). Thus although a loop may have a very high capacity, practical implementation will sometimes limit the data rate on the line.

Figure 5.6 shows that a modulation scheme using such a channel must have a very flexible way to heavily use parts of the channel that are "good" and can transmit many bits/Hz. Likewise, the modulation scheme should be able to ignore the parts of the channel that are "bad" and support very close to zero bits/Hz. Later chapters discuss how the various modulation schemes proposed for ADSL and VDSL accomplish this goal.

Figure 5.6 Information-carrying ability as a function of frequency for the case study loop.

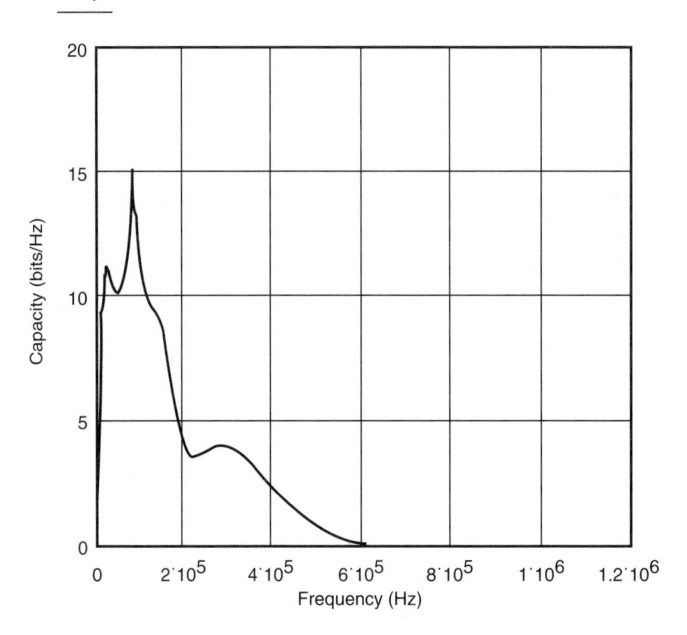


## *Summary*

This chapter dealt with the maximum theoretical data rate that can reliably be sent through a channel—often called *channel capacity*. Although practical systems do not operate at capacity, it is a good benchmark for evaluating a modulation and detection system. A derivation of channel capacity, applicable to DSL over twisted pairs, was given. The main constraints for this example include a bandwidth constraint and an average PSD constraint. (A more general derivation is given in Appendix 5A at the end of this chapter.) This chapter also evaluated the capacities of different ADSL and VDSL loops with common crosstalk scenarios and examined the dependence of capacity on SNR over frequency via a case study.

## *Exercises*

1. Find the entropy of a random variable X given that X can take on the value 0 with probability p and the value of 1 with probability 1-p. Graph H with respect to p. Note the point at which X has the greatest entropy.

2. As discussed in Chapter 4, a 2B1Q symbol can have one of four values: -3, -1, +1, and +3. If the four levels are equally likely, find the entropy of a 2B1Q symbol.

How many bits of uncertainty can be cleared up if the 2B1Q symbol is properly received and decoded?

3. Repeat exercise 2 for an 8-PAM signal with equally likely levels.
4. Graph the channel capacity of an AWGN channel according to Eqtn. 5.14 with respect to SNR. Repeat the procedure assuming that the channel has a bandwidth of 1.1 MHz.
5. Find the SNR and bits/Hz with respect to frequency of T1.601 Loop #13. Assume that the signal is downstream ADSL and that 24 ISDN NEXT disturbers exist. Find the channel capacity of this loop.
6. Evaluate the channel capacity for the case in exercise 5 assuming a 6 dB margin. Again assume that the input is an ADSL downstream signal.
7. For this chapter's case study, evaluate the channel capacity assuming a downstream ADSL signal. For the same loop, find the PSD that would achieve capacity assuming no PSD constraint, but rather a total power constraint of 20 dBm into 100 ohms. Find the resultant capacity and compare it to the PSD-constrained case.

## *Endnotes*


1. James Aslanis and John Cioffi, Achievable Information Rates on Digital Subscriber Loops: Limiting Information Rates with Crosstalk. *IEEE Transactions on Communications,* Vol. 40, No. 2, Feb. 1992.
2. Cover, et al, Asynchronous Multiple-Access Channel Capacity. *IEEE Transactions on Information Theory*, Vol. IT-27, No. 4, July 1981.
3. Robert G. Gallager, *Information Theory and Reliable Communication.* John Wiley & Sons, New York, 1968.
4. Thomas Cover and Joy Thomas, *Elements of Information Theory.* John Wiley & Sons, New York, 1991.
5. Roger Cheng and Sergio Verdu, Gaussian Multiaccess Channels with ISI: Capacity Region and Multiuser Water-Filling. *IEEE Transactions on Information Theory*, Vol. 39, No. 3, May 1993.
6. Roman Kuc, *Introduction to Digital Signal Processing.* McGraw-Hill, New York, 1988.
7. Walter Hirt and James Massey, Capacity of the Discrete-Time Gaussian Channel with Intersymbol Interference. *IEEE Transactions on Information Theory*, May 1998.

# *Appendix 5A*

In this chapter, the capacity of a channel was derived with two constraints on the input to the channel:

- The input power is finite and less than some W.
- The input PSD is flat over frequency.

In this appendix, the second constraint is lifted, and the capacity is derived with only the finite power constraint. Thus the treatment of capacity in the appendix is more general than the corresponding treatment in the main chapter.

In Eqtn. 5A.1, the capacity or maximum rate of the channel is given as a summation of the maximum rates of L subchannels. All quantities are given except for the power in each subchannel $S_i$.

Eqtn. 5A.1

$$R_{max} = \sum_{i=1}^{L} \frac{B}{L} \log_2 \left( 1 + \frac{|G_i|^2 S_i}{|H_i Q_i + Z_i|^2} \right)$$

This discussion assumes that in the original untransformed system, a power constraint W was placed on x such that the average power (normalized to 1 ohm) was Eqtn. 5A.2.

Eqtn. 5A.2

$$\sum_{i=0}^{N-1} |x_i|^2 \le W$$

Using Parseval's relation, the power constraints of the transformed sequence becomes Eqtn. 5A.3.

Eqtn. 5A.3

$$\sum_{i=0}^{N-1} |X_i|^2 \le N^2 W$$

Or stated in terms of power, Eqtn. 5A.4.

Eqtn. 5A.4

$$\sum_{i=0}^{N-1} |X_i|^2 \le N^2 W$$

For notational ease, the substitution Eqtn. 5A.5 will be made:

Eqtn. 5A.5

$$T_i = \frac{|G_i|^2}{|H_i Q_i + Z_i|^2}$$

Thus the expression for capacity, including the substitution of Eqtn. 5A.5, becomes Eqtn. 5A.6.

Eqtn. 5A.6

$$R_{max} = \sum_{i=1}^{L} \frac{B}{L} \log_2 \left(1 + T_i S_i\right)$$

As N approaches infinity, the summation can be converted to an integration and the rate of the total system can be bounded by Eqtn. 5A.7.

Eqtn. 5A.7

$$R \leq \int_0^B \log_2(1 + S(f)T(f))df$$

The constraints on the power then become Eqtn. 5A.8 and Eqtn. 5A.9

Eqtn. 5A.8

$$S(f) > 0$$

Eqtn. 5A.9

$$\int_0^B S(f)df \leq W$$

The first constraint is intuitive, as power cannot be negative, and the second constraint reflects the total power constraint.

To find an S that maximizes Eqtn. 5A.7 under the constraints of Eqtn. 5A.8 and Eqtn. 5A.9, you can use the Lagrange multiplier technique and maximize Eqtn. 5A.10.

Eqtn. 5A.10

$$\log_2(1 + S(f)T(f)) + \lambda S(f)$$

With respect to S where $\lambda$ is a Lagrange multiplier. Continuing by differentiating gives Eqtn. 5A.11.

Eqtn. 5A.11

$$0 = \frac{T(f)}{1 + S(f)T(f)} + \lambda$$

Eqtn. 5A.11 can be solved for S as Eqtn. 5A.12.

Eqtn. 5A.12

$$\begin{aligned} S(f) &= \frac{T(f) + \lambda}{\lambda T(f)} \\ &= \frac{1}{\lambda} + \frac{1}{T(f)} \\ &= \left[c - T^{-1}(f)\right]^{+} \end{aligned}$$

Here c is a constant to be determined by Eqtn. 5A.13.

Eqtn. 5A.13

$$\left[c - T^{-1}(f)\right]^{+} = \begin{cases} c - T^{-1}(f), & c - T^{-1}(f) \geq 0 \\ 0, & c - T^{-1}(f) < 0 \end{cases}$$

Eqtn. 5A.12 specifies the power spectral density necessary on the channel to achieve capacity. When the power spectral density is known, Eqtn. 5A.7 can be used to find the actual capacity. If $T^{-1}$ is plotted as shown in Figure 5A.1, the power spectral density needed to achieve capacity is analogous to the distribution that water would take if poured into an imaginary container with $T^{-1}$ as the container's bottom. For this reason, Eqtn. 5A.12 is often referred to as the "water-filling" solution. The constant c then becomes the level of the "water" such that the total power distributed in the container obeys the power constraints of the signal.

Figure 5A.1 Illustration of the water-filling solution for the capacity-achieving PSD of a channel.

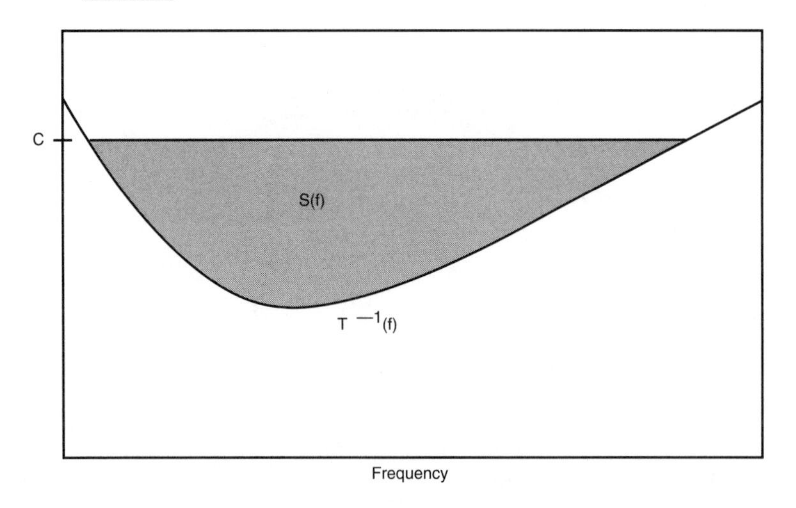

CHAPTER 6

# *DSL Modulation Basics*

In this chapter:

- Descriptions of the different blocks used in DSL transmitters and receivers
- Basic concepts of single carrier and multicarrier modulation schemes
- Discussion of forward error correction coding and decoding
- Overview of equalization methods and the performance bounds on each

This chapter presents some of the basics of modulation systems as well as other blocks used by DSL communications systems. Subsequent chapters cover ADSL and VDSL in detail and show how the specific implementations of the basic blocks are put together to create a full DSL transmitter and receiver. Many of the topics and blocks in this chapter are also relevant to communication systems other than those used on DSLs, including voice-band modems, cable modems, and magnetic storage channels.

Figure 6.1 shows a generic block diagram of a transmitter and a receiver that is used for reference throughout this chapter. This figure is not specific to either a Carrierless Amplitude and Phase (CAP)/Quadrature Amplitude Modulation (QAM) or a discrete multitone (DMT) system, but rather it illustrates where some of the relevant blocks fit in the "big picture." Notice that most blocks in the transmitter have corresponding sister blocks in the receiver.

Figure 6.1 Transmitter and receiver block diagram for a communications system.

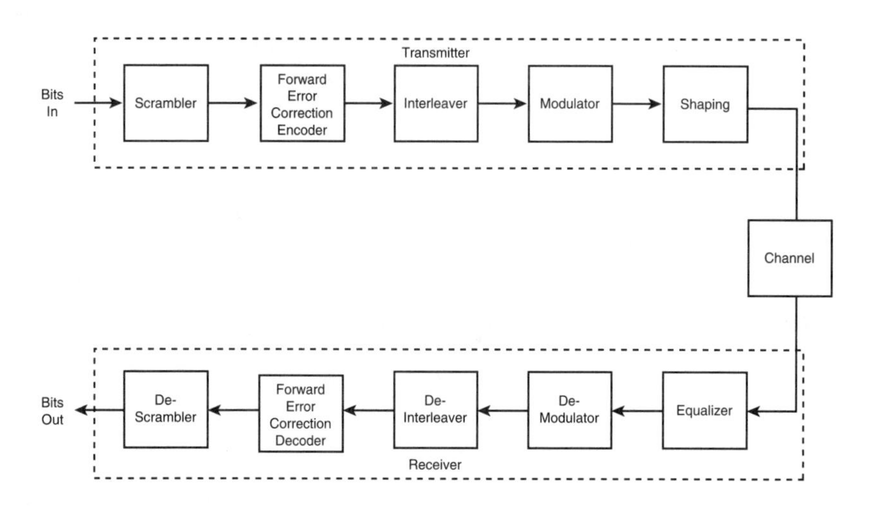


# Modulation/Demodulation Blocks

A modulation block typically converts input bits into waveforms to be sent over the channel. A demodulation block in the receiver maps received waveforms back to the bits that most likely generated the waveform. The main idea is to re-create the bits at the demodulator output that were present at the modulator input. The following sections describe three types of modulation common in DSLs: QAM, CAP, and multitone.

## QAM Modulation

QAM utilizes a sine wave and a cosine wave with the same frequency component to convey information. The waves are sent over a single channel simultaneously, and the amplitude (including sign and magnitude) of each wave conveys the information (bits) being sent. At least one period (and sometimes more) of the waves is sent to convey a set of bits before a new set of bits is to be sent (resulting in new magnitudes for the sine and cosine waves). QAM modulation has been used for many years and is the basis for many of the voice-band modem specifications, including V.34.

### A Simple QAM Example

An example of a QAM modulation system that can send four bits of information per QAM symbol appears in Figure 6.2.

Figure 6.2 An example of a QAM system sending four bits per symbol over a channel.

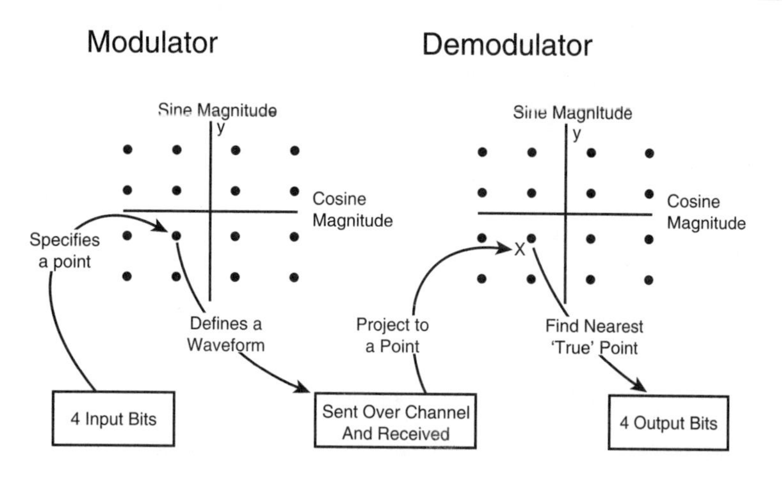


The four bits of information to be transmitted are mapped to 1 of 16 points on a QAM *constellation*. Note that for four bits of information, 16 points allow a unique point for any combination of bits. The x and y components of the point to which the bits were mapped specify the amplitude of the cosine wave and sine wave to be sent over the channel. Both the transmitter and receiver know the predetermined method of mapping between the bits and a point.

After cosine and sine waves are sent over the channel, the receiver recovers and estimates the amplitude of each wave (which normally requires significant equalization and processing as described later in the chapter). These magnitudes are projected on a constellation identical to that used at the transmitter. Again, the cosine wave amplitude is the x component, and the sine magnitude is the y component. Normally, noise and distortion in the channel and compromised electronics in the transmitter and receiver prevent the point projected onto the receiver's constellation from falling directly on top of a "true" point. The receiver, however, selects the closest "true" point to the projected point as the one the transmitter is most likely to use to generate the QAM symbol. This point is then mapped into four bits using the same mapping method employed in the transmitter (but mapping in the opposite direction). If too much noise is present at the receiver, the point projected on the constellation may be closer to the wrong known point than the accurate known point, resulting in an error estimating the recovered symbol.

This example is sometimes called 16-QAM because the constellation has 16 points. In general, other numbers of bits can be transmitted per QAM symbol. For example, if two bits per symbol were transmitted, a four-point constellation would be necessary and the modulation would be called 4-QAM. Figure 6.3 shows a 4-QAM constellation overlaid on a 16-QAM constellation. The average energy of each signal set is approximately the same. Note that the distance between the points in the 4-QAM constellation (labeled $d_4$) is greater than the distance between the points in the 16-QAM constellation (labeled $d_{16}$). Consequently, on the same channel more noise is necessary to cause an error when using 4-QAM than when using 16-QAM. Said differently, 16-QAM requires a greater SNR than 4-QAM requires.

Figure 6.3 An example of a 4-QAM constellation overlaid on a 16-QAM constellation with the same average energy.

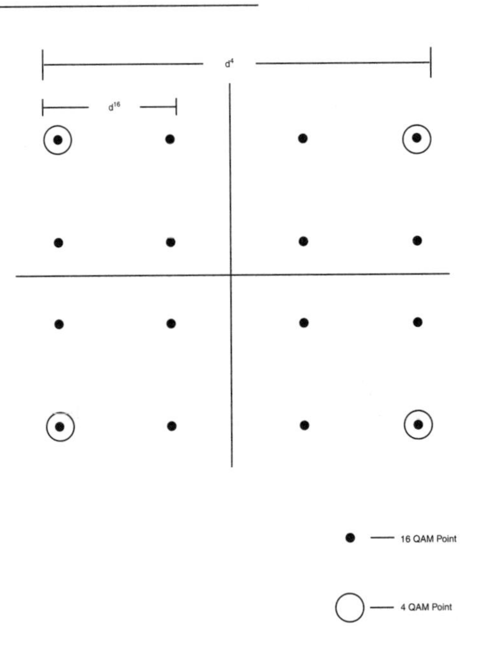


A block diagram of a QAM modulator appears in Figure 6.4. The branch of the modulator that carries the cosine wave amplitude is sometimes called the *in phase* branch. The cosine amplitude is called the in phase, or the I, component. The branch of the modulator that carries the sine wave amplitude is sometimes called the *quadrature* branch. The sine amplitude is called the quadrature, or the Q, component.

Figure 6.4 QAM modulator block diagram.

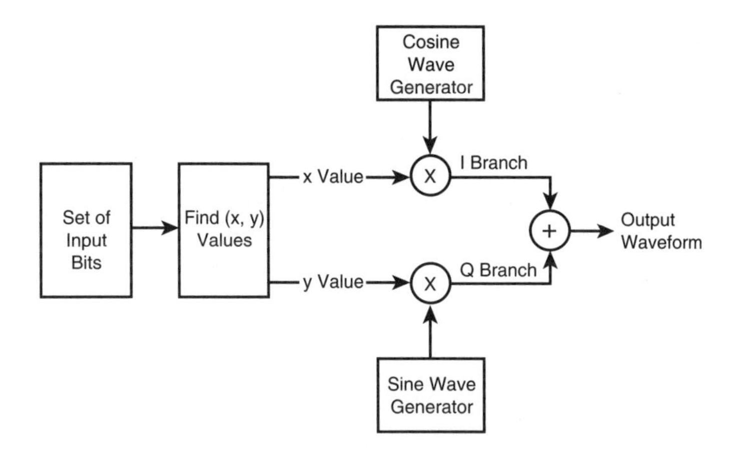


## *A Mathematical Treatment of QAM*

Orthogonality between a sine wave and a cosine wave allows them to transmit data simultaneously over a channel. Considering a single period of each wave, the orthogonality principle can be illustrated by Eqtn. 6.1.

Eqtn. 6.1

$$\int_0^{\tau} \sin\left(\tfrac{2\pi t}{\tau}\right)\cos\left(\tfrac{2\pi t}{\tau}\right)dt = 0$$

In Eqtn. 6.1, $\tau$ is the period of the sine and cosine waves. Because of the orthogonality property, the sine and cosine functions are sometimes called *basis functions*.

An example of a QAM demodulator appears in Figure 6.5.

The input to this block comes from the channel, and the output projects onto the receiver's constellation. If one assumes that the channel has no loss and that the receiver has perfect phase timing, then it is not difficult to write the equations for each point in Figure 6.5 for some symbol, i. Eqtn. 6.2 gives the input signal at point A.

Eqtn. 6.2

$$V_A(t) = X_i \cos(\omega t) + Y_i \sin(\omega t)$$

Figure 6.5 (a) Logical view of a QAM demodulator. (b) Signal-processing view of a QAM demodulator.

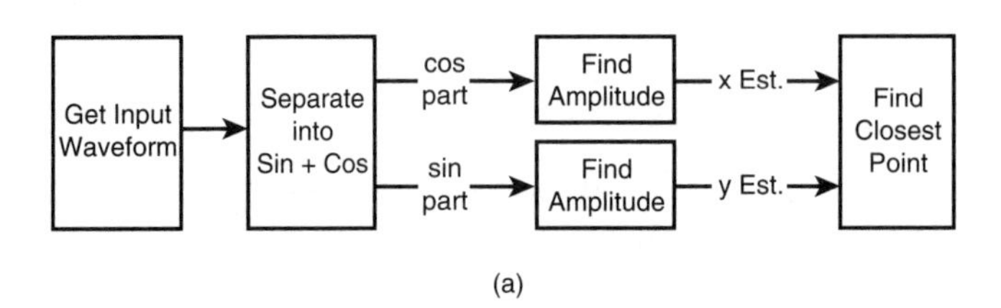


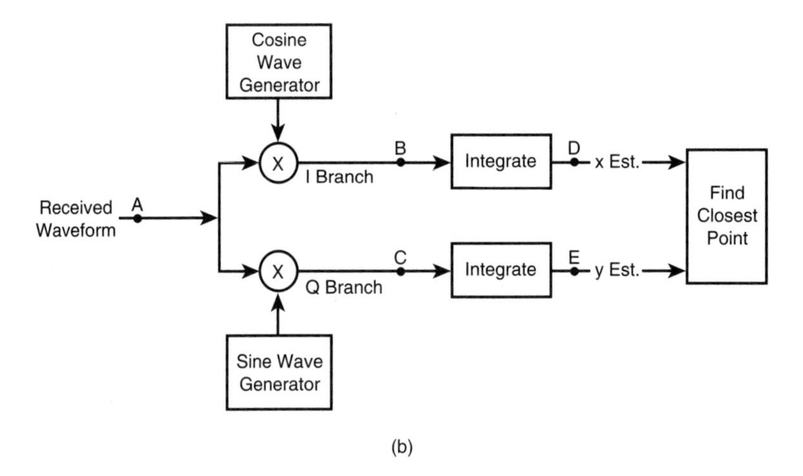


In Eqtn. 6.2, $X_i$ is the amplitude (sign and magnitude) of the cosine wave encoded at the transmitter, and $Y_i$ is the amplitude (sign and magnitude) of the sine wave encoded at the transmitter. After passing through the multiplication blocks at points B and C, the signals are given by Eqtn. 6.3 and Eqtn. 6.4.

Eqtn. 6.3

$$V_B(t) = X_i \cos^2(\omega t) + Y_i \sin(\omega t)\cos(\omega t)$$

Eqtn. 6.4

$$V_C(t) = X_i \cos(\omega t)\sin(\omega t) + Y_i \sin^2(\omega t)$$

The signals at points B and C then independently pass through integration blocks. These blocks integrate over a single period and reset after each symbol. The switches in Figure

6.5 sample the output of the integrator at the end of each integration period. The signal at points D and E can thus be written as Eqtn. 6.5 and Eqtn. 6.6.

Eqtn. 6.5

$$\begin{aligned}
V_D(t) &= \int_0^{\tau} V_B(t)dt \\
&= \int_0^{\tau} X_i \cos^2(\omega t) + Y_i \sin(\omega t)\cos(\omega t)dt \\
&= \int_0^{\tau} X_i \cos^2(\omega t)dt + \int_0^{\tau} Y_i \sin(\omega t)\cos(\omega t)dt \\
&= \frac{\tau X_i}{2} + 0 = \frac{\tau X_i}{2}
\end{aligned}$$

Eqtn. 6.6

$$\begin{aligned}
V_E(t) &= \int_0^{\tau} V_C(t)dt \\
&= \int_0^{\tau} X_i \cos(\omega t)\sin(\omega t) + Y_i \sin^2(\omega t)dt \\
&= \int_0^{\tau} X_i \cos(\omega t)\sin(\omega t)dt + \int_0^{\tau} Y_i \sin^2(\omega t)dt \\
&= 0 + \frac{\tau Y_i}{2} = \frac{\tau Y_i}{2}
\end{aligned}$$

Note that the terms going to zero in Eqtn. 6.5 and Eqtn. 6.6 do so because of the orthogonality property of sines and cosines. The values at the output of the demodulator then contain a scaled estimate of the magnitudes of the received sine and cosine waves to project onto the receiver's constellation map.

**Note**

The scale factor can be removed before integrating by scaling the multiplication cosine and sine waves.

This analysis assumes that the baseband pulses in the modulator used to shape the symbols before multiplying by the sine and cosine waves were simple pulses with unity amplitude during the entire symbol period. In general, a shaping filter may be used to shape the pulse before the mixing function. (The purpose of this step is to reduce the bandwidth of the transmitted signal by filtering out higher-frequency components.) If the shaping filter has a response of p(t), then the transmitted signal for symbol i will be of the form shown in Eqtn. 6.7.

Eqtn. 6.7

$$V(t) = X_i p(t - i\tau)\cos(\omega t) + Y_i p(t - i\tau)\sin(\omega t)$$

Though the shaping pulse is not very important to the basics of QAM, including the shaping term is more complete and allows comparison to CAP in the next section.

Thus I have shown mathematically how a QAM signal can be received and demodulated. Other methods of demodulation are possible as well, and most take advantage of the orthogonality property discussed in this section.

## *CAP Modulation*

Similar to a QAM modulator, a carrierless amplitude and phase (CAP) modulator uses a constellation to encode bits at the transmitter and decode bits at the receiver. The x and y values resulting from the encoding process are then used to excite a digital filter. A CAP modulator appears in Figure 6.6.

Figure 6.6 A CAP modulator.

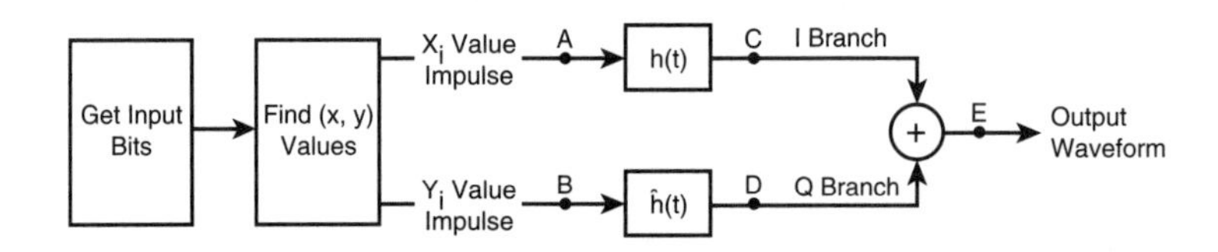


The modulator has two branches—one in-phase branch and one quadrature branch. The impulse responses of the digital filters are *hilbert transform pairs* or more simply, a *hilbert pair*. Two functions that form a hilbert pair are orthogonal to one another. In general, any valid hilbert pair can be used to make a CAP modulator; however, today's implementation of CAP uses a cosine wave and a sine wave shaped with a transmit pulse. Typically, CAP modulation is performed with digital filters instead of in-phase and quadrature multipliers.

When comparing a CAP modulator to a QAM modulator, consider voltages at the different points in Figure 6.6. Assume that this modulator uses the same constellation encoding method, the same size constellation, and the same symbol rate as the QAM modulator discussed earlier. The signal at each point can then be written as Eqtn. 6.8 and Eqtn. 6.9.

Eqtn. 6.8

$$V_A(t) = X_i \delta(i\tau)$$

Eqtn. 6.9

$$V_B(t) = Y_i \delta(i\tau)$$

Note that both quantities are simply discrete impulses that will excite the digital filters in the modulator.

Given the impulse responses of the filters in the modulator, the output of the two filters—found by simple convolution—due to the inputs is shown by Eqtn. 6.10 and Eqtn. 6.11.

Eqtn. 6.10

$$\begin{aligned} V_C(t) &= X_i \delta(i\tau) * h(t) \\ &= \int_0^{\infty} X_i \delta(i\alpha) h(t-\alpha) d\alpha \\ &= X_i h(t - i\tau) \end{aligned}$$

Eqtn. 6.11

$$\begin{aligned} V_D(t) &= Y_i \delta(i\tau) * \hat{h}(t) \\ &= \int_0^{\infty} Y_i \delta(i\alpha) \hat{h}(t-\alpha) d\alpha \\ &= Y_i \hat{h}(t - i\tau) \end{aligned}$$

In these equations, the sifting property was used to arrive at both results.

The output of the modulator due to symbol i is then given by Eqtn. 6.12.

Eqtn. 6.12

$$V_E(t) = X_i h(t - i\tau) + Y_i \hat{h}(t - i\tau)$$

Comparing Eqtn. 6.12 to the output of a QAM modulator with shaping given in Eqtn. 6.7, note that the two would be identical if Eqtn. 6.13 and Eqtn. 6.14 hold.

Eqtn. 6.13

$$h(t - i\tau) = p(t - i\tau)\cos(\omega t)$$

Eqtn. 6.14

$$\hat{h}(t-i\tau)=p(t-i\tau)\sin(\omega t)$$

If one further designed the system such that $\omega\tau=\pi$, then Eqtn. 6.13 and Eqtn. 6.14 can be written as Eqtn. 6.15 and Eqtn. 6.16.

Eqtn. 6.15

$$h(t-i\tau)=p(t-i\tau)\cos(\omega(t-i\tau))$$

Eqtn. 6.16

$$\hat{h}(t-i\tau)=p(t-i\tau)\sin(\omega(t-i\tau))$$

Both of these equations are valid because i is an integer. The time impulse responses of the filters with no dependence on i then become Eqtn. 6.17 and Eqtn. 6.18.

Eqtn. 6.17

$$h(t)=p(t)\cos(\omega t)$$

Eqtn. 6.18

$$\hat{h}(t)=p(t)\sin(\omega t)$$

These results are important because they show that properly relating the symbol rate and center frequency of a CAP and QAM system allows the time-domain waveforms to be identical. For other relationships of $\omega$ and $\tau$—assuming that they are the same across a CAP and QAM system—it can be shown that the modulation systems differ by a rotation. That is, the waveforms will be identical if one of the sets of constellations used for encoding and decoding is rotated about the origin of the plane.[1]

## *Multitone Modulation*

Multitone modulation is the basis of the DMT version of ADSL as well as some multicarrier versions of VDSL. This type of modulation is sometimes called *orthogonal frequency division multiplexing* (OFDM). For simplicity, this book simply refers to multitone modulation as DMT.

DMT builds on some of the ideas of QAM. Imagine having more than one constellation encoder. Each encoder receives a set of bits that are encoded using a constellation encoder as described in the previous sections. The output values from the constellation encoder are again the amplitudes of cosine and sine waves; however, a different sine and cosine frequency is used for each constellation encoder. All the sine and cosine waves are then summed together and sent over the channel. This waveform is a simple DMT symbol, represented by the diagram in Figure 6.7.

Figure 6.7 Conceptual view of a DMT modulator.

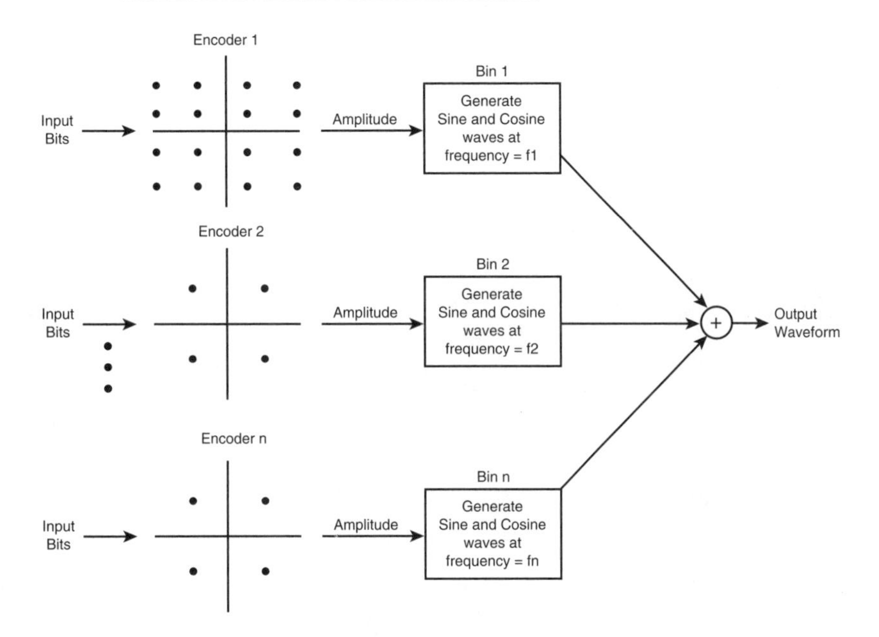


Assuming that one can separate the sines and cosines at different frequencies at the receiver, each set of waveforms can be independently decoded in the same way that a QAM signal can be decoded and the resultant bits output by the constellation decoders.

Note that the idea of using different frequencies to transmit information is not unique to DMT. Television and radio have readily employed such techniques. The difference here is that DMT has a "receiver" tuned to all of the channels at once, whereas the others typically tune to only one channel. Some names used for frequency channels in DMT are *frequency bins* (or *bins*), tones or DMT tones, and subchannels.

Perhaps not obvious in the previous discussion is the assumption that the waveforms in each bin are completely separable from one another. If this is not the case, then decoding for each bin is difficult because the sine and cosine waves in each bin can be corrupted by signals from other bins. One key to DMT is that the sine and cosine frequencies used in each bin should be integer multiples of a common frequency and that the symbol period, $\tau$, be the inverse of that common frequency. (It can also be an integer multiple of the reciprocal of that frequency.)

This common frequency will be referred to as the *fundamental frequency*. From the analysis done for QAM, one can already claim that the sine and cosine waves at the fundamental frequency form basis functions. To ensure that interference does not exist between bins, one must show that sine and cosine waves from any bin are orthogonal to sine and cosine waves in any other bin. Mathematically, this orthogonality can be written as Eqtns. 6.19, 6.20, and 6.21.

Eqtn. 6.19

$$\int_0^{\tau} \cos(n\omega_f t)\cos(m\omega_f t) = 0$$

Eqtn. 6.20

$$\int_0^{\tau} \cos(n\omega_f t)\sin(m\omega_f t) = 0$$

Eqtn. 6.21

$$\int_0^{\tau} \sin(n\omega_f t)\sin(m\omega_f t) = 0$$

Here n and m are different integers, and $\omega_f$ is the radian fundamental frequency. Performing the integration in Eqtn. 6.19 yields Eqtn. 6.22.

Note that if n=m, then the first term will converge to $1/2\ \tau$, which follows from the result of Eqtn. 6.5. The relationships in Eqtn. 6.20 and Eqtn. 6.21 can be likewise shown except that in Eqtn. 6.20 the orthogonality holds even for the case of n=m, similar to the discussion around Eqtn. 6.1. Thus the demodulation of a DMT symbol relies on the orthogonality of sine and cosine waves at different frequencies as well as between a sine and cosine wave at the same frequency.

The modulation and demodulation procedures in Figure 6.7 are brute-force methods of creating and detecting a DMT symbol. These methods would be very monotonous on an actual implementation and are generally not the way DMT systems are realized. To better understand how an implementation could be simplified, consider the summation of a sine and cosine of duration $\tau$. Such a waveform could be written as Eqtn. 6.23.

Eqtn. 6.22

$$\begin{aligned}
&\int_0^{\tau} \cos(n\omega_f t)\cos(m\omega_f t) \\
&= \int_0^{\tau} \frac{1}{2}\cos((n-m)\omega_f t) + \frac{1}{2}\cos((n+m)\omega_f t)dt \\
&= \left[\frac{\sin((n-m)\omega_f t)}{2\omega_f (n-m)} + \frac{\sin((n+m)\omega_f t)}{2\omega_f (n+m)}\right]_0^{\tau} \\
&= \frac{\sin\left((n-m)\frac{2\pi}{\tau}\tau\right)}{2\omega_f (n-m)} + \frac{\sin\left((n+m)\frac{2\pi}{\tau}\tau\right)}{2\omega_f (n+m)} \\
&= \frac{\sin((n-m)2\pi)}{2\omega_f (n-m)} + \frac{\sin((n+m)2\pi)}{2\omega_f (n+m)} \\
&= 0 \text{ for n, m integers and n} \neq \text{m}
\end{aligned}$$

Eqtn. 6.23

$$s(t) = \begin{cases} X_n \cos(n\omega_f t) + Y_n \sin(n\omega_f t) & \text{for } 0 < \text{t} \leq \tau \\ 0 & \text{otherwise} \end{cases}$$

Such a signal is representative of the contribution of a single bin, or simply bin n, to a single DMT symbol. If s(t) is sampled at a rate of $2*N*f_f$, the resulting nonzero values of the signal are given in Eqtn. 6.24.

Eqtn. 6.24

$$\begin{aligned}
s_k &= X_n \cos\left(n\omega_f \frac{k}{2Nf_f}\right) + Y_n \sin\left(n\omega_f \frac{k}{2Nf_f}\right) \\
&= X_n \cos\left(\frac{\pi nk}{N}\right) + Y_n \sin\left(\frac{\pi nk}{N}\right) \quad \text{for } 0 < \text{k} \leq 2\text{N}
\end{aligned}$$

In a DMT system, N would represent the largest bin transmitting a signal. This signal is at a frequency of $Nf_f$. Because the Nyquist theorem states that the sampling rate for a system must be twice the largest frequency in the system, the rate of $2Nf_f$ was chosen. If we take the *discrete fourier transform* (DFT) of $s_k$ using 2N points in the transform, the result is Eqtn. 6.25.

Eqtn. 6.25

$$S_m = \sum_{k=0}^{2N} \left( X_n \cos\left(\frac{\pi nk}{N}\right) + Y_n \sin\left(\frac{\pi nk}{N}\right) \right) e^{\frac{-j2\pi km}{2N}}$$
$$= \begin{cases} N(X_n - jY_n) & \text{for m = n} \\ N(X_n + jY_n) & \text{for m = 2N - n} \\ 0 \text{ otherwise} \end{cases}$$

Because the DFT decomposes a signal into its frequency-domain components, the result of Eqtn. 6.25 should not be surprising. Simply put, over the 2N points being transformed, the signal had energy at only one frequency. Nonzero values at two points in the frequency domain arise because, in the frequency domain the frequency represented after the transform represents a two-sided spectrum. For all real values in the time domain (consistent with $s_k$), the 2N points will exhibit complex conjugate symmetry about the center of the points. This is analogous to having positive and negative frequency components resulting from a fourier transform. The zero values in Eqtn. 6.25 result from no energy being present at the frequencies represented by those points. In essence, Eqtn. 6.25 illustrates orthogonality of sines and cosines at different frequencies, as well as a sine and cosine wave at the same frequency.

The result of Eqtn. 6.25 suggests another way to produce a DMT symbol. Instead of mapping the output of a constellation encoder into a cosine and sine amplitude, the output can be mapped into a complex number in a vector. The value from the X-, or cosine, axis would represent the real part of the complex number, and the Y value, or sine magnitude, would represent the imaginary part. If the outputs of all the constellation encoders are ordered in the vector, then each vector point represents one of the DMT bins. If N bins existed in the DMT system, the complex vector would have N entries. A suffix containing the complex conjugates of the vector's original entries can then be added to this vector such that the new vector has complex conjugate symmetry. An *inverse discrete fourier transform* (IDFT) on this new vector would then produce a real-valued time-domain sequence equivalent to the original DMT modulator in Figure 6.7. Figure 6.8 shows this new method of performing DMT modulation.

Figure 6.8 also shows a method of DMT demodulation. Note that it is basically the reverse of the modulator except that a DFT is used instead of an IDFT. This approach should make sense because the DFT goes from the time domain to the frequency domain. Because the time-domain values are real, the output of the DFT block has complex conjugate symmetry. Only half of the output is then necessary to drive the constellation decoders. Implementations often use the fast fourier transform (FFT) and inverse fast fourier transform (IFFT) to realize the modulator and demodulator. These transform algorithms realize the DFT and IDFT with reduced computational complexity.

Figure 6.8 DMT modulation using an IDFT.

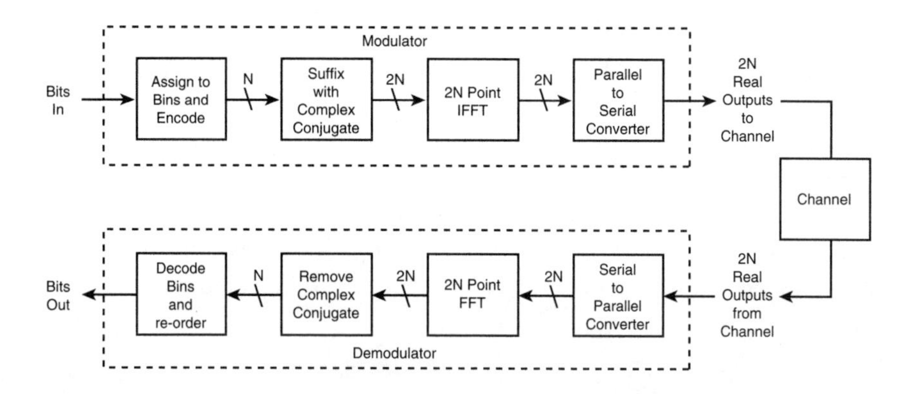


DMT allows a communications system to be very flexible and optimally utilize a channel. For example, consider the channel discussed as a case study in Chapter 5, "DSL Theoretical Capacity in Crosstalk Environments." The SNR of the channel, assuming an ADSL downstream PSD at the transmitter, appears in Figure 6.9.

Figure 6.9 SNR with respect to frequency of a sample channel; DMT bins in the areas where the SNR is high can use more dense QAM constellations.

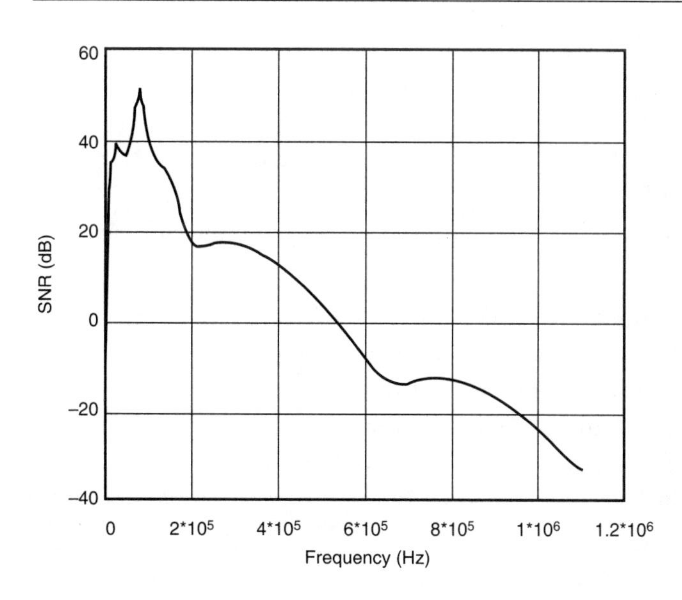

Compared to the bins when the SNR is low, the bins occupying parts of the channel where the SNR is high can be used to transmit more bits. The process involves increasing the number of points used in the constellations of the "good" bins. If you plot the information sent as a function of frequency for this channel using DMT, the resulting curve takes on a shape similar to that shown in Figure 5.6, thus showing the optimal utilization of the channel.

DMT also allows a simple method to increase or decrease the output PSD of a transmitter in selected frequency regions. The process involves scaling the complex valued vector for the bin where power adjustment is desired before performing the IFFT. Such adjustment can boost power in regions where the channel insertion loss is poor, or reduce power in regions where interference with other systems must be avoided. Note that power boosts and reductions can be done in a CAP/QAM system also with external shaping filters, although these tend to be complex and can be difficult to implement for arbitrary shaping requirements.

**Note**

A classic example is the reduction of VDSL's PSD in certain bands to avoid interference with amateur radio or emergency bands.

## *Forward Error Correction*

A forward error correction (FEC) block adds redundancy to the data to be transmitted. Usually this redundancy is a small fraction of the actual transmitted payload. The value of adding the redundancy is that the FEC block might be able to correct bits that the demodulator decodes incorrectly. The power of a forward error correction technique is often described using the term *coding gain*. A reasonable coding gain for a DSL forward error correction technique may be 3 dB at a bit error rate (BER) of $10^{-7}$. Stated another way, adding the FEC to the system makes the system perform as well as a system without coding but with twice the transmit power (3 dB more transmit power). This attribute is indeed formidable.

Two types of forward error correction are common in DSL systems. The first type is a cyclic block code known as a *Reed-Solomon coding*. The second is a *convolution code* known as *trellis-coded modulation* (TCM) and ties in tightly with the constellation encoding used in both CAP/QAM and DMT modulation. The following sections discuss the basics of these two techniques. Detailed treatment of these subjects can be found in various books, including *Digital Modulation and Coding* and *Error Control Coding: Fundamentals and Applications*.[2,3]

## *Reed-Solomon Coding*

Reed-Solomon coding was first described around 1960.[4] This type of coding is a subset of cyclic coding, which in turn is a subset of block coding. Reed-Solomon codes are also a special case of nonbinary BCH codes named for Bose and Ray-Chaudhuri and Hocquenghem who independently discovered them around 1960.[5,6] Many readers may be familiar with *cyclic redundancy check codes* (CRC codes) used in data transmission. Because CRC codes are simple and popular, they will be used as a starting point in describing Reed-Solomon codes that are typical in DSL systems.

CRC codes are typical in almost all types of layer 2 protocols including high-level data link control (HDLC)-based schemes and Ethernet. Normally, these codes operate on bits and are used for error detection but not error correction. Additionally, the number of bits being covered by the check bits is variable.

**Note**

Operating on bits infers that modulo 2 arithmetic is used for coding and checking. It can also be said that the arithmetic is done in GF(2).

Reed-Solomon codes used in DSL systems use *galois field arithmetic*. Specifically, they operate on GF(256), meaning that each symbol can take on one of 256 values (0 through 255). Thus each symbol is made up of eight bits—one byte long. The arithmetic rules in GF(256) define the results when two symbols are added, subtracted, multiplied, or divided. It turns out that in GF(256) carrying is not done in the same way that it is done in the base-10 decimal system.

A Reed-Solomon code produces codewords made up of a fixed number of data bytes and a fixed number of check bytes. When operating in GF(256), the total number of bytes in a codeword must be less than 255. Using different codes can vary the number of data bytes and check bytes making up the codeword. A typical DSL implementation might use a codeword size of 240 bytes, 224 of which are data bytes and 16 of which are check bytes. Figure 6.10 shows a Reed-Solomon codeword with these properties.

**Figure 6.10** Example of a Reed-Solomon codeword in GF(256).

| DATA PAYLOAD<br>224 bytes | CHECK BYTES<br>16 bytes |
|---|---|

Reed-Solomon codes allow for error correction in the receiver. Note that in the receiver, the Reed-Solomon decoder block in the previous example would receive 240 bytes from the demodulator and output 224 bytes (only the data bytes) to the rest of the receiver. In general, the number of symbols that a Reed-Solomon decoder can correct is half the number of check bytes used. Thus for the example of 16 check bytes, eight byte errors would be correctable even if these bytes all occur in a row (a burst error). This correction is an extremely powerful benefit for very little overhead (6.6% in this example).

## Convolution Codes

Convolutional coding differs from block coding in that the former does not have codewords made up of distinct data sections and block sections. Instead, redundancy is distributed throughout the coded data. Convolutional codes are sometimes referred to as *trellis codes*. Normally, convolutional encoding is simple, and decoding is a bit more difficult.

**Note**

Trellis codes can be linear or nonlinear. Convolutional codes are a subset of linear trellis codes.

Figure 6.11 shows a simple convolutional encoder.

Figure 6.11 A simple, rate 1/2 convolutional encoder; note that shifting in one bit produces two output bits.

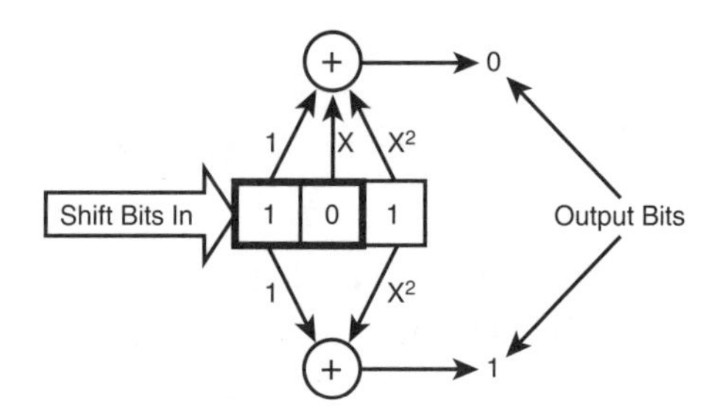


Data bits are shifted into the encoder from the left. Each data bit shifted in produces two data output bits defined by the modulo 2 additions shown. The outputs are defined by generator polynomials. This simple convolutional encoder is said to be a rate 1/2 encoder with a constraint length 2. The rate refers to the ratio of input bits to output bits, and the constraint length refers to the number of delays in the encoder. These parameters, along with the generator polynomials, determine the processing gain of the code.

Note that the first two positions in Figure 6.11 have a darkened outline. The values in these positions are called the *state of the encoder*. The convolutional encoder can be viewed as a finite state machine, changing state based on an input condition and producing an output when it changes state.

Figure 6.12 shows a simple example of using the encoder in Figure 6.11 to encode a set of bits. Note that the initial condition of the encoder is assumed to be all zeros and that the top output is taken before the bottom output.

**Figure 6.12** Input and output of the sample convolutional encoder.

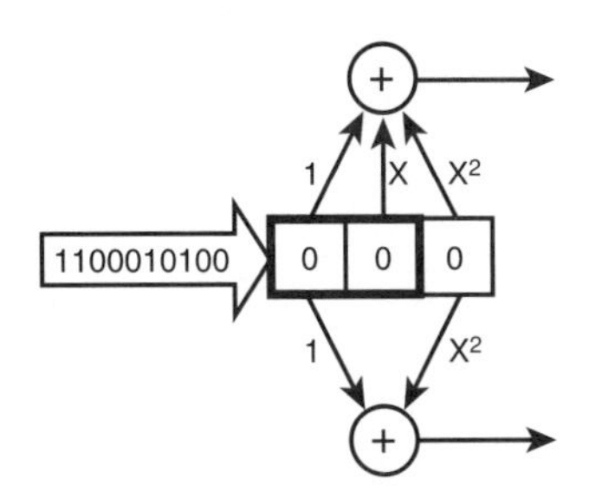


<table>
<tr><td>Input</td><td colspan="2">0</td><td colspan="2">0</td><td colspan="2">1</td><td colspan="2">0</td><td colspan="2">1</td><td colspan="2">0</td><td colspan="2">0</td><td colspan="2">0</td><td colspan="2">1</td><td colspan="2">1</td></tr>
<tr><td>Output</td><td>0</td><td>0</td><td>0</td><td>0</td><td>1</td><td>1</td><td>1</td><td>0</td><td>0</td><td>0</td><td>1</td><td>0</td><td>1</td><td>1</td><td>0</td><td>0</td><td>1</td><td>1</td><td>0</td><td>1</td></tr>
</table>

Each node in the trellis represents a state. Each line represents a pair of received bits. Note that each state has two lines leading to the state and two lines leaving it. When two bits are received, a transition occurs between the current state and the new state. The type of line connecting the current and new state (solid or dotted) determines whether the decoder will output a zero or a one in response to the two bits received. Note that two bits in produce one bit out—just the opposite of the encoder.

Also in Figure 6.13, note that the numbers on each line represent the input to the receiver that will cause the state transition. The type of line (solid or dotted) represents the decoder output.

Decoding of a convolutional code can be done with a trellis and the *Viterbi algorithm*. The trellis in Figure 6.13 can be used for this encoder example.

If no errors occur, the bits received while in a state correspond to one of the paths leaving the state. When errors occur on the line, the received bits may correspond to neither of the lines leaving the state and knowing which line to follow is unclear. The decoding algorithm allows for multiple paths through the trellis to be kept active. By tallying the number of mismatches (errors) that have occurred along each path, the decoding algorithm terminates those with a high error count. The error count is called a *metric*. As time passes and the decoding process continues, a path with a low metric develops, and all others are terminated. This path is then used to produce the output of the decoder.

Figure 6.13 A trellis that can be used to decode the sample convolutional encoder: each node represents one of the four states.

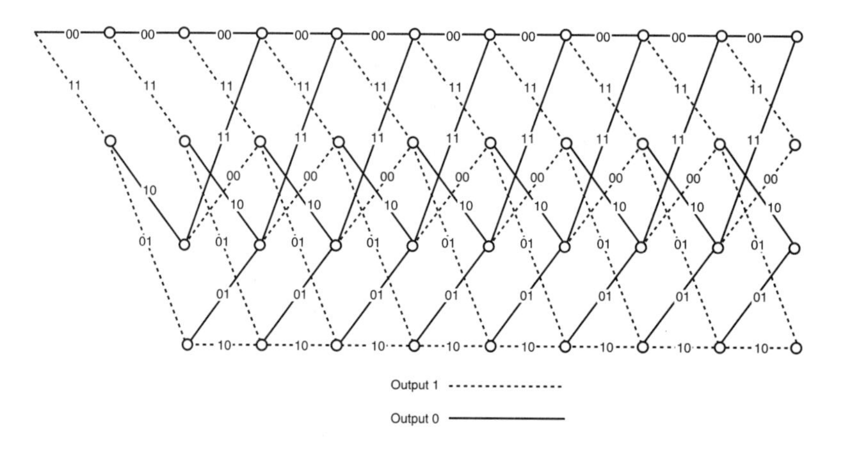


An example of decoding is the best way to understand the decoding algorithm. Consider the coded output from Figure 6.12. The decoding algorithm for the first four bits is shown in Figure 6.14.

Figure 6.14 Trellis decoding example; the first four received bits are zeros and begin the decoding process.

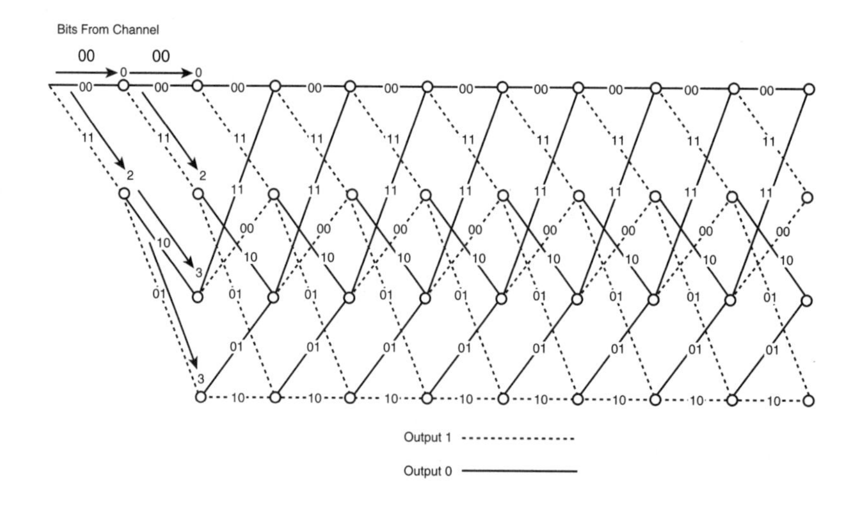

The received bits are shown above the trellis (00, 00). For these four bits, all paths in the trellis are kept alive (this step is just to get the algorithm started). At each node, a metric is computed that counts the mismatches between the received bits and the path bits. (Note that the metric is a running sum for each path.) At this point, the paths are extended using the following rules:

1. Two lines will lead to each node in the next set of states.
2. For each node, follow only one line to each node, specifically, the line having the lowest metric.

> **Note**
>
> The example being discussed uses hard-decision decoding. Another method, using soft decisions, does not completely decode the received bits before the trellis, but instead assigns a metric to each received bit, indicating how close to a one or zero it is. Soft decoding can outperform hard decoding by several dB.

3. Terminate the other path that would arrive at that node.

Note that four paths always remain active. Ideally, one main path develops through the trellis, and all others terminate. In actual practice, some delay is necessary to identify the correct path.

Turning back to the example, consider the next two bits being received. This situation is shown in Figure 6.15.

**Figure 6.15** Example of moving along the trellis—only one of the two lines leading each state is followed.

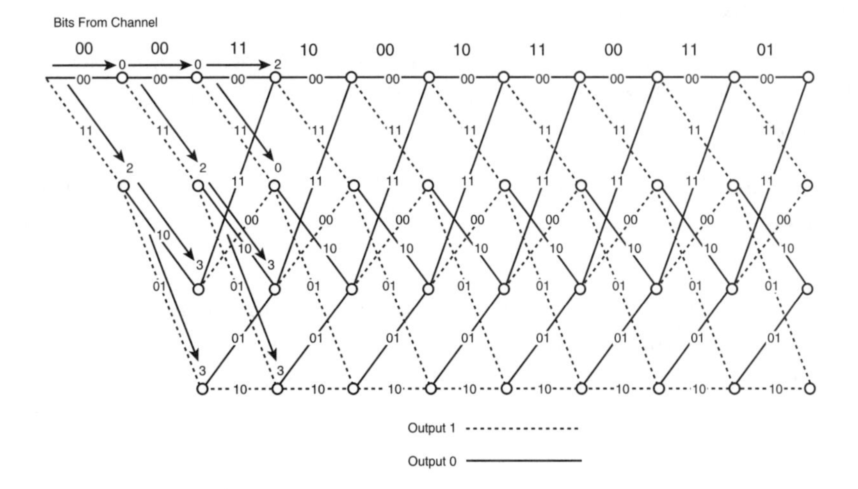

The previous rules are now used to decide which paths will go forward. For example, for the top node, the path through the line marked 00 was followed instead of the path through the line marked 11. The former produced a metric of two, whereas the latter would have produced a metric of three. Thus the first is followed, and the second terminated.

Figure 6.16 shows the next step in the decoding process.

Figure 6.16 An example of moving along the trellis; note that one path (which we will eventually find to be the correct path) still has a metric of zero.

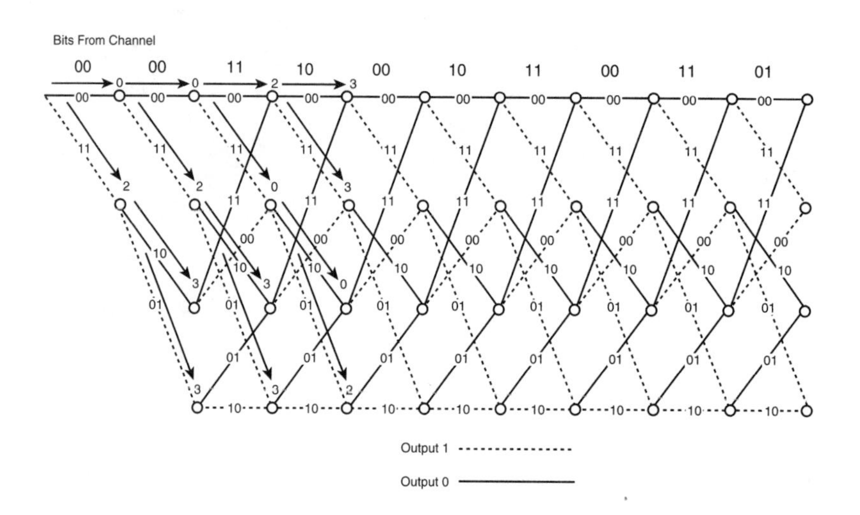


Notice that the new metrics are simply the previous metrics with a number of new mismatches added. The best metric still has a value of zero, indicating that it is a good candidate for the correct path. Figure 6.17 shows the result of several more iterations.

Note that many paths along the way have been terminated and that only one main path gives rise to all the existing paths. This main path is highlighted in Figure 6.18, and the decoded output for each set of inputs appears at the bottom of the trellis.

These outputs match the first few bits at the input to the decoder. As time passes and decoding continues, the main path will extend farther into the trellis and more outputs are decoded.

In Figure 6.18, the path can be decoded only up to the point where the main path diverges. When a surviving path is chosen up to a certain point, the "bad" paths up to that point are all discarded. Future input bits will be necessary to decode beyond this point.

Figure 6.17 Trellis appearance after several iterations from the example.

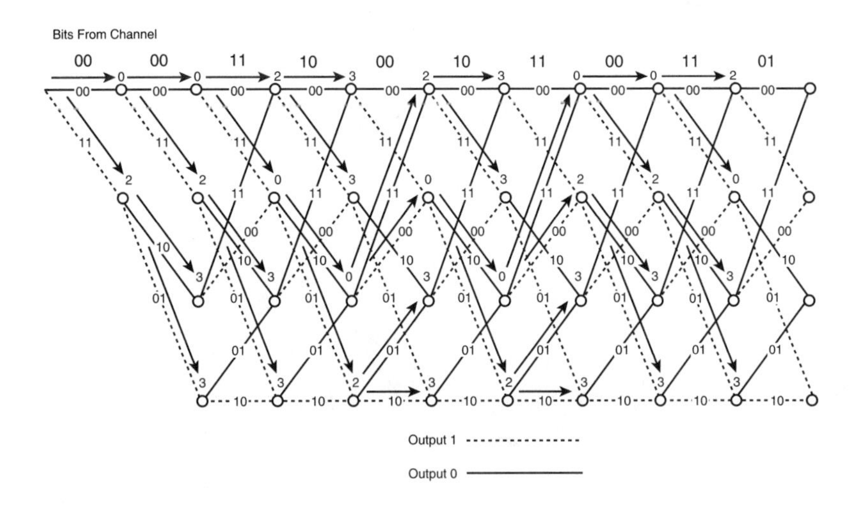


Figure 6.18 The decoded output of the trellis.

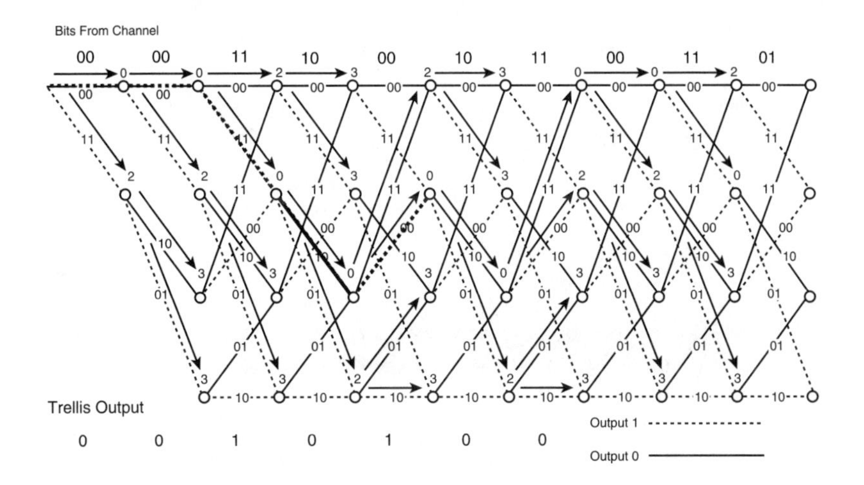


The preceding example considered decoding when no errors were made on the channel; thus error correction was not necessary. Consider now the case in which an error is made on the channel and one of the bits is incorrectly received. Figure 6.19 shows the first three steps of this decoding process.

Although none of the metrics is zero, the same rules are used to move through the trellis. The example is continued for several more iterations in Figure 6.20.

Figure 6.19 First few iterations for the sample channel with an error in the received bits; note that none of the metrics is zero in this case.

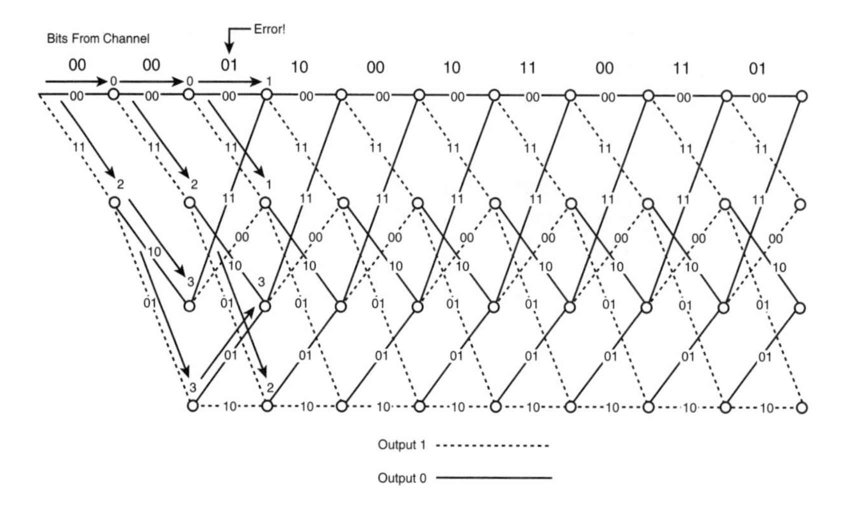


Figure 6.20 More iterations for a channel with received bit errors; note the paths that have terminated due to high metrics.

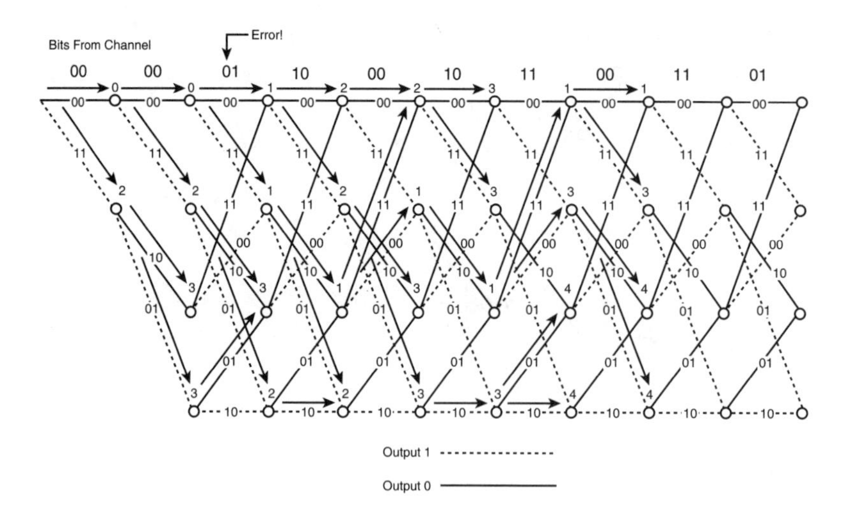


In Figure 6.20, many of the metrics have grown, but one is still small. Also, note that up to a certain point in the trellis all paths have terminated except one (the one with the lowest metric). This path is darkened in Figure 6.21, and the output shown at the bottom of the trellis.

**Figure 6.21** Decoding of the trellis; note that even with the error, only one main path survives and decodes with the proper values.

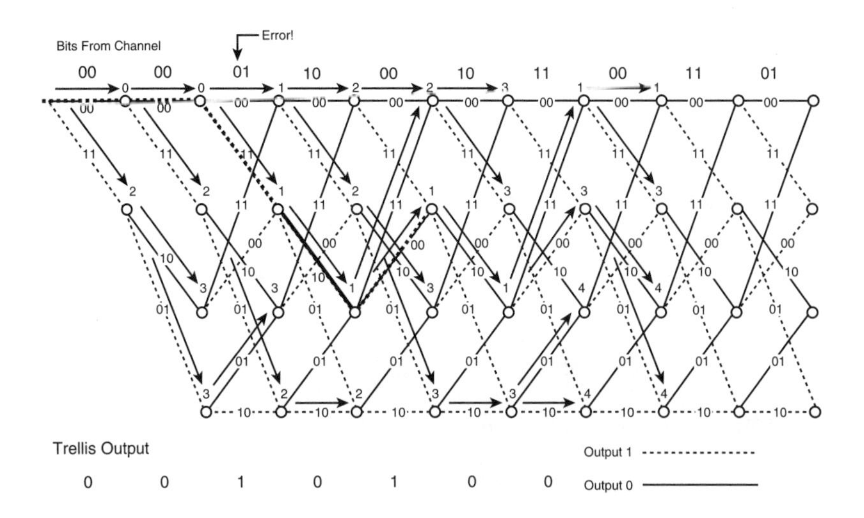


Even with the bit error that occurred on the line, the output of this example produced the correct sequence (identical to the preceding example), illustrating the forward error correcting mechanism of convolutional codes. In all, it is fair to say that bad paths are eventually terminated and a surviving path, even when some errors occur, is eventually apparent.

## Note

Trellis decoding can also be compared to following directions to someone's house. If you make a wrong turn at some point, the rest of the directions most likely do not make sense; however, it is possible to retrace your route to the point at which the directions stopped making sense.

Convolution coding may also be combined with modulation. When combined with QAM (as well as other types of modulation, such as pulse amplitude modulation [PAM]), the result is *trellis-coded modulation.* Normally, a subset of the bits (one or two) making up a QAM symbol is passed through an encoder. The uncoded bits usually determine the quadrant or quadrant portion a point will be in. The coded bits are the least significant bits distinguishing a point from neighboring points. Coding the least significant bits adds to the likelihood that a constellation point will not be incorrectly decoded as a neighboring point, the most common type of error occurrence in QAM. TCM is analyzed in many publications, most notably G. Ungerboeck's 1987 article in *IEEE Communications Magazine,* vol. 25, no. 2, entitled "Trellis Coded Modulation with Redundant Signal Sets—Part I: Introduction."[7]

# *Interleaving*

Many FEC blocks have trouble when it comes to correcting long strings of errors. Such errors are common on channels and can occur on DSLs (for example, when a transient spike is incident upon the twisted pair). The purpose of interleaving is to spread out a codeword such that channel burst errors are also spread out. Interleaving often occurs between a block forward error correction module and a modulation module in a transmitter. The corresponding block in the receiver is a "de-interleaving" block. The following sections describe two types of interleavers: a block interleaver and a convolutional interleaver. Block interleavers are simple and easy to understand; however, convolutional interleavers are more common in DSLs, as they have delay and memory requirement advantages.

**Note**

A poorly filtered ringing burst or ground bounce due to lightning or other disturbances can produce such a transient spike.

## *Block Interleaving*

Figure 6.22 shows an example of a block interleaver—specifically, a block interleaver with depth D=3 and span N=7.

Figure 6.22 A block interleaver with depth 3 and span 7.

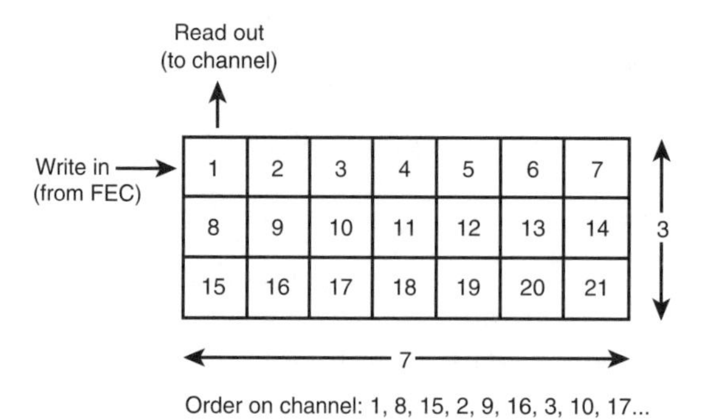


The numbers in the blocks represent the order in which bits would enter the interleaver. In general, bits are written in rows and read out in columns. Normally, a single row contains a complete FEC codeword. Thus the codeword length for this example would be 7. Figure 6.23 shows a de-interleaver with D=3 and N=7.

Figure 6.23 A de-interleaver with depth 3 and span 7; the output of the de-interleaver is the proper order of the symbols.

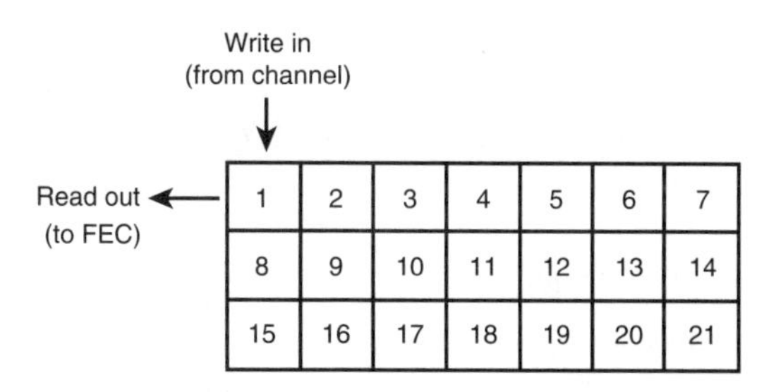


Here the bits coming from the channel are written into the de-interleaver in columns and read out in rows. The rows are then properly ordered bits and are normally destined for an FEC block.

Table 6.1 shows the value of interleaving.

Table 6.1 Comparison on burst errors with and without interleaving

| Bit Order (No Interleaving) | 1 | 2 | 3 | 4 | 5 | 6 | 7 | 8 | 9 | 10 | 11 | 12 | 13 | 14 | 15 | 16 |
|---|---|---|---|---|---|---|---|---|---|---|---|---|---|---|---|---|
| Bit Order (With Interleaving) | 1 | 8 | 15 | 2 | 9 | 16 | 3 | 10 | 17 | 4 | 11 | 18 | 5 | 12 | 19 | 6 |
| Error Burst | | | | | | | | | | | | | | | | |
| Received Bits (No Interleaving) | 1 | 2 | | | | 6 | 7 | 8 | 9 | 10 | 11 | 12 | 13 | 14 | 15 | 16 |
| Received Bits (De-interleaved) | 1 | | 3 | 4 | 5 | 6 | 7 | 8 | | 10 | 11 | 12 | 13 | 14 | | 16 |

The first two rows of Table 6.1 illustrate the order in which these bits will be sent over the channel both without interleaving and with the interleaver from Figure 6.22. If a burst error occurred in the channel as shown in the third row of Table 6.1, note the bit numbers being corrupted in each case. The last two rows of Table 6.1 show the bits that would be sent to the FEC block in the receiver on the sample channel. Note that when interleaving is used, the bit errors are broken up, giving the FEC block a better chance to correct the errors. This example might be used on a channel where burst errors are not expected to be longer than three bit periods. A more practical interleaver would normally have larger D and N parameters.

Interleaving adds delay to the end-to-end transport of data and also requires memory buffers at both the transmitter and receiver. In general, both the transmitter and receiver need about DxN bits of memory space to support block interleaving and a delay of about 2DN bits is inherent. For a trivial example like the one in Table 6.1, such negative aspects may seem insignificant, but for larger interleaver depths and spans, they are very significant.

**Note**

For higher-layer protocols that use acknowledgments, such as TCP, the added delay can cause the protocol to stall, drastically reducing the data throughput.

## *Convolutional Interleaving*

Convolutional interleaving is more efficient than block interleaving with respect to memory requirements and end-to-end delay. Note that convolutional interleaving does not imply anything about the FEC type, and it can function with either block or convolutional FEC techniques. A convolutional interleaver with a codeword size N=7 and a depth D=3 appears in Figure 6.24.

Figure 6.24 A convolutional interleaver with depth D=3 and codeword size N=7.

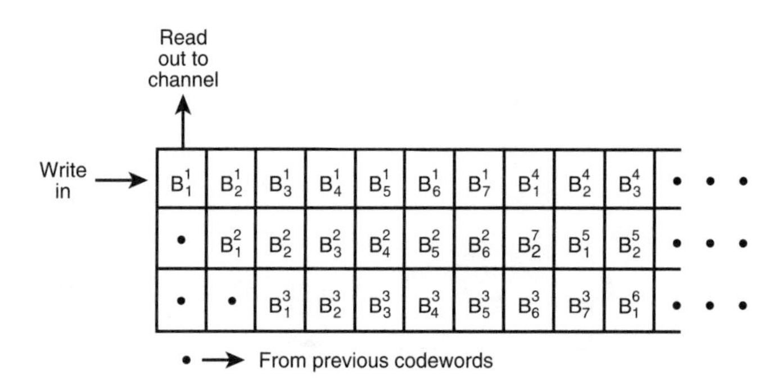


In this example, the superscripts of B represent the codeword to which B belongs. The subscripts represent the bit number within that codeword. The codewords are again written into the interleaver in rows and read out in columns. The difference between this scheme and the block interleaver scheme is that codewords do not always start in the same column in the convolutional interleaver as they do in the block interleaver. In addition, the rows do not end. The depth and length of the interleaver decide whether the next codeword should be written into the next row or in the top row immediately following the codeword already written there.

Figure 6.25 shows a convolutional de-interleaver for use in a receiver when a transmitter uses a convolutional interleaver.

Figure 6.25 A convolutional de-interleaver with depth D=3 and length N=7; reading values out of the de-interleaver is not as straightforward as for the block de-interleaver.

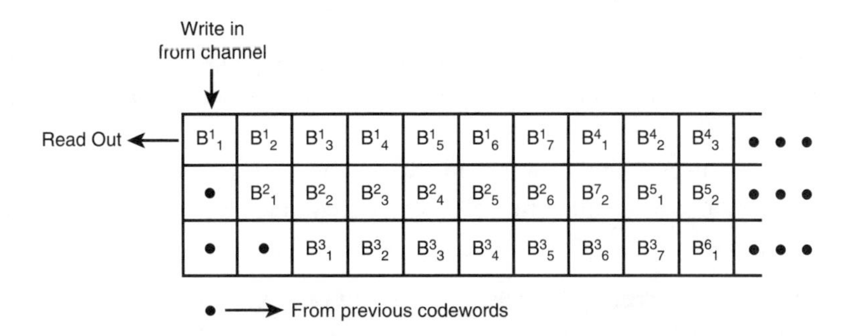


Here bits are written in columns and read out in rows for further processing by an FEC block. The de-interleaver must read only a single codeword from each row and then proceed to the next row until the last row is read. After reading the last row, the de-interleaver returns to the first row and reads from the next unread position.

Convolutional interleaving can distribute a codeword of length N over an interval of ND, incurring an end-to-end delay of ND bits.

Note that although the interleaving examples given are in terms of bits (the depth and codeword size are in terms of bits), many coding schemes work at the byte level (for example, Reed-Solomon codes operating in GF(256)), or even more generally, at the symbol level. Interleaving schemes employed along with FEC systems operating at the byte or symbol level normally operate at the byte or symbol level as well. For example, assuming a codeword size of seven bytes, a payload stream is interleaved at the byte level, and the end-to-end delay of a convolutional interleaver and de-interleaver is ND bytes.

Note that other types of interleaving are also possible. For example, helical interleaving is a different interleaving scheme.[8]

## *Scrambling*

Most DSLs include a scrambler in the transmitter and a descrambler in the receiver. A scrambler decreases the probability that a long string of ones or zeros are passed to the modulator. Such strings may occur in either a packet-based system or an ATM system when no packets or cells exist to transmit and the input to the transmitter is held either high or low. Often it is said that a scrambler randomizes data and a descrambler removes the randomness.

**Note**

T1 lines do not include scramblers; however, ADSL, VDSL, HDSL, and ISDN all use scrambling in the transmitter and descrambling in the receiver.

Scrambling has two positive benefits in a communications system. First, it helps ensure that the transmitted power spectral density observed at the output of the transmitter is predictable. When deriving equations for PSD in Chapter 4, "Power Spectral Densities and Crosstalk Models," an assumption was made about the statistics of the transmitter (for example, for ISDN it was assumed that all quats were equally probable). Scrambling helps to make such assumptions valid.

Second, many types of circuits or algorithms depend on a mix of ones and zeros to operate properly. Timing recovery phase locked loops and some adaptive equalization algorithms are good examples of these. Scrambling helps ensure that such blocks function properly. Note that scrambling in a transmitter is typically not used for data encryption or security purposes. The means of descrambling received data at the receiver are much too simple to be effective for security applications.

Figure 6.26 illustrates a scrambler and a descrambler.

Figure 6.26 An example of a scrambler and descrambler; the $z^{-1}$ blocks represent a one-bit delay, and all addition is modulo 2.

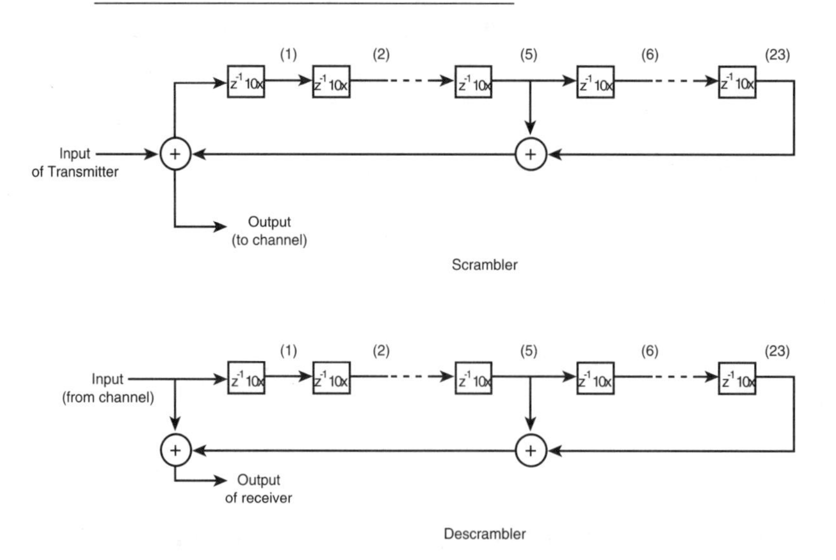


Note that the scrambler is a recursive device, whereas the descrambler does not include feedback. The pair shown are characterized by a generating polynomial g(x), which in the case of Figure 6.26 is given by Eqtn. 6.26.

Eqtn. 6.26

$$g(x) = 1 + x^{-5} + x^{-23}$$

This popular scrambler is used in several communications standards. The coefficients in Eqtn. 6.26 indicate the tap locations at which the scrambler performs an exclusive OR function. The generating polynomial of a scrambler is usually chosen to produce a *maximum length sequence* when the input to the scrambler is constant ones or constant zeros. Assuming the proper coefficients are chosen, longer scramblers normally produce longer maximum length sequences.

**Note**

Assuming the initial taps of a scrambler are not all zero, the maximum length sequence of a scrambler is given by 2n–1, where n is the order of the scrambler's generating polynomial.

The scrambler shown in Figure 6.26 is called a *self-synchronizing scrambler*. This name reflects the fact that the descrambler can begin in an arbitrary state (the initial values inside the scrambler) and, after a finite amount of time, begin to output the correctly unscrambled data. In addition, a bit error on the channel will not catastrophically affect the descrambler or cause it to significantly multiply the number of errors produced. The lack of feedback in the descrambler is directly related to these properties, as feedback can cause a bit error to infinitely affect the operation of the block.

A bit error on the channel is somewhat enhanced by a descrambler. In general, a single bit error is increased by a factor equal to the number of nonzero taps in the descrambler. In the example shown in Figure 6.26, a single bit error at the input produces three bit errors at the output. (More precisely, a bit error at the input produces an output error as it propagates to all of the nonzero tap positions.) This event is often termed the *error propagation factor*. Such increases in errors are often deemed acceptable when traded off against the many benefits of scrambling.

# *Equalization*

Most modern communications systems that operate near theoretical limits employ equalization in the transmitter, receiver, or both to optimize or nearly optimize transmission. Often the equalization is done digitally by adaptive digital filters. This approach provides a very flexible way to accommodate different types of channels and different types of noise environments. In most DSLs (as well as in voice-band modems), the adaptive filters converge to optimal initial settings during a training period and then can be updated during normal runtime operation of the system. Training can also be done without a predetermined training phase in what is called *blind equalization*.[9]

**Note**

Updating is necessary as many channels and noise conditions change over the lifetime of operation. A good example is the change in twisted pair characteristics due to the difference of temperature during the day and night.

Any time a channel's frequency response is not flat over the range of frequencies being transmitted, *intersymbol interference* (ISI) can occur. A channel with a frequency response that is not flat is sometimes called a *nonflat channel*, a *dispersive channel*, a channel with memory, or simply a channel with ISI. Additionally, a nonflat channel has an impulse response that is nonzero at more than one point.

**Note**

A flat channel will have an impulse response that is simply an impulse. Such a channel is often called a *memoryless* channel.

Figure 6.27 shows an impulse response of a channel with ISI. Note that an impulse in at time 0 produces an output with components over a range of times.

Figure 6.27 Response of a channel with ISI to an impulse.

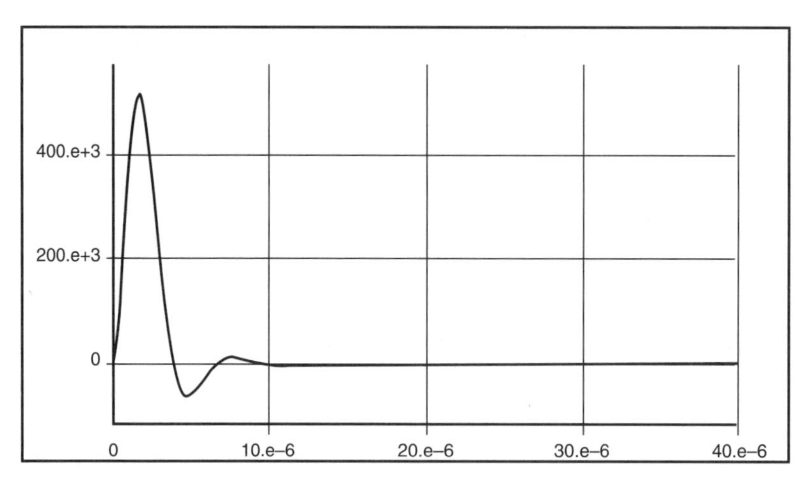


Figure 6.28 shows a single pulse transmitted through this channel.

Figure 6.28 Output of a channel with ISI when excited by a pulse at the input.

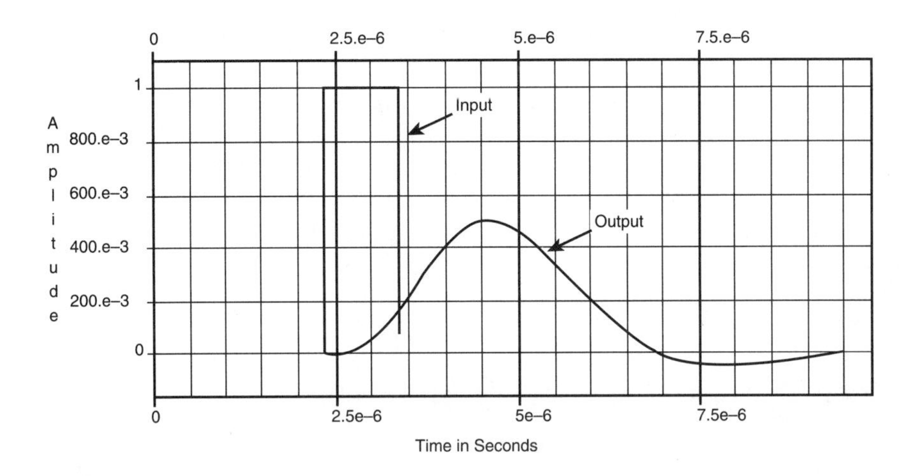


Note that the pulse is spread out and distorted at the output of the channel. Such spreading would interfere with the next pulse transmitted. Also, a previously transmitted pulse would interfere with the one shown. A channel can have even more memory than the one shown, causing pulses or symbols to be spread out and interfere not only with immediately adjacent symbols but also with several or many of them. The different types of equalization discussed in the next few sections are mainly intended to remove some or all of this ISI so that optimal decisions can be made on incoming symbols in the receiver. The following analysis is done in various domains, depending on which domain is more intuitive for the explanation. Keep in mind that transforms can be used to find equivalent duals in different domains.

## *Folded Spectrums*

Before analyzing different types of equalizers, a quick discussion of signal spectrums and the effect of sampling on signal spectrums is in order. First, consider a signal s(t) with a frequency spectrum of S(f) passing through an LTI system h(t) with spectrum H(f). This is shown in Figure 6.29.

Figure 6.29 A channel with sampling at the output, illustrating spectrum folding.

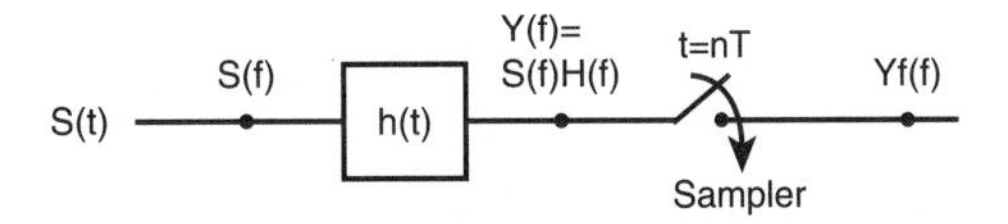

Working in the frequency domain, the output of the system Y(f) is simply given by the multiplication of the input spectrum and transfer function section as shown by Eqtn. 6.27.

Eqtn. 6.27

$$Y(f) = S(f)H(f)$$

When the output is sampled at some frequency $f_s$, the output samples are denoted by $y_i$ and the output frequency spectrum is denoted as $Y_f(f)$. This spectrum is a ***folded spectrum***. When working with two-sided PSDs, the equation defining the spectrum $Y_f(f)$ appears as Eqtn. 6.28.

Eqtn. 6.28

$$Y_f(f) = \sum_{n=-\infty}^{\infty} Y(f + nf_s) \quad \text{for } |f| < \frac{f_s}{2}$$

Note that if Y(f) does not have frequency components at frequencies higher than $f_s/2$, then the spectrum of $Y_f(f)$ will be the same as that for Y(f). Normally, when a folded spectrum is computed, it is computed only for small n values (–2, –1...2) because these components dominate in the calculation.

**Note**

Put another way, in this case aliasing does not occur because the signal was sampled at a rate at least two times higher than the highest frequency component it contains.

Folded spectrums are sometimes normalized with respect to frequency in discussions when the frequencies being discussed are not known or are generalized. This step is done by recognizing that in the frequency domain, the radian-sampling frequency $2\pi f_s$ is equal to a complete period. The frequency variable f is then replaced by the quantity $\theta f_s/2\pi$ where $\theta$ is the normalized frequency variable. The normalization of a folded frequency spectrum appears in Eqtn. 6.29.

Because f was defined on $(-f_s,f_s)$, $\theta$ is defined on $(-\pi/2,\pi/2)$. Folded spectrums are revisited in later sections on equalization.

Eqtn. 6.29

$$
\begin{aligned}
Y_f(f) &= \sum_{n=-\infty}^{\infty} Y(f + nf_s) \\
&= \sum_{n=-\infty}^{\infty} Y\left(\frac{\theta f_s}{2\pi} + nf_s\right) \\
&= \sum_{n=-\infty}^{\infty} Y\left(\frac{\theta}{2\pi\tau} + \frac{n}{\tau}\right) \\
&= \sum_{n=-\infty}^{\infty} Y\left(\frac{\theta + 2\pi n}{2\pi\tau}\right) \quad |\theta| < \pi
\end{aligned}
$$

## *Linear Equalization*

A linear equalizer (LE) is typically placed in a receiver in front of the demodulator. To analyze an LE, consider the system shown in Figure 6.30.

Figure 6.30 Simple block diagram for LE analysis.

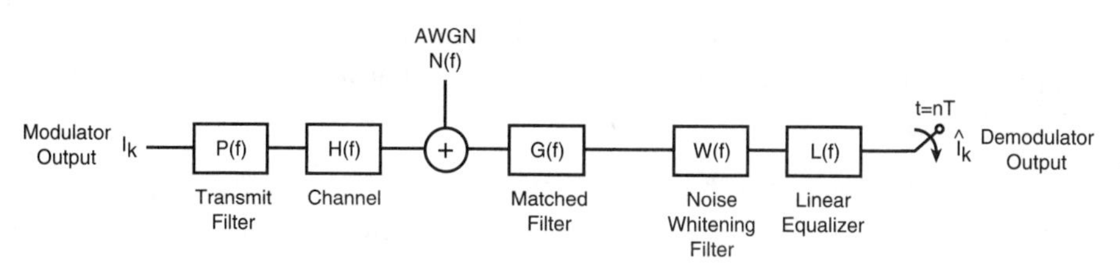


In the description that follows, some of the filtering functions may be combined, resulting in less complex implementations than actually implementing several filters in cascade. However, for explanatory purposes, it is convenient to discuss the filters as separate entities. In the following discussion, assume that all filters may be implemented digitally and that the sample frequencies are large enough to prevent folding (the appreciable frequencies of all signals are much less than $f_s/2$). However, folding may take place when sampling the output of the system.

It is well known that the front end of any optimal receiver is made up of a *matched filter* followed by some type of processing. The matched filter M(f) is typically matched to the convolution of the channel response and the excitation pulse response (the shaping filter). Eqtn. 6.30 gives the response of the matched filter in Figure 6.30.

Eqtn. 6.30

$$M(f) = [H(f)P(f)]^*$$

The filter following the matched filter in Figure 6.30 designated by W(f) is called a *noise-whitening filter*. The purpose of this filter is to remove any memory in the noise output of the matched filter before further processing. Eqtn. 6.31 gives the typical frequency response of this filter.

Eqtn. 6.31

$$W(f) = \frac{1}{G^*(f)}$$

G(f) and $G^*(f)$ are factors of the transfer function between the modulator and noise-whitening filter input including the channel, the shaping filter, and the matched filter. This is reflected in Figure 6.30. G(f) and $G^*(f)$ are chosen such that $G^*(f)$ has minimum phase so that W(f) is realizable.

The symbol * in Eqtn. 6.30 and Eqtn. 6.31 denotes the complex conjugate operator. A linear equalizer would then follow the noise-whitening filter.

You can use two techniques to optimize the response of a linear equalizer. The first optimization technique results in the *zero-forcing* (ZF) linear equalizer. The second optimization technique results in the *mean squared error* (MSE) linear equalizer.

The zero-forcing LE is optimized with respect to the *peak distortion criterion*. This criterion attempts to force the output of the equalizer to have no residual ISI. For a sufficiently long equalizer, the result of the peak distortion criterion is an equalizer with a response given by Eqtn. 6.32.

Eqtn. 6.32

$$L(f) = \frac{1}{G(f)}$$

Note that the net result of the noise-whitening filter and the ZF equalizer is the inverse of the channel up to the input of the noise-whitening filter (an added processing delay is not discussed but would also be present). Assuming the baud rate of the system is $\tau$, the SNR resulting from a zero-forcing linear equalizer is found by Eqtn. 6.33.

Eqtn. 6.33

$$SNR_{ZFLE} = \frac{Signal_{ZFLE}}{Noise_{ZFLE}} = \frac{\int_{-\frac{1}{\tau}}^{\frac{1}{\tau}} S_f(f)df}{\int_{-\frac{1}{\tau}}^{\frac{1}{\tau}} \frac{N_o}{X_f(f)} df}$$

In this equation $N_o$ is the PSD of the noise process N(f), and S(f) is the PSD of the signal power at the output of the modulator. Note that the spectrums shown are folded spectrums sampled at the baud rate of the system (denoted by the subscript *f*). The folded spectrum $X_f(f)$ is given by Eqtn. 6.34.

Eqtn. 6.34

$$X_f(f) = \sum_{n=-\infty}^{\infty} \left| P\left(f + \frac{n}{\tau}\right) H\left(f + \frac{n}{\tau}\right) \right|^2$$

Note from Eqtn. 6.33 that when the folded spectrum $X_f(f)$ contains a null or some value close to zero, the SNR of the zero-forcing linear equalizer becomes very small. This artifact of the ZF equalizer occurs because, at frequencies with significant channel loss, the equalizer tries to compensate by introducing high gain. This gain enhances the background noise, causing the SNR to approach zero.

The MSE linear equalizer minimizes the mean square error energy at its output. This type of equalization can trade off the amount of error caused by ISI with the amount of error caused by background noise. For a sufficiently long equalizer, the optimal response of the MSE case is shown in Eqtn. 6.35.

Eqtn. 6.35

$$L(f) = \frac{G^*(f)}{G(f)G^*(f) + N_o} = \frac{G^*(f)}{P(f)H(f)M(f) + N_o}$$

In this equation, $N_o$ is the variance of the white noise process N(f). Note that when the noise is negligible, the MSE linear equalizer becomes identical to the ZF linear equalizer (given by Eqtn. 6.32). The residual mean squared error resulting from the MSE linear equalizer is shown in Eqtn. 6.36.

Eqtn. 6.36

$$J_{MSE} = \int_{-\frac{1}{\tau}}^{\frac{1}{\tau}} \frac{N_o}{X_f(f) + N_o} df$$

The resulting SNR at the output of the MSE equalizer is shown in Eqtn. 6.37.

Eqtn. 6.37

$$SNR_{MSE} = \frac{1 - J_{MSE}}{J_{MSE}}$$

It should be evident from Eqtn. 6.36 and Eqtn. 6.37 that the MSE linear equalizer does not perform well in environments where the folded spectrum of the channel up to the receiver has nulls, because the error increases as the SNR decreases. Such nulls are common in many channels, including twisted pairs. The performance problems of linear equalizers on these channels provide motivation to discuss other types of equalization in subsequent sections. However, linear equalizers do play a role in these other types of equalization and are used in many different receivers.

## Decision Feedback Equalization

Performance problems of the linear equalizer have motivated the search for some type of nonlinear equalization technique. The *decision feedback equalizer* (DFE) is one such technique. A DFE is made up of a *feed-forward filter* and a *feedback filter*. Note that each of these filters is normally a linear filter, but they are used together in a nonlinear fashion. The DFE is illustrated in Figure 6.31.

Figure 6.31 A decision feedback equalizer on a communications channel.

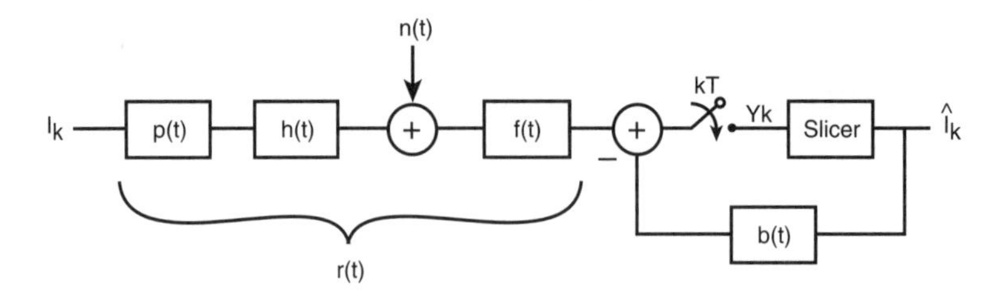

The slicer shown in Figure 6.31 is the decision-making device. Thus the input to the slicer is an estimate of the symbol to be decoded, and the output is the actual signal nearest to the estimate. The slicer is the nonlinear component of a DFE. Note that the output of the slicer goes into the feedback filter. The output of the feedback filter is subtracted from the received signal. In effect, the feedback section of the DFE removes ISI from a symbol due to previously transmitted symbols. This procedure is sometimes termed postcursor ISI. Assuming that the slicer makes the correct decisions, the feedback filter provides a way to remove the postcursor ISI without significantly amplifying noise (a major shortcoming of linear equalizers).

### Note

Of course, making the assumption that the decisions are made by slicer is not always correct. However, when the probability of error is low (less than $10^{-7}$), this assumption allows for a simpler analysis of the performance of a DFE. The result stemming from this ideal assumption is called the *ideal DFE bound.*

To mathematically illustrate the operation of the feedback filter, consider an optimization technique minimizing the mean squared error at the output of the slicer. The expected value of the error can be written as in Eqtn. 6.38.

Eqtn. 6.38

$$J_{DFE} = E\left|Y_k - I_k\right|^2$$

If we let r(t) represent the combined response of the pulse-shaping filter, the channel, and the feed-forward filter, then the input to the slicer is made up of three terms. These terms are the signal component exiting the feed-forward filter, the noise component exiting the feed-forward filter, and the signal from the feedback filter. Thus we can write $Y_k$ as in Eqtn. 6.39.

Eqtn. 6.39

$$\begin{aligned} Y_k &= I_k * r(t)\big|_{t=kT} + n(t) * f(t)\big|_{t=kT} - \sum_{n=1}^{\infty} b_n I_{k-n} \\ &= \sum_{n=-\infty}^{\infty} r_n I_{k-n} - \sum_{n=1}^{\infty} b_n I_{k-n} + n(t) * f(t)\big|_{t=kT} \\ &= \sum_{n=-\infty}^{-1} r_n I_{k-n} + \sum_{n=1}^{\infty} (r_n - b_n) I_{k-n} + r_0 I_k + n(t) * f(t)\big|_{t=kT} \end{aligned}$$

In Eqtn. 6.39, the third line simply results from distributing the first summation into the second and third terms. If we then substitute Eqtn. 6.39 into the mean squared error equation, we get the result shown in Eqtn. 6.40.

Eqtn. 6.40

$$\begin{aligned} J_{MSE} &= E\left|\sum_{n=-\infty}^{-1} r_n I_{k-n} + \sum_{n=1}^{\infty}(r_n - b_n)I_{k-n} + r_0 I_k + n(t) * f(t)\Big|_{t=kT} - I_k\right|^2 \\ &= E\left|\sum_{n=-\infty}^{-1} r_n I_{k-n} + \sum_{n=1}^{\infty}(r_n - b_n)I_{k-n} + I_k(r_0 - 1) + n(t) * f(t)\Big|_{t=kT}\right|^2 \\ &= E\left|\sum_{n=-\infty}^{-1} r_n I_{k-n}\right|^2 + E\left|\sum_{n=1}^{\infty}(r_n - b_n)I_{k-n}\right|^2 + E\left|I_k(r_0 - 1)\right|^2 + E\left|n(t) * f(t)\Big|_{t=kT}\right|^2 \end{aligned}$$

Note in Eqtn. 6.40 that the expected value of all multiplication cross-terms goes to zero due to independence, and thus, the multiplication cross-terms are not shown on the last line. Assuming that $\sigma_I^2$ is the variance of $I_n$ and $N_o/2$ is the variance of the noise term, Eqtn. 6.40 can be simplified as shown in Eqtn. 6.41.

Eqtn. 6.41

$$J_{MSE} = \sigma_I^2 \sum_{n=-\infty}^{-1} r_n^2 + \sigma_I^2 \sum_{n=1}^{\infty}(r_n - b_n)^2 + \sigma_I^2 (r_0 - 1)^2 + \frac{N_o}{2}\int_{-\infty}^{\infty} f^2(t)dt$$

Note that all terms in Eqtn. 6.41 are positive and that only the second term involves coefficients from the feedback filter. You should easily see that $b_n = r_n$ for $n > 0$ are the optimal feedback filter coefficients. This result would force the second term of Eqtn. 6.41 to zero and minimize $J_{MSE}$ with respect to the feedback filter. Logically, this value follows because $r_n$ for $n > 0$ defines the interference from past symbols into the current slicer input. Mimicking this operation in the feedback filter and subtracting its output cancels this postcursor ISI.

You could continue to minimize the mean squared error, $J_{MSE}$, by standard optimization techniques and solve for the optimal feed-forward filter.[10,11] In addition, optimization can be done for the transmitter pulse-shaping filter. The result is that the feed-forward filter is a matched filter followed by a tapped delay line.

**Note**

If the matched filter is implemented in front of the feed-forward filter, then it is simply just the tapped delay line.

Eqtn. 6.42 gives the MSE result after minimization.

Eqtn. 6.42

$$J_{MSE} = \exp\left( -2\tau \int_0^{\frac{1}{\tau}} \ln\left(1 + SNR_f(f) df\right) \right)$$

$SNR_f$ in this equation represents the folded SNR at the input to the receiver given by Eqtn. 6.43.

Eqtn. 6.43

$$SNR_f(f) = \sum_{n=-\infty}^{\infty} \frac{S\left(f + \frac{n}{\tau}\right) \left| C\left(f + \frac{n}{\tau}\right) \right|^2}{N\left(f + \frac{n}{\tau}\right)} \quad 0 < f < \frac{1}{2\tau}$$

In this equation, C(f) is the combined response of the pulse shaping filter and the channel. The resulting SNR at the input to the slicer is then shown in Eqtn. 6.44.

Eqtn. 6.44

$$\begin{aligned} SNR_s &= \frac{1 - J_{MSE}}{J_{MSE}} \\ &\approx \frac{1}{J_{MSE}} \\ &= \frac{10}{\ln(10)} \left( 2\tau \int_0^{\frac{1}{2\tau}} \ln\left(1 + SNR_f(f) df\right) \right) \end{aligned}$$

The approximation in the second line of Eqtn. 6.44 holds for all $J_{MSE} < 1$.

Note the difference between $SNR_f$ and $SNR_s$. The first is a function of frequency and represents the SNR of signal power to noise power at the front end of the receiver at a given frequency. The signal power in this case may be made up of signal energy from more than one symbol due to ISI. The second term, $SNR_s$, is the SNR at the input to the decision device. The signal power in this case is only due to energy of the symbol entering the slicer. The noise power is due to the filtered background noise and any residual ISI.

The analysis of DFEs up to this point has assumed a PAM-based system. A DFE can be applied to a CAP/QAM-based system to decode received symbols as well. Figure 6.32 shows the basic structure of a DFE in this environment.

Figure 6.32 A decision feedback equalizer for a CAP/QAM system.

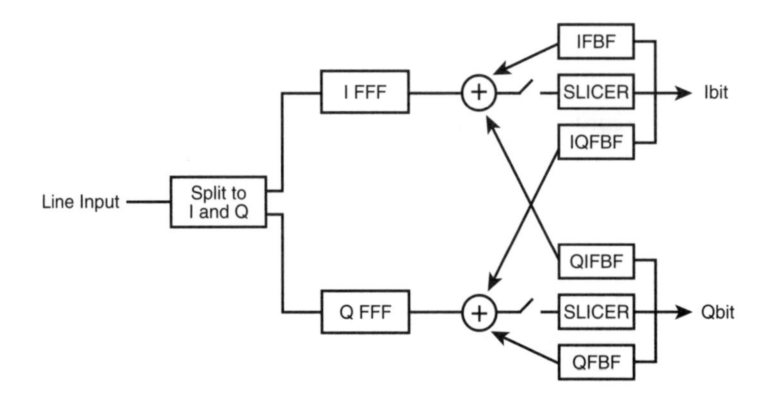


Note that both the I and Q paths have a feed-forward filter. In addition, four feedback filters exist, allowing interference to be eliminated on the two branches due to both previous I components and previous Q components. Eqtn. 6.45 gives the SNR at the input to either slicer for the CAP/QAM DFE.

Eqtn. 6.45

$$
\begin{aligned}
SNR_s &= \frac{1 - J_{MSE}}{J_{MSE}} \\
&\approx \frac{1}{J_{MSE}} \\
&= \frac{10}{\ln(10)}\left(\tau \int_{f_c - \frac{1}{2\tau}}^{f_c + \frac{1}{2\tau}} \ln\left(1 + SNR_f(f)df\right)\right)
\end{aligned}
$$

The folded SNR, $SNR_f$, is given by Eqtn. 6.46.

Eqtn. 6.46

$$SNR_f(f) = \sum_{n=0}^{\infty} \frac{S\left(f+\frac{n}{\tau}\right)C\left(f+\frac{n}{\tau}\right)}{N\left(f+\frac{n}{\tau}\right)} \quad \text{for } f_c - \frac{1}{2\tau} < f < f_c + \frac{1}{2\tau}$$

## *Tomlinson-Harashima Precoding*

Although not obvious at first, Tomlinson precoding, sometimes called *Tomlinson-Harashima precoding* (THP) is related to decision feedback equalization. A main difference between the two is that for THP the feedback filter is not present in the receiver, but rather a precoding feedback filter is present in the transmitter. Figure 6.33 shows a simple THP system.

Figure 6.33 A Tomlinson-Harashima precoder in a transmitter along with a channel and a receiver.

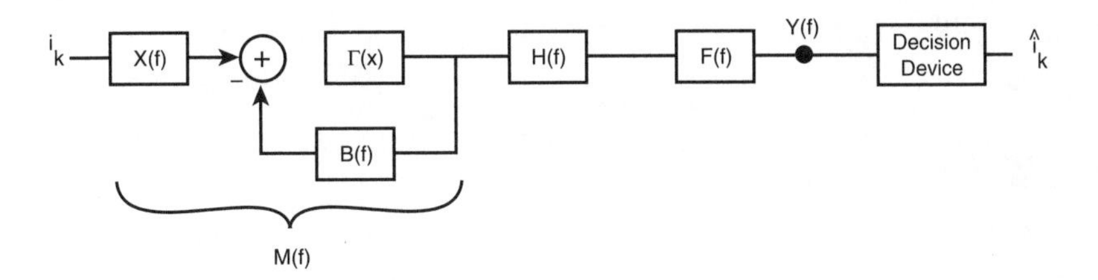


The operator Γ(x) in Figure 6.33 is a modulo operator. Ignoring this operator, the output of the receiver (input to the decision devise) is given by Eqtn. 6.47.

Eqtn. 6.47

$$\begin{aligned} Y(f) &= M(f)H(f)F(f) \\ &= \left(\frac{X(f)}{1-B(f)}\right)H(f)F(f) \\ &= \frac{X(f)H(f)F(f)}{1-B(f)} \end{aligned}$$

Figure 6.34 shows a DFE with similar notation.

Figure 6.34 Block diagram of a decision feedback equalizer.

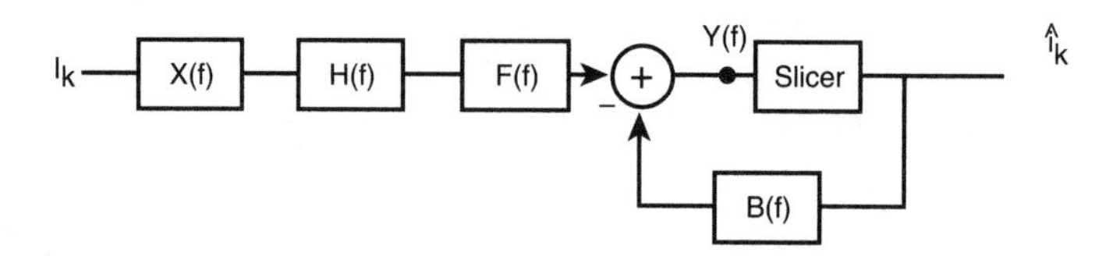


Assuming ideal decisions, Eqtn. 6.48 gives the input of the slicer.

Eqtn. 6.48

$$\begin{aligned} Y(f) &= X(f)H(f)F(f) - \hat{X}(f)B(f) \\ &= X(f)H(f)F(f) - X(f)B(f) \\ &= X(f)\big(H(f)F(f) - B(f)\big) \end{aligned}$$

In Eqtn. 6.48, the second line reflects the ideal decision assumption.

For the zero-forcing case, no ISI should exist at the input to the decision device. For the THP, Eqtn. 6.49 must hold.

Eqtn. 6.49

$$\frac{H(f)F(F)}{1+B(f)} = 1$$

For the DFE, the zero-forcing condition requires that Eqtn. 6.50 be true.

Eqtn. 6.50

$$H(f)F(f) - B(f) = 1$$

Note that Eqtn. 6.50 can be rewritten as Eqtn. 6.51.

Eqtn. 6.51

$$H(f)F(f) = 1 + B(f)$$

*or*

$$\frac{H(f)F(f)}{1+B(f)} = 1$$

Because this result matches Eqtn. 6.49, it is clear that the ZF constraint placed on the THP filters results in the same requirements being placed on the DFE filters. Thus the optimal F(f) and B(f) are the same for both systems.

The MSE condition applied to the THP does not result in a similar intuitively convenient comparison with the DFE. However, it can be shown that the optimum F(f) and B(f) for the MSE THP are identical to those for the MSE DFE.[12,13]

## *Frequency-Domain Equalization*

Equalization can be done in the frequency domain instead of in the time domain. Given a signal s(t) and a linear, time-invariant filter with impulse response s(t), the basic relationship between filtering (convolution) in the time domain and filtering in the frequency domain is given by Eqtn. 6.52.

Eqtn. 6.52

$$s(t) * h(t) \Leftrightarrow S(f)H(f)$$

The arrow in this equation represents the fourier transform. In other words, convolution in the time domain is equivalent to multiplication in the frequency domain, as shown in Figure 6.35.

Figure 6.35 The duality of time-domain and frequency-domain representations of filtering.

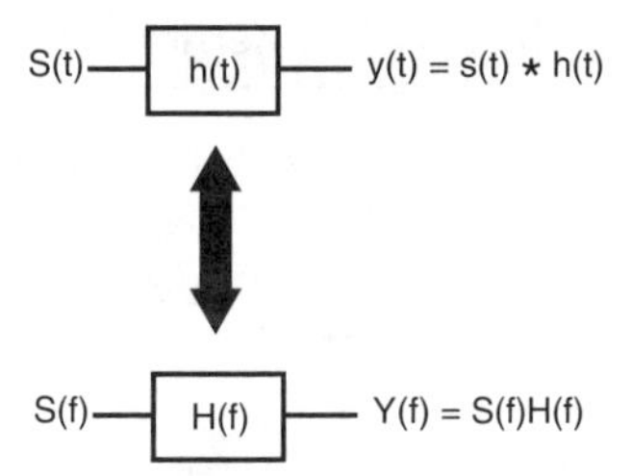


It may be apparent that a modulation technique such as DMT, which involves a transformation between the time domain and frequency domain at the receiver using a discrete fourier transform, may lend itself nicely to frequency-domain equalization. In this case, a single complex multiplication to the received value in each DMT bin would sufficiently equalize the channel. However, a small caveat must be considered before this technique will work.

### Note

Such equalization is sometimes much more flexible than equalization in the time domain. Additionally, long filters can realize significant savings in the number of multiplications needed.

Assume that two signals s(t) and h(t) are sampled and an N-point discrete fourier transform taken. Multiplication of the resulting frequency-domain components is related to the sampled time-domain values as shown in Eqtn. 6.53.

Eqtn. 6.53

$$S^{\langle N\rangle}(kf_o)H^{\langle N\rangle}(kf_o) = s^{\langle N\rangle} \otimes h^{\langle N\rangle}$$

The superscripts in Eqtn. 6.53 represent the block or window size of the samples. The fundamental difference between Eqtn. 6.52 and Eqtn. 6.53 is the circular convolution in the latter as opposed to a linear convolution in Eqtn. 6.52. With discrete domains, multiplication in the frequency domain is equivalent to circular convolution in the time domain. The obvious problem with directly implementing frequency-domain equalization is that the output of a twisted pair channel looks like a linear convolution of the input and the channel response, not a circular convolution.

To make the output look like a circular convolution, a *cyclic prefix* is added to each symbol to be transmitted over the channel. The process involves appending the last L samples in the symbol to the beginning of the symbol before sending it over the channel. If the original symbol is N samples long, the resulting symbol will be N + L samples long, as illustrated in Figure 6.36.

Figure 6.36 An example of a cyclic prefix insertion block.

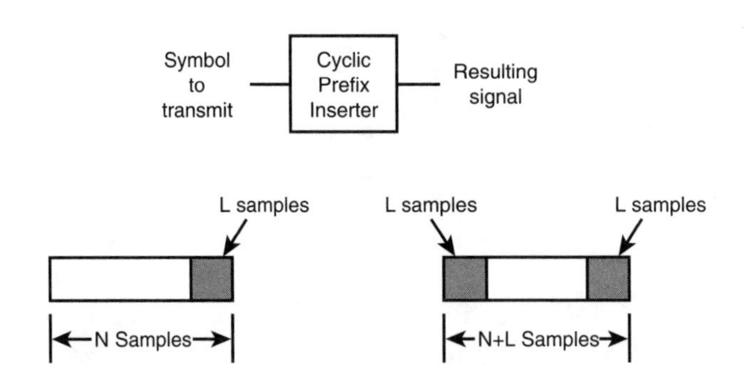


At the receiver, only N samples are processed, and the L samples added are dropped. However, the N-length block appears to have been circularly convolved with the channel. In other words, the channel has been made a *circular channel.*

Note that adding a cyclic prefix does add overhead and reduces the amount of information that can be sent over the channel. For long blocks of data (large N), this overhead is usually a very small fraction of the payload and can be ignored. Also, if the impulse response of the channel is larger than the length of added cyclic prefix, energy from the previous symbol will leak into the N saved samples and the channel will no longer look circular. This occurrence is common. The solution is to combine some front-end time-domain equalization before removing the N samples at the receiver. The front-end equalization can effectively shorten the channel response to less than L samples, thus restoring the circular property desired.

## *Shaping*

Shaping filters are often used at the output of a modulator. Analog or digital filters or both are used. Most often, shaping is used to ensure that the output waveforms of the transmitter adhere to pertinent spectral requirement. For DSLs, some of these requirements are discussed in Chapter 4. Shaping is often challenging, as the out-of-band energy must be properly attenuated while the in-band energy should experience minimal distortion.

## *Summary*

This chapter presented an overview of DSL modulation basics and discussed many of the transmitter and receiver blocks used in a DSL system. CAP and QAM modulation and demodulation were discussed and the similarities and differences between them explained. DMT modulation and demodulation was also discussed, along with the method of coupling a frequency-domain equalizer into the system.

The benefits and methods of forward error correction were discussed. The differences between block and convolutional coding and decoding were explained. Block and convolutional interleaving were also discussed. The method of each was explained, as well as the delay and memory requirements of the two schemes.

Equalization in general and various forms of equalization, including a linear equalizer, a decision feedback equalizer, and a Tomlinson-Harashima precoder, were discussed. Performance equations for the equalizers were given, and the pros and cons of each discussed.

This chapter introduced the various modulation and demodulation blocks used in DSL systems. The next two chapters discuss how DSL systems actually implement some of these blocks to produce a realizable, high-performance transmitter and receiver.

## *Exercises*

1. For a given background noise power, derive and plot the probability of error as a function of average signal power for a 4-QAM, 16-QAM, and 64-QAM system.
2. For a 4-QAM system, assume that the background noise power is 1/10 of the average signal power. Through simulation, plot the location of a point projected at the receiver onto the constellation decoder if the constellation point from the first

quadrant is sent. Repeat this procedure on the same constellation plot 1,000 times. Do the same assuming the other three constellation points were sent.

3. Repeat exercise 2 assuming that the background noise power is 1/5 of the average signal power. Repeat again assuming 1/3. Discuss the locus of received points and how they relate to error probability.
4. Simulate a DMT transmitter with eight tones by using uniform iid random variables on [–A,A] as real and imaginary inputs for the IFFT. Over many symbols, record the peak time-domain power value (output sample squared) and keep a running average of the square of the samples. Repeat and compare for 32 and 64 tone DMT symbols and for a QAM symbol with the same uniform input. Comment on the peak-to-average ratio (PAR) of the squared time-domain samples.
5. Find the output sequence for a rate 1/2 convolutional encoder with generator polynomials $G1(x)=1+x+x^2$ and $G2(x)=x$. Assume the same input sequence given in the text.
6. Using the convolutional decoder example given in the text, decode the received signal assuming that the seventh bit was received in error.
7. Write the input and output sequence of a block interleaver with a depth of 5 and a span of 9. Repeat for a convolutional interleaver with the same properties. Calculate the time that it would take for a single codeword to reach the receiver with both schemes as well as if no interleaver were used.
8. Find the folded spectrum of an ISDN signal sent over the mid-CSA Loop and sampled at the receiver at 80 kHz. Compare this spectrum to the spectrum received assuming no folding.
9. Find the SNR of an HDSL signal at the input to the decision device using CSA Loop #4 if a ZF linear equalizer is used. Repeat assuming that a MSE linear equalizer is used. Assume that additive white Gaussian noise (AWGN) is present with a PSD of –140 dBm/Hz along with 24 self-NEXT disturbers.
10. Repeat exercise 9 assuming that a DFE is used.
11. Create two arbitrary vectors of length 8. Verify that the FFT of the circular convolution of the vectors is equal to the frequency-domain multiplication of the FFT of each vector.

## *Endnotes*

1. Ken Kerpez, CAP and QAM System Interoperability, T1E1.4 Contribution 96-242, Nashua, NH, September 5, 1996.

2. Stephen G. Wilson, *Digital Modulation and Coding*. NJ: Prentice Hall, 1996.
3. Shu Lin and Daniel J. Costello Jr., *Error Control Coding: Fundamentals and Applications*. Englewood Cliffs, NJ: Prentice Hall, 1983.
4. I. S. Reed and G. Solomon, "Polynomial Codes over Certain Finite Fields," *J. Soc. Ind. Appl. Math*, 9, June 1960, pp. 300–304.
5. R. C. Bose and D. K. Ray-Cyaudhuri, "On a Class of Error Correcting Binary Group Codes," *Inf. Control*, 3, March 1960, pp. 68–79.
6. A. Hocquenghem, "Codes Corecteurs d'Erreurs," *Chiffres*, 2, 1959, pp. 147–156.
7. G. Ungerboeck, "Trellis Coded Modulation with Redundant Signal Sets—Part I: Introduction," *IEEE Communications Magazine*, vol. 25, no. 2, February 1987.
8. S. Benedetto, M. Mondin, and G. Montorsi, "Performance Evaluation of Trellis Codes," *IEEE Transactions on Information Theory*, vol. IT-32, March 1987.
9. J.J. Werner contribution on blind equalization.
10. J. Salz, "Optimum Mean Square Decision Feedback Equalization," *Bell System Technical Journal*, vol. 52, no. 8, October 1973.
11. Carlos A. Belfiore and John H. Park, "Decision Feedback Equalization," *Proceedings of the IEEE*, vol. 67, no. 8, August 1979.
12. Richard D. Wesel and John Cioffi, "Achievable Rates for Tomlinson-Harashima Precoding," submitted to *IEEE Transactions on Information Theory*, 1996.
13. Richard D. Wesel and John Cioffi, "Precoding and the MMSE-DFE," *Proceedings of the 28th Asilomar Conference on Signals, Systems, & Computers*, 1995, pp.1144–1149.

CHAPTER 7

# ADSL Modulation Specifics

In this chapter:

- Details of a DMT ADSL transmitter and receiver implementation
- Discussion of data and overhead channels for DMT ADSL
- Explanation of training and initialization procedures for DMT ADSL
- Discussion of a CAP/QAM ADSL transmitter and receiver
- Discussion of special ADSL functions to accommodate ATM

The preceding chapter discussed many of the common blocks used in DSL transmitters and receivers. This chapter describes how these blocks are implemented to create an ADSL transceiver. Specifically, blocks will be discussed for ADSL end modems: the ADSL transceiver unit–remote (ATU-R) normally located at a customer premise and the ADSL transceiver unit–central office (ATU-C) normally located at the central office aggregation point. Many of the ADSL options are described along with the more common implementations of them. The first part of the chapter deals with a discrete multitone (DMT) ADSL implementation, and the second part deals with a Carrier Amplitude and Phase/Quadrature Amplitude Modulation (CAP/QAM) implementation. The DMT implementation is according to the specification in the ANSI document T1E1.4/98-007R4,[1] and the CAP/QAM implementation is according to the specification in ANSI document T1E1.4/97-104R2a.[2]

## DMT ADSL Systems

Figures 7.1a and 7.1b show a block diagram of a DMT ADSL transmitter. Figure 7.2 shows a block diagram of an upstream and downstream DMT ADSL receiver.

Figure 7.1a Block diagram for an upstream ADSL transmitter.

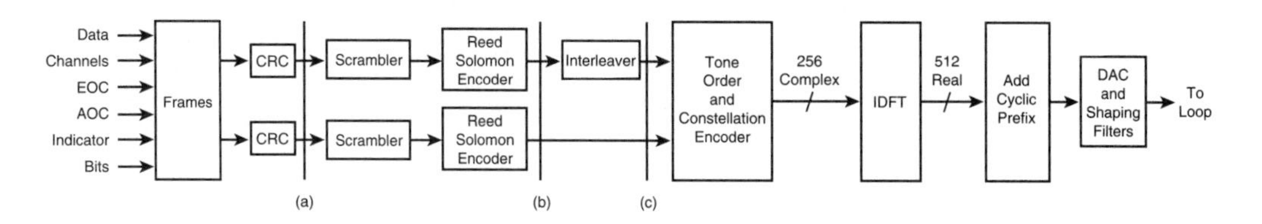


Figure 7.1b Block diagram for a downstream ADSL transmitter.

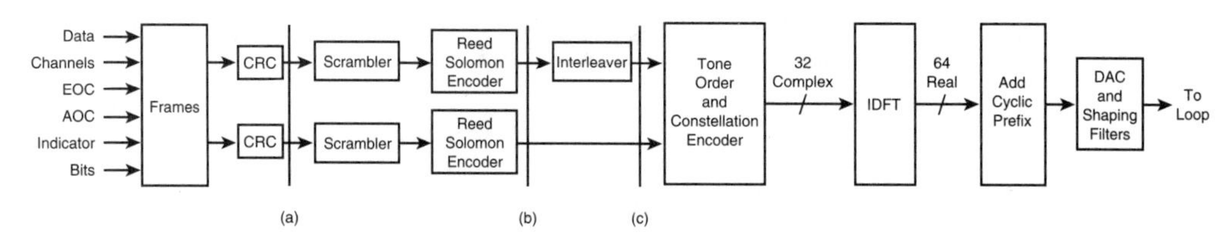


Figure 7.2 shows a DMT ADSL receiver.

Figure 7.2 Block diagram for an upstream (a) and downstream (b) ADSL receiver.

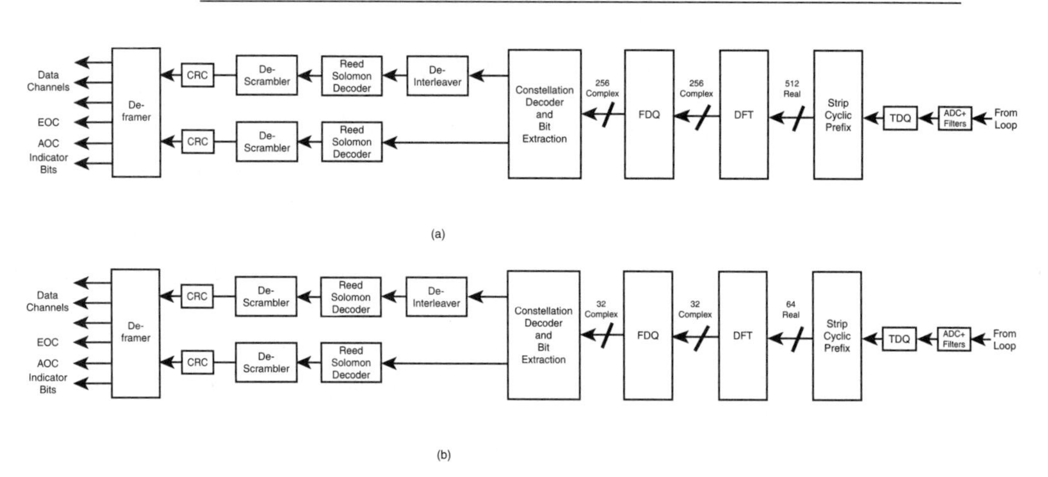


Both upstream and downstream transmitters and receivers are shown in these figures. As discussed previously, DMT ADSL is the standard for ADSL set forth by ANSI and ETSI[1,3]. The specifications allow for high-speed transmission greater than 6 Mbps in the downstream direction and lower-speed transmission up to about 640 kbps in the upstream direction. The actual rates depend on line and noise conditions as discussed in Chapter 5, "DSL Theoretical Capacity in Crosstalk Environments." Also, the company providing the ADSL service (often called the access provider) may limit the actual rate. In such cases, a line that could potentially support higher rates is configured to support a lower rate.

**Note**

This situation is typical in many service offerings. An access provider may offer customers different service tier options with some tiers providing higher rates (and presumably costing more) than others.

## *Transmitter and Receiver Blocks*

The following sections discuss the various blocks shown in Figure 7.1a, 7.1b, and 7.2 for DMT ADSL. Note that many of the blocks were discussed in general terms in Chapter 6, "DSL Modulation Basics." Because of the asymmetric nature of ADSL, some of the blocks may have a different implementation at the ATU-C and the ATU-R.

It is worth mentioning up front that the transceivers may carry more than one data channel and also carry an embedded operations channel (EOC), an ADSL overhead channel (AOC), and also various other bits to be described later in this chapter. All but the data is often simply termed *overhead*. It is said that a single *physical channel*, the physical ADSL link or ADSL channel, carries more than one *logical channel*, the logical channels being all data and overhead channels. Combining and separating the logical channels at the transmitter and receiver is done using a frame structure that is described later in this chapter.

The logical data channels on an ADSL link consist of four possible simplex downstream channels and three possible full-duplex channels. The logical channels are sometimes called *bearer channels*. The simplex channels are named *AS0*, *AS1*, *AS2*, and *AS3*. The duplex channels are named *LS0*, *LS1*, and *LS2*. The duplex channels may have different rates upstream and downstream including a rate of zero in one or both directions. When a duplex channel rate is zero in one direction and nonzero in the other direction, it is analogous to a simplex channel in the nonzero direction. The allowable rates of the different bearer channels are listed in Table 7.1. For all logical data channels, the actual rates must be multiples of 32 kbps (the reason is explained later). In many implementations, only one of the simplex data channels is used in the downstream direction, and a duplex channel is used in simplex mode for the upstream direction. Typically, the channels used would be AS0 and LS0.

Table 7.1 ADSL Logical Data Channels and Associated Rates

| Channel | Type | Allowable Rates | Comments |
|---|---|---|---|
| AS0 | Downstream Simplex | 0-8.192 Mbps | This channel is most commonly used as the lone downstream channel. |
| AS1 | Downstream Simplex | 0-4.608 Mbps | |
| AS2 | Downstream Simplex | 0-3.072 Mbps | |
| AS3 | Downstream Simplex | 0-1.536 Mbps | |
| LS0 | Duplex | 0-640 kbps | Can have different rates in each direction. This channel is most commonly configured as an upstream simplex channel. |
| LS1 | Duplex | 0-640 kbps | Can have different rates in each direction. |
| LS2 | Duplex | 0-640 kbps | Can have different rates in each direction. |

The different logical channels allow ADSL to be flexible enough to support many different applications. For example, one of the first drivers of ADSL was video on demand over a telephone line. In this case, a downstream simplex channel, AS0, for example, can be used to send streaming digital video and audio while a duplex channel, LS0, for example, can be used to send and receive control information, such as pause or rewind commands. In any case, the total bandwidth available on the physical channel must be greater than the sum of the bandwidths used by each logical channel. (One may argue that these should be equal; however, some of the physical channel bandwidth is used for other purposes such as synchronization. This topic is discussed later.)

Note also that the ADSL transmitter and receiver each have two associated paths. One path is called the *fast path,* and one path is called the *interleaved path.* Cyclic redundancy check (CRC) generation, scrambling, and forward error correction are independent on each path. The main difference between the two is that the interleaved path contains an interleaving function in the transmitter (and de-interleaving function in the receiver), whereas the fast path does not. The name fast path comes from the fact that interleaving, and thus the delay caused by interleaving, is not present so that data gets through the transmitter and receiver faster than in the other path. Simply put, bits transmitted on the fast path get through the transmitter and receiver with less delay than bits transmitted through the interleaved path. You will see later that a logical data channel is assigned to either the fast or interleaved channel, but never to both. However, both paths can be active simultaneously (each with one or more logical data channels assigned to it). Many implementations, however, use only one of the paths.

## *Framing*

The framing structure of ADSL is key to understanding how the different logical channels are combined or separated when transmitted on the physical ADSL channel. ADSL has a framing structure as well as a *superframe structure.* The superframe structure is 17 ms long and is made up of 68 frames, which are numbered 0 through 67, and a synchronization symbol. Figure 7.3 shows this structure.

Figure 7.3 The structure of an ADSL superframe.

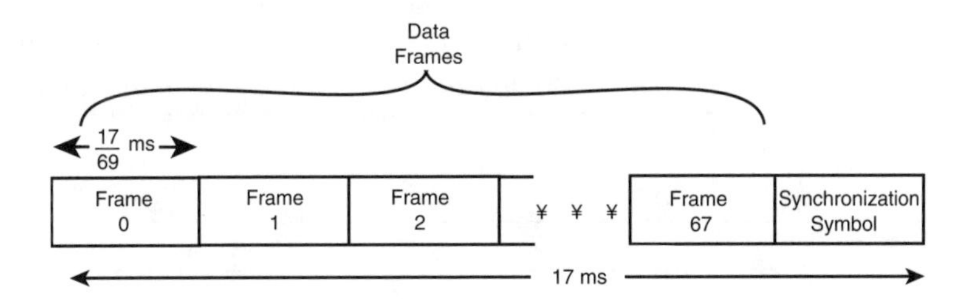


The basic superframe structure is the same for both the upstream and downstream directions. The synchronization symbol has the same duration as a data frame, and it can be thought of as a frame that carries no user data. For short, I refer to the *synchronization symbol* as simply the *sync symbol* or *sync frame.*

Each ADSL frame, including the sync frame, corresponds to a single ADSL symbol. Each symbol has a duration of just over 246 us (specifically 250×68/69 us). Adjusting for the time taken by the sync frame, the data symbols have a rate of 4 kHz. This relationship can be seen in Eqtn. 7.1.

Eqtn. 7.1

$$f_{data\_symbol} = \frac{68 \text{ data frames}}{17 \text{ ms}} = 4000 \frac{\text{data frames}}{\text{s}} = 4 \text{ kHz}$$

Each data frame within a superframe has a similar structure. Each data frame contains a specified number of data bytes for each of the active logical data channels in both the fast and interleaved paths. Overhead bytes are also contained in each data frame; however, the overhead bytes contained are different for different frame numbers. Figure 7.4 shows a general data frame for an ADSL link.

Figure 7.4 The structure of a data frame including both the fast path and interleaved path components.

| Fast Overhead | Fast Data | Fast Overhead | Interleaved Overhead | Interleaved Data | Interleaved Overhead |
|---|---|---|---|---|---|

Each data frame is divided into a fast buffer part and an interleaved buffer part. As one might expect, the fast part contains the data to be sent through the fast path while the interleaved part contains data to be sent through the interleaved path. Figure 7.5 shows a more detailed description for both the upstream and downstream directions of the fast part of point A in Figure 7.1.

Figure 7.5 Detailed structure of the fast path portion of an ADSL data frame for (top) the downstream direction and (bottom) the upstream direction.

| Fast Byte | AS0 Bytes | AS1 Bytes | AS2 Bytes | AS3 Bytes | LS0 Bytes | LS1 Bytes | LS2 Bytes | AEX Byte | LEX Byte |
|---|---|---|---|---|---|---|---|---|---|

| Fast Byte | LS0 Bytes | LS1 Bytes | LS2 Bytes | LEX Byte |
|---|---|---|---|---|

In Figure 7.5, you can see that each logical channel is allotted a certain number of bytes. Note that the downstream data frame has room for bytes from all seven bearer channels, whereas the upstream frame has space only for the LS channels because the AS channels are not transmitted in the upstream direction. If a logical channel is not using the fast path, the number of bytes allotted to that channel is zero (thus it would take up no space in the frame). Note that the allocation of bytes (eight bits), rather than fractions of bytes, to a data channel in a data frame combined with the 4 kHz data frame rate gives rise to the 32 kbps granularity of a channel's data rate. In fact, the data rate of a data channel can be found by Eqtn. 7.2:

Eqtn. 7.2

$$R_{channel} = \#\text{ data bytes per frame} \times 8\frac{\text{bits}}{\text{byte}} \times 4000\frac{\text{data frames}}{s}$$

Present in the data frame for the fast buffer are several overhead bytes, the *fast byte*, the *AEX byte*, and the *LEX byte* in the downstream direction and the fast byte and LEX byte in the upstream direction. The description of these bytes is given in the following paragraphs.

The fast byte is short for the *fast path synchronization byte*. The fast byte is always present even if no bearer channels are using the fast path. It has four different uses depending on the frame number (0 through 67). Its four uses are

- Carrying the CRC check for a superframe
- Carrying the indicator bits
- Carrying the embedded operations channel (EOC)
- Carrying synchronization control information to allow stuffing and robbing of bytes for synchronization

In frame 0, the fast byte carries superframe CRC information. This information is the CRC computed on the entire preceding superframe. In frames 1, 34, and 35, the fast byte carries *indicator bits (ibs)*. An indicator bit communicates some type of condition or primitive indication to the peer transceiver. There are 24 ibs in all; each is explained later in this chapter. In data frames 2 through 33 and 36 through 67, the fast byte carries either EOC data or synchronization control information. Either type of payload is two bytes long and must occupy an even-numbered-frame fast byte followed by an odd-numbered-frame fast byte.

An EOC message is distinguished from a synchronization control message because all bytes in an EOC message have a Least Significant Bit (LSB) of one, whereas all bytes in a synch control message have an LSB of zero. EOC messages and synch control messages are discussed later in this chapter. In summary, Figure 7.6 gives the makeup of the fast byte for both the upstream and downstream directions.

Figure 7.6 Makeup of the fast byte for all frames within the ADSL superframe.

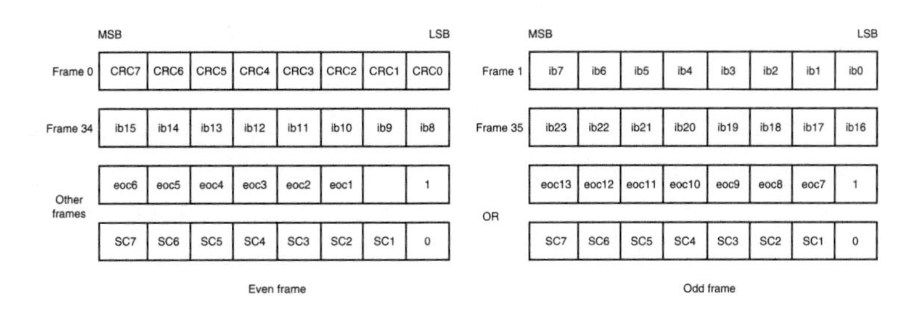

The AEX and LEX bytes are used to stuff an extra byte into an AS and LS channel. The AEX byte is present in the fast path only if at least one of the AS bearer channels uses the fast path. Likewise, LEX is present on the path only if an LS channel is present. Whether or not a byte is added to a bearer channel during a frame is controlled by synchronization control information in the fast byte. As stated previously, the synchronization control information can be present in a fast byte during frames 2 through 33 and 36 through 67. Table 7.2 gives the format of an eight-bit synchronization message.

Table 7.2 Format of a Fast Channel Synchronization Message (Fast Byte)

| Bits | Function | Code |
|---|---|---|
| 7,6 | Designates the AS channel to stuff or rob | 00—channel AS0.<br>01—channel AS1.<br>10—channel AS2.<br>11—channel AS3. |
| 5,4 | Controls the action to be taken on the designated AS channel | 00—do nothing.<br>01—add AEX byte.<br>10—add AEX and LEX bytes.<br>11—delete last byte. |
| 3,2 | Designates the LS channel to stuff or rob | 00—channel LS0.<br>01—channel LS1.<br>10—channel LS2.<br>11—no synchronization action. |
| 1 | Controls the action to be taken on the designated LS channel | 0—add LEX byte.<br>1—delete last byte. |
| 0 | Sync/EOC indicator | 0—indicates byte is a fast control byte.<br>1—In downstream direction, indicates that LEX byte is an EOC byte, but synchronization on AS channels may be performed as indicated. In upstream direction, indicates that the Fast byte is an EOC byte. |

Quite simply, the synchronization messages define what action is to be taken on a specified AS channel and a specified LS channel for that frame. Note that for an AS channel, up to two bytes can be stuffed (AEX and LEX) or one byte deleted. A deleted byte is sometimes called a *robbed byte*. For the LS channels, only one byte can be stuffed or robbed. Note that for the upstream, the upper four bits of the fast byte are unused because the upstream channel does not use AS bearers.

Byte stuffing and robbing is used to equalize the rate of a logical channel over the ADSL link with the rate needed to transport the channel. A good example of this is if a logical channel is transparently transporting a T1 line (1.544 Mbps). If the logical channel is running at a rate slightly lower than 1.544 Mbps, every so often the channel must stuff a byte or its input buffers will overflow. Likewise, if the logical channel were running slightly faster than 1.544 Mbps, every so often it must rob a byte so that its input buffers do not underflow. In each case, stuffing and robbing allow the logical channel, on average, to run at 1.544 Mbps.

Figures 7.7a and 7.7b show the fast path data buffer at points B and C in Figure 7.1a and Figure 7.1b for both the upstream and downstream directions, respectively. The only difference at this point is the addition of the Forward Error Correction (FEC) bytes from the encoder. Note that at these points, the data payload is scrambled.

Figure 7.7a The structure of the fast path portion of an ADSL data frame after the FEC block for the upstream direction.

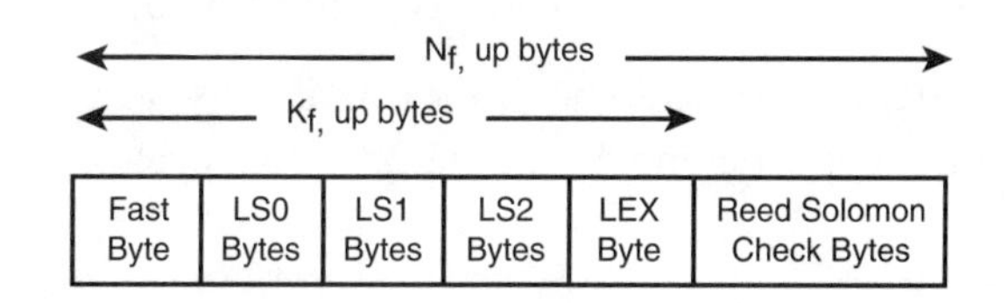


Figure 7.7b The structure of the fast path portion of an ADSL data frame after the FEC block for the downstream direction.

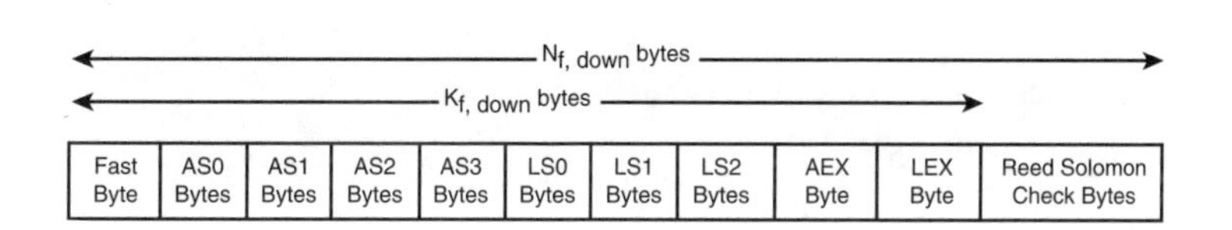


Figure 7.8a and Figure 7.8b show the data buffer at point A in Figure 7.1a and Figure 7.1b for the interleaved path.

Figure 7.8a The structure of an interleaved mux data frame before the FEC block for the upstream direction.

| Sync Byte | LS0 Bytes | LS1 Bytes | LS2 Bytes | LEX Byte |
|---|---|---|---|---|

Figure 7.8b The structure of an interleaved mux data frame before the FEC block for the downstream direction.

| Sync Byte | AS0 Bytes | AS1 Bytes | AS2 Bytes | AS3 Bytes | LS0 Bytes | LS1 Bytes | LS2 Bytes | AEX Byte | LEX Byte |
|---|---|---|---|---|---|---|---|---|---|

At this point, the data frame on the interleaved path is called a *mux data frame*. Note that at point A, interleaving has not yet been done. Similar to the fast path, the interleaved buffer also contains an AEX and a LEX byte (only LEX in the upstream direction). Again, the AEX byte is present only if an AS bearer channel uses the interleaved path. The LEX byte is present unless no channels are using the interleaved path. Analogous to and in lieu of the fast byte, however, the interleaved path contains a *sync byte*.

The sync byte has several functions. Some of these functions are analogous to fast bytes' functions, but at least one is different. In general, the sync byte performs the following actions:

- Carries CRC information for the interleaved channel in the preceding superframe
- Carries synchronization control information for stuffing and robbing bytes from bearer channels using the interleaved path
- Signals when the LEX byte of the interleaved path is used to carry a byte for the ADSL overhead channel (AOC) channel
- Carries the AOC channel when no bearer channels are using the interleaved path

Similar to the fast byte, in frame zero, the sync byte carries eight bits of CRC information. These cover the bearer channels assigned to the interleaved path over the preceding superframe. In all frames 1 through 67, when at least one bearer channel is using the interleaved path, the sync byte carries eight bits decoded as shown in Table 7.3.

Table 7.3 Format of an Interleaved Channel Synchronization Message (Sync Byte)

| Bits | Function | Code |
|---|---|---|
| 7,6 | Designates the AS channel to stuff or rob | 00—channel AS0.<br>01—channel AS1.<br>10—channel AS2.<br>11—channel AS3. |
| 5,4 | Controls the action to be taken on the designated AS channel | 00—do nothing.<br>01—add AEX byte.<br>11—add AEX and LEX bytes.<br>10—delete last byte. |
| 3,2 | Designates the LS channel to stuff or rob | 00—channel LS0.<br>01—channel LS1.<br>10—channel LS2.<br>11—no synchronization action. |
| 1 | Controls the action to be taken on the designated LS channel | 1—add LEX byte.<br>0—delete last byte. |
| 0 | LEX Byte Sync/AOC indicator | 0—indicates the LEX byte is to be used for synchronization as specified in bits 1–7.<br>1—In downstream direction, indicates that the sync byte contains AOC data and limited synchronization can be done. In upstream direction, indicates that the LEX contains AOC data. |

As shown in Table 7.3, a mechanism to add or delete a byte from any channel, almost identical to that for the fast byte, is provided by the sync byte. Note that for the upstream, no mechanism is present to add or stuff bytes in the AS channels because no AS channels run in the upstream direction. In the downstream direction, either one or two bytes can be appended to an AS channel, whereas only one byte can be appended to an LS channel. Note that the LSB of the sync byte indicates the contents of the LEX byte. If the LSB is one, the LEX byte contains AOC information. In this case, bits 4 through 7 of the sync byte still define synchronization action to perform to the AS channel (although only a single byte may be added). Additionally, a delete may still be performed to the LS channels. However, because LEX is already carrying AOC information, neither a one-byte stuff can be performed on the LS channels nor can a two-byte stuff be performed on the AS channels.

When no channels are using the interleaved path, no interleaved path bearer channel synchronization is necessary. In this case, both the AEX and LEX bytes are absent, and the sync byte is used to carry the AOC channel in frames 1–67.

On the interleaved path, multiple mux data frames can make up an FEC code word. This situation contrasts with the fast path, where a one-to-one correspondence between symbols and FEC codewords occurs. The actual number of mux data frames in an FEC codeword is decided during initialization. If this number is labeled as S, then at point B in Figure 7.1a and Figure 7.1b, the interleaved path buffer is shown in Figure 7.9. In this case, note that the FEC codeword spans multiple frames but is always an integer number of frames.

Figure 7.9 An FEC codeword on the interleaved path spanning S data frames.

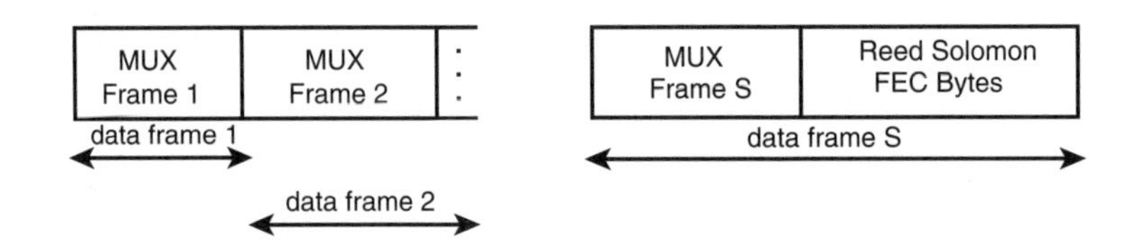


At point C in Figure 7.1a and Figure 7.1b, the FEC codewords are interleaved. Thus, the data to be modulated is made up of bytes from several FEC codewords. If a codeword is made up of S mux data frames for a total of N bytes and has K check bytes, then each interleaved path frame will consist of (N+K) bytes.

## *CRC Generator/Detector Blocks*

For both the fast and interleaved paths, an eight-bit CRC is generated for each superframe and sent during the first frame of the next superframe. On the fast path, the CRC is sent via the fast byte. On the interleaved path, the CRC is set via the sync byte. The CRC for a path covers all bits sent through that path during a superframe except for the FEC bytes, the previous CRC, and the synchronization symbol (frame 68). The CRC is meant to help keep track of how many superframes were received that contained an error uncorrectable by the FEC blocks. Both CRC generators and receivers are based on the generator polynomial shown in Eqtn. 7.3.

Eqtn. 7.3

$$G(x) = 1 + x^2 + x^3 + x^4 + x^8$$

## *Scrambler/Descrambler*

For both the ATU-R and the ATU-C, the generator polynomial for the scramblers and descramblers is given in Eqtn. 7.4.

Eqtn. 7.4

$$g(x) = 1 + x^{18} + x^{23}$$

As mentioned previously, the fast and interleaved paths are scrambled and descrambled independently. All bits from all logical channels are scrambled and descrambled. FEC bytes are not scrambled because they are added after the scrambling block. At the receiver, they are removed before the descrambling block. The only frame that does not pass through the scrambler is the sync frame.

## *Forward Error Correction Block*

After being scrambled, the data in both the fast and interleaved paths enter independent FEC blocks. Both paths utilize the Reed-Solomon technique in Galois field (256). The number of check bytes K and the number of payload bytes N that make up a codeword are decided during initialization of the ADSL link. For the fast path, a single codeword is sent for each data frame. For the interleaved path, a codeword can consist of an integer number of frames S, as long as S is a power of two less than or equal to 16 (S=1, 2, 4, 8, or 16). For both paths, the number of check bytes may be any even number between 0 and 16, inclusive. As a rule of thumb, the ratio of the number of check bytes to the number of data payload bytes should be around 0:1. This ratio gives a reasonable coding gain without a significant amount of overhead.

## *Interleaver/De-interleaver*

After passing through the forward error correction block, data in the interleaved path enters the interleaver in the transmitter, in both the ATU-R and ATU-C. DMT ADSL uses convolutional interleavering. The interleaver depth is expressed in Reed-Solomon codewords and may be any power of two up to and including 64. The interleaver works at the byte level because the Reed-Solomon code works in GF(256).

At the receiver, data enters the de-interleaver after constellation decoding and bit extraction (explained later in the chapter) are performed. The de-interleaver essentially reorders the bytes back into Reed-Solomon codewords for processing by the FEC decoder. The fast and interleaved paths have independent de-interleavers. Each de-interleaver uses the same parameters as its respective interleaver in the transmitter agreed upon during initialization.

## *Tone Ordering and Constellation Encoder*

Before entering the DMT modulation block, the payload to be transmitted consists of a buffer of data from the fast channel and a buffer of data from the interleaved channel. To modulate this data, you must first extract the proper number of bits for each DMT bin and code these bits into a complex $Z_i$ value (one complex value per bin). The process of extracting the bits and assigning them to tones is called *tone ordering*. The process of encoding the extracted bits into complex values for each bin is called *constellation encoding*. Constellation encoding for a DMT modulator is discussed in Chapter 6.

For tone ordering, the number of bits assigned to each bin must be known. This information is decided during startup training when the bins are analyzed. After analysis, the ATU-C and ATU-R calculate and agree to the number of bits to be handled by each bin.

Commonly, the value that represents calculated bits per bin is called the bin's *bi value*. The number of bits per bin for ADSL must be between 2 and 15, inclusive, or be 0. The sum of the number bits carried by all subcarriers directly correlates to the number of bits in each fast and interleaved path data frame. If the number of bits per bin is known, tone ordering follows this process:

1. Extract the N bits to be sent from the fast and interleaved paths.
2. Order these bits from 1 to N, putting the fast buffer bits first and the interleaved buffer bits second.
3. Execute the following algorithm on the ordered bits:

```
For k=2 to 15
{
While an unused tone remains that has b_i=k bits
{
Find the lowest numbered tone that that has b_I=k bits
Assign the next k bits from the data buffer to that tone
}
}
```

Essentially, the preceding algorithm assigns bits from the fast path to the tones carrying fewer bits. Bits from the interleaved buffer are, in general, assigned to tones carrying more bits. The reason is that the algorithm begins filling lightly loaded bins first and the fast bits are ordered before the interleaved bits (step 2). Note that one of the tones may carry bits from both the fast and interleaved paths.

### Note

Another option defined for DMT ADSL uses trellis-coded modulation on each bin. In this case, bit ordering is done slightly differently.[1]

The tone-ordering algorithm was designed to minimize error occurrences. Impulses during a symbol (whether from an outside influence or from a clip) will cause an almost even amount of noise in each DMT bin. In this case, tones carrying a larger number of bits are more likely to make a decoding error than are tones carrying a smaller number of bits. Thus, in general, an impulse will cause more bit errors on the interleaved path and less on the fast path. In the receiver, the bit errors on the fast path will likely be corrected by the forward error correction. The bit errors from the interleaved path, although more numerous than the fast path, will be split up and, because they will most probably belong to different FEC codewords (due to interleaving), will also be corrected. Note that the main assumption here is that impulses do not occur regularly.

After bits are properly assigned to each subchannel, constellation encoding takes place. The encoding technique is independent for each subchannel. The number of points in each bin's constellations depends on the number bits assigned to the bin. In the range of 2 to 15 bits per bin, the constellation size range is 4 points ($2^2$) to 32,768 points ($2^{15}$).

The output of each constellation encoder is a complex value. As discussed in the preceding chapter, this complex value represents one of the input points to the DMT modulator's Inverse Fast Fourier Transform (IFFT). For the downstream direction, a total of 256 bins exists. Thus the output of the tone order and constellation encoder at the ATU-C is 256 complex values. In the upstream direction, a total of 32 bins exists, and so the output of the block at the ATU-R is 32 complex values.

The constellation encoder block also includes a fine-gain-adjustment function. Apart from the $b_i$ value, each tone has assigned to it a fine-gain-adjustment value called $g_i$. The constellation output values are scaled by this $g_i$ value. Fine-gain adjustment equalizes the probability of error on subchannels by adjusting the output signal power (and thus the received signal power) by a factor in the range between -1.25 and +2 (approximately 0.75 to 1.33). For subchannel i, the constellation encoder's output complex value is referred to as $Z_i$.

At first the reason for reducing the output signal power and increasing the error probability on a subchannel may not seem obvious. The reason lies in the *peak-to-average ratio* (PAR) of a DMT symbol. PAR is simply the ratio of the peak possible power of a DMT sample (a time domain's value after the IFFT) to the average power of a DMT sample. A large PAR is undesirable because the analog front end of a modem with a large PAR must have a very large dynamic range. This condition tends to make the front-end designs more complex and also have higher power consumption. It turns out that DMT has a very high PAR (about 15 dB). The high peak power results from the situation when the outputs of all the individual bins add constructively to a high value. In many cases, this value actually goes outside the linear region of the analog front end and clipping results. Reducing power on tones that do not need the extra power to maintain a reasonable probability of error reduces not only the overall output power of the modem but also the probability that a clip will occur.

**Note**

Several algorithms are under development to reduce the PAR value for DMT.[4,5,6]

In both the upstream and downstream direction, it is common for some of the bins to carry zero bits of information. The constellation encoder will output a value of zero for these bins. The $g_i$ values for these bins are normally set to zero.

### *Modulation Block*

The heart of the transmitter is the DMT modulation block. The ATU-C and ATU-R, although both employing DMT, differ in the number of points being used in the IFFT. In the ATU-C, the modulator has 256 bins and uses 255 complex frequency points (and their complex conjugates) along with two special points to generate 512 real time-domain points. Thus the downstream signal is made up of 256 bins or tones. The spacing between DMT bin centers is 4.3125 kHz with the first center at 4.3125 kHz and the highest center at 1.104 MHz. One of the special points described previously is the tone at 1.104 MHz. This tone is called the *Nyquist tone,* is not used to carry data, and normally is set to zero. The other special point is a DC term that is also set to zero. Eqtn. 7.5 gives the basic equation governing the IFFT.

Eqtn. 7.5

$$x_k = \sum_{i=0}^{511} Z_i e^{\frac{j\pi k i}{256}} \quad \text{for } k = 0..511$$

The $Z_i$ value i=0 (in Eqtn. 7.4) is the DC term set to zero. The next 256 $Z_i$ values are the complex values coded from the QAM constellations for each subcarrier (including the Nyquist value of zero). The final 255 $Z_i$ values are the complex conjugates of the values from i=1 through i=255 with symmetry about the Nyquist tone. An equation for these values is given in Eqtn. 7.6.

Eqtn. 7.6

$$Z_m = conj(Z_{512-m}) \text{ for m} = 257..511$$

For the special case of the synchronization symbol, all the $Z_i$ are set to real values with nominal amplitude.

The DMT modulator in the ATU-R uses 32 bins. The 32 complex values (including a zero for tone 32) representing the constellation encoding outputs for each subchannel, a DC term, and the complex conjugates of tones 1 through 31 produce 64 real time-domain values after the IFFT operation. Again, the tone spacing is 4.3125 kHz, which gives a tone range from 4.3125 kHz to 138 kHz. The basic equation for the ATU-R IFFT is shown in Eqtn. 7.7.

Eqtn. 7.7

$$x_k = \sum_{i=0}^{63} Z_i e^{\frac{j\pi ki}{64}} \text{ for k} = 0..63$$

Here the first value of $Z_i$ is the DC term, and the next 31 are the complex QAM encoder outputs for each subchannel (including the value of zero at the Nyquist tone). The values of $Z_i$ for i=33 to i=63 are complex conjugates of tones 1 through 31 with complex conjugate symmetry about the Nyquist tone. An equation for this is shown in Eqtn. 7.8.

Eqtn. 7.8

$$Z_m = conj(Z_{64-m}) \text{ for m} = 33..63$$

For the case of the synchronization symbol, all $Z_i$ are set to real values with nominal amplitude.

The flexibility of DMT modulation can be seen by considering the three areas in the frequency spectrum shown in Figure 7.10.

Figure 7.10 Spectral placement of the downstream, upstream, and POTS regions for frequency division multiplex (FDM) mode ADSL.

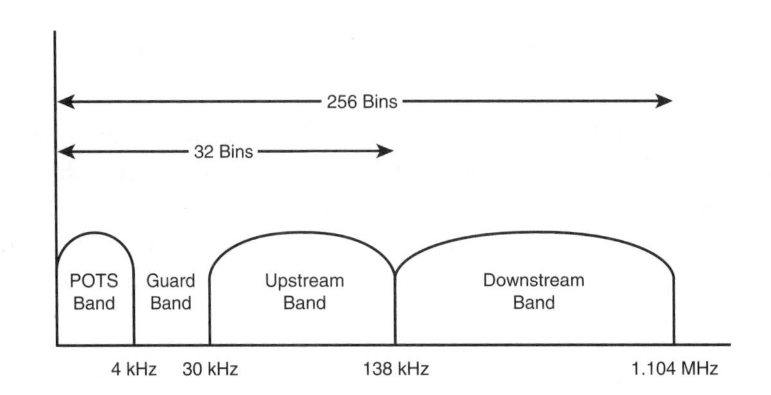


The lower-frequency area is the POTS band used for traditional voice transmission. This area is left open by the ADSL upstream and downstream signals by simply not using the first seven bins in both directions. Reserving the first seven bins in both directions is accomplished by assigning the bins in this region a $Z_i$=0 as well as using analog filtering on the front end of the transmitter to filter these frequencies. Setting the complex constellation encoder output to zero "turns off" the bin or tone. The middle frequency range in Figure 7.10 is used by the upstream signal. The downstream signal can leave this area free if it turns off tones 8 through 32. The upper-frequency region in Figure 7.10 is used exclusively by the downstream channel. The upstream channel does not need to do anything special to leave this region open because that channel transmits only on bins up to 138 kHz.

This type of DMT operation is called *frequency division multiplex* (FDM) DMT or *frequency division duplex* (FDD) DMT. The names stem from the fact that the duplexing function of the line is realized by using separate upstream and downstream frequencies. Note in this case, for multiple ADSL lines in a binder group, self-FEXT exists. However, the more damaging self-NEXT does not exist.

Another type of operating mode exists and is called *echo-canceled* (EC) DMT. For EC mode, the downstream also transmits on all tones above 30 kHz, causing it to overlap the ATU-R output in the 30 kHz to 138 kHz region. EC operation allows the downstream channel to achieve higher rates than does FDM mode.

EC operation has a couple of downsides. First, self-NEXT exists at the ATU-C and the ATU-R because local transmitters are using the 30 kHz to 138 kHz region. Note that at the ATU-R, NEXT exists only in the 30 kHz to 138 kHz; the upper part of the downstream band is self-NEXT free. Second, both the analog and digital portions of the modem design become more complex. The analog portion must handle a more dynamic range, and the digital portion must implement a very precise echo canceler. Though some ADSL implementations do employ an EC mode, many network implementers prefer FDM mode because of the absence of self-NEXT.

## *Pilot Tones*

In both the upstream and downstream directions, one of the DMT tones is reserved for a pilot signal. In the downstream direction, the pilot tone is at 276 kHz, or equivalently tone 64. In the upstream direction, the pilot tone is at 69 kHz, or equivalently tone 16. The pilot tones can be used to resolve sample timing at the receiver. The pilot tones will never be used for data, but will always send bits (0,0) from a 4QAM constellation (the constellation point in the first quadrant).

## *Cyclic Prefix Block*

As discussed in Chapter 6, a cyclic prefix can help to make a channel look circular so that equalization can occur more easily in the frequency domain. In the downstream direction, a 32-sample cyclic prefix is appended to the 512 time-domain samples. Basically, the prefix consists of samples 480 through 511. The output sequence of the DMT transmitter consists of 544 samples given by $x_{480}, x_{481}, \ldots, x_{511}, x_0, x_1, \ldots, x_{511}$. The cyclic prefix for the upstream channel is four samples long and contains samples $x_{60}$ through $x_{63}$ from the output of the upstream IFFT. Thus the output sequence in the upstream direction consists of 68 samples given by $x_{60}, x_{61}, \ldots, x_{63}, x_0, x_1, \ldots, x_{63}$.

It is important to understand and distinguish the data frame rate and the frame rate of a DMT symbol as well as to understand why the bin spacing is 4.3125 kHz. The cyclic prefix as well as the synchronization frame play a role in these rates and frequencies. The data frame rate, as explained earlier, is exactly 4 kHz (or 4 kBaud). Because every 69th frame is a synchronization frame, the frame rate must be slightly higher than the data frame rate and is given by Eqtn. 7.9.

Eqtn. 7.9

$$
\begin{aligned}
R_{frame} &= \frac{69}{68} R_{data} \\
&= \frac{69}{68} 4000 \\
&\approx 4.0588 \text{ kHz}
\end{aligned}
$$

Given the number of samples in the downstream direction (544) per frame, Eqtn. 7.10 gives the sample rate leaving the ATU-C (and arriving at the ATU-R).

Eqtn. 7.10

$$R_{\text{Sample, DS}} = \frac{69}{68} 4000 \frac{\text{frames}}{\text{s}} \left( 544 \frac{\text{samples}}{\text{frame}} \right)$$
$$= 2.208 \frac{\text{Msample}}{\text{s}}$$

In the upstream direction, the sample rate leaving the ATU-R (and arriving at the ATU-C) is similarly given by Eqtn. 7.11.

Eqtn. 7.11

$$R_{\text{Sample, DS}} = \frac{69}{68} 4000 \frac{\text{frames}}{\text{s}} \left( 68 \frac{\text{samples}}{\text{frame}} \right)$$
$$= 276 \frac{\text{ksample}}{\text{s}}$$

The frequency-domain resolution of an FFT (the spacing of points after the transform) with 2N real time-domain samples is given by Eqtn. 7.12.

Eqtn. 7.12

$$\text{Res}_{FFT} = \frac{R_{Sample}}{2N}$$

For the downstream direction, a simple calculation with 2N=512 and a sample frequency of 2.208 MHz give a bin spacing in the frequency domain of 4.3125 kHz. Likewise, for the upstream direction 2N=64 and a sample rate of 276 kHz, the resulting bin spacing is also 4.3125 kHz.

## *Time-Domain Equalization Block*

Though a DMT system can do equalization most efficiently in the frequency domain, a time-domain equalizer is generally present at the front end of the receiver. This equalizer generally serves two purposes. First, the equalizer removes any intersymbol interference from the channel that is longer than the cyclic prefix. Such interference would otherwise cause a symbol to interfere with the next symbol in time, and the channel would not look circular. Second, the time-domain equalizer can be used to partially bandpass the incoming signal and filter out-of-band energy. For example, the time-domain equalizer at the input to an ATU-C might contain a low pass section to reduce energy that is below tone number 33 and a high pass section to reduce energy that is above 138 kHz. In general, the implementor decides on the length and type of the time-domain equalizer used in an ADSL receiver. The filters are generally adaptive in nature and are trained to initial optimal values during initialization.

## *Demodulation Block with FDQ*

The demodulator in an ADSL receiver very much follows from the modulator. At the ATU-R, the heart of the demodulator is basically an FFT operating on 512 real-valued points. Note that before the FFT is performed, the 32-sample cyclic prefix is removed. After the FFT is performed, only the lower 256 complex FFT outputs are kept, and the other 256, which are complex conjugates, are dropped. Eqtn. 7.13 describes the demodulation mathematically.

Eqtn. 7.13

$$Z'_k = \sum_{i=0}^{511} x_i e^{-\frac{j\pi ki}{256}} \quad \text{for k = 0..255}$$

Here the $Z'_k$ are the unequalized values to be decoded by each bin's QAM constellation decoder.

At the ATU-C, the heart of the demodulator is an FFT using 64 real time-domain samples. Again, the four-sample cyclic prefix is discarded before the FFT block. Only the first 32 FFT outputs are kept, and their complex conjugates are dropped. Eqtn. 7.14 describes the FFT.

Eqtn. 7.14

$$Z'_k = \sum_{i=0}^{63} x_i e^{-\frac{j\pi ki}{32}} \quad \text{for k = 0..31}$$

At both the ATU-R and ATU-C, frequency-domain equalization is done after the FFT operation. As discussed in Chapter 6, this process is accomplished by simply using one complex multiply for each bin. The multiplier values are determined during initialization of the ADSL link and are normally updated either during the synchronization frame or on a frame-by-frame basis.

### *Constellation Decoding and Bit Extraction*

After the demodulation and equalization, the values for each bin are individually decoded using a QAM constellation decoder. Decoding involves the same constellation mapping on each bin as used in the transmitter. Bins carrying zero bits do not have to be decoded.

The decoded bits from each bin are sent to either the fast or interleaved buffer, using the reverse procedure of how they were distributed. This procedure is known as *bit extraction* and is essentially the reverse of tone ordering.

The rest of the receiver process is the reverse of the transmitter process. Specifically, de-interleaving is done on the interleaved path, and Reed-Solomon decoding and descrambling are done independently on each path. The resulting bit streams are passed through the CRC check block. Note that this block does not change any of the payload bits. Deframing is then performed, at which point the fast and sync bytes are decoded. If necessary, the AEX or LEX bytes are stuffed onto the appropriate channels, or a byte is removed from the end of the bytes of an indicated AS or LS channel. At this point, all the incoming data are passed out onto the appropriate interface (including all LS and AS channels as well as AOC and EOC information).

### *Timing Recovery*

For proper operation of most systems, the sampling clock at the transmitter and receiver must be exactly the same frequency. Put another way, the transmitter and receiver clocks must be locked. Typically, locking is done by one end being the master clock and the other end recovering the clock. In an ADSL system, either end can be the master, and the decision is made during link initialization. Most often the ATU-C is the master, and the ATU-R does clock recovery. In this case, we say that the ATU-R is *loop timed*. Loop timing can be done in many ways but most often involves a *phase-locked loop* (PLL). The ATU-R basically needs to recover the 2.208 MHz sampling clock. All other clocks and timing can be found from that clock, including the frame rate, the superframe rate, and the upstream transmit clock. In addition, the upstream sampling clock of 276 kHz can be derived from the 2.208 MHz sampling clock. A modem may have to use the synchronization frame to find superframe boundaries.

## *DMT ADSL Overhead*

As discussed in the framing sections, ADSL has several overhead channels. The overhead includes the EOC, AOC, and the ibs. The uses of the overhead are discussed in the next few sections.

## *EOC Channel*

As discussed previously, the EOC is carried in the fast byte of the fast channel. The ATU-C uses the EOC to retrieve information from and write information to the ATU-R during runtime. An EOC message contains five fields spanning 13 bits, as shown in Table 7.4. An EOC message always starts with a fast byte in an even frame number and continues into the fast byte from the following odd frame.

Table 7.4 Format of an EOC Message

| Bits | Description | Notes |
|---|---|---|
| 1,2 | Address Field | 00—ATU-R address<br>01—unused<br>10—unused<br>11—ATU-C |
| 3 | Data/Opcode | 0—Data (used for reads and writes)<br>1—Opcode |
| 4 | Odd/Even Byte | Used for multibyte transfers to delineate successive bytes |
| 5 | Autonomous Message Field | Set to 0 by ATU-R to send 'dying gasp' message |
| 6–13 | Information Field | Data or op codes |

Information that can be read from the ATU-R includes the ATU-R vendor ID, the version number, serial number, results of a self-test, line attenuation, SNR margin, and configuration information. This information is held at the ATU-R in registers that the ATU-C can read via the EOC. Additionally, the ATU-C can use the EOC to command the ATU-R to perform a self-test, send corrupt CRC values, and prepare to receive corrupt CRC values.

**Note**

Often corrupt CRCs are generated to make sure that other performance information (the indicator bits) are properly logging erroneous frame information.

The EOC protocol is very simple. The ATU-C is the master of the channel, and the ATU-R is the slave, responding to commands from the master. The protocol uses a message/echo-response scheme. To send a message to the ATU-R, the ATU-C sends the message three consecutive times. These three messages represent three outstanding messages and one outstanding command. Upon receiving three identical EOC messages, the ATU-R echoes the messages by sending them back to the ATU-C. Upon receiving three echoes, the ATU-C is assured that the ATU-R properly received the command. If the command instructs the ATU-R to take some action (perform a self-test, for instance), that action begins. If the command is a read or write, the next read or write message can be sent. These messages would follow a similar message/echo-response scheme.

One exception to the normal EOC protocol is the *dying gasp* that can be autonomously set by the ATU-R and sent to the ATU-C. Besides sending the dying gasp op-code, the ATU-R also sets bit 5 of the EOC message to zero, indicating an autonomous message. Such a message is sent if power is lost at the ATU-R, but for some reason the ATU-R transceiver is still powered on.

## *AOC Channel*

As discussed previously, during training the twisted pair channel is analyzed, and the number of bits that each DMT bin can support is calculated. In general, bins with a better-received SNR can support more bits and denser constellations than bins with lower-received SNRs. Over time, the characteristics of the channel may change, resulting in some bins not being able to still support the same amount of bits and other bins being able to support more. Such changes may be the result of other services being added to the binder group, causing crosstalk, or even from changes in temperature, causing the twisted pair channel's characteristics to change. The AOC allows for a process known as *bit swapping* in which a bit to be removed from one bin (causing its constellation to become smaller) is placed into another bin (causing its constellation to become larger).

AOC messages can be carried in either the sync byte of the interleaved channel or the LEX byte of the interleaved channel. An AOC message consists of an eight-bit message header followed by several more bytes, depending on the type of message. Table 7.5 describes the possible message headers.

Table 7.5 AOC Message Headers

| Value | Message | Notes |
|---|---|---|
| 00000000 | Idle Mode | |
| 00001111 | Reconfiguration | |
| 1100xxxx | Reserved | Vendor-specific commands are allowable. |
| 11110000 | Unable to comply | Reply to a bit swap request. |
| 11111100 | Extended Bit Swap Request | Used to move two bits from a carrier. This header allows for six message fields instead of four. |
| 11111111 | Bit Swap Request | Normal bit swap to move one bit from one bin to another. |

In general, when a bit swap is desired, a bit swap request message is sent. If the bit swap is approved, a bit swap acknowledgment message is returned. Note that the receiver sends the request and the transmitter returns the reply. (This arrangement makes sense because the receiver has primary knowledge of the received SNR of each bin, so the receiver knows when the channel characteristics have changed.) Bit swap requests are sent five times and indicate from which bin a bit is to be removed and to which bin a bit is to be added. Additionally, power adjustments to the bins may be made. Figure 7.11 shows the format of a bit swap request message.

Figure 7.11 The format of a bit swap request message.

| 1 byte | 2 bytes | | 2 bytes | | 2 bytes | | 2 bytes | |
|---|---|---|---|---|---|---|---|---|
| Message Header | Subcommand 1 | | Subcommand 2 | | Subcommand 3 | | Subcommand 4 | |
| 0 x FF | Command | Subchannel Index | Command | Subchannel Index | Command | Subchannel Index | Command | Subchannel Index |

The command basically allows four subcommands to be sent as shown in Table 7.6. In a single bit swap request, all information to change a bit from one bin to another and the appropriate power changes can be made.

Table 7.6 Bit Swap Request Commands

| Value | Meaning |
|---|---|
| 00000000 | Do nothing |
| 00000001 | Increase the number of bits in this bin by one |
| 00000010 | Decrease the number of bits in this bin by one |
| 00000011 | Adjust the transmitted power in this bin by +1 dB |
| 00000100 | Adjust the transmitted power in this bin by +2 dB |
| 00000101 | Adjust the transmitted power in this bin by +3 dB |
| 00000110 | Adjust the transmitted power in this bin by -1 dB |
| 00000111 | Adjust the transmitted power in this bin by -2 dB |
| 00001xxx | Reserved for vendor-specific commands |

A bit swap acknowledgment consists of three bytes as shown in Figure 7.12.

Figure 7.12 The structure of a bit swap acknowledgment message.

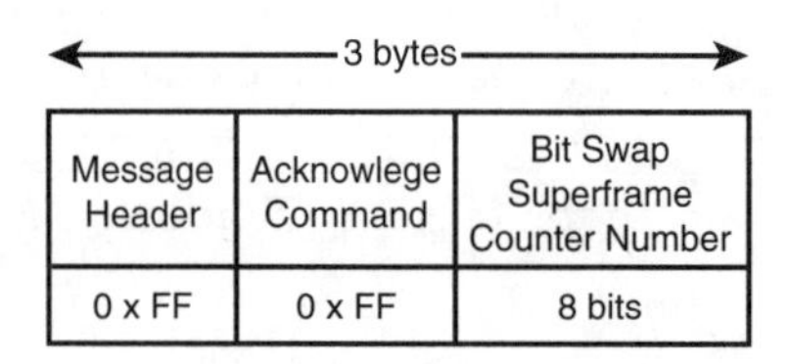


The third byte of the acknowledgment contains a superframe counter number, indicating the superframe after which the bit swap will take place. The number must be at least 47 superframes greater than when the request was received. Beginning with frame 0 of the superframe following the number specified in this field, the bit and power changes take place at the transmitter and receiver. Note that this change affects constellation size as well as the tone-ordering algorithm described previously. The bit swap acknowledgment is sent five times. A bit swap will not take place if an acknowledgment is not received within 450 ms of the request or if an unable-to-comply message is received after a request.

## *Indicator Bits*

For each superframe, 24 *indicator bits* are sent using an overhead byte in frame 1, frame 34, and frame 35. Some of these indicator bits are defined, and others are reserved for future use. These bits indicate to the peer ATU anomalies that have occurred during the preceding superframe. These anomalies include whether or not CRC errors occurred in the fast and interleaved paths, whether or not the FEC had to correct any bytes during the previous superframe, whether or not a loss of signal (LOS) occurred during the last superframe, and whether a remote default indication is present. (A remote default indication (rdi) is present when two consecutive synchronization symbols are incorrectly received.) The indicator bits have additional functionality when ATM cells are being carried over ADSL. Specifically, indicator bits are reserved to indicate whether an ATM cell header error occurred during the previous superframe and whether a cell delineation is properly achieved on each path. In addition, four indicator bits are reserved to transfer a *network timing reference* (NTR) from the ATU-C to the ATU-R. This topic is discussed later in this chapter. Table 7.7 lists and defines the indicator bits for an ADSL system.

Table 7.7 Indicator Bit Definitions

| Indicator Bits | Name | Function |
|---|---|---|
| ib0-ib7 | N/A | Reserved for future use (set to 1) |
| ib8 | febe-i | Indicates whether a CRC error occurred during the preceding superframe on the interleaved path |
| ib9 | fecc-i | Indicates whether errors were detected and corrected by the Reed-Solomon decoder during the previous superframe on the interleaved path |
| ib10 | febe-f | Indicates whether a CRC error occurred during the previous superframe on the fast path |
| ib11 | fecc-f | Indicates whether errors were detected and corrected by the Reed-Solomon decoder during the previous superframe on the fast path |
| ib12 | los | Indicates that a loss of received signal has been detected |
| ib13 | rdi | Indicates that two consecutive synchronization symbols were incorrectly received |
| ib14 | ncd-i | For ATM operation, indicates whether cell delineation has been achieved on the interleaved path |
| ib15 | ncd-f | For ATM operation, indicates whether cell delineation has been achieved on the fast path |
| ib16 | hec-i | For ATM operation, indicates whether a cell header error was detected during the previous superframe on the interleaved channel |
| ib17 | hec-f | For ATM operation, indicates whether a cell header error was detected during the previous superframe on the fast channel |
| ib18-19 | N/A | Reserved for future use |
| ib20-23 | NTR0-3 | Transmits the network timing reference offset |

## *ADSL Reduced Overhead Modes*

When ADSL was being developed, one of the main applications was video on demand (VOD). VOD over an ADSL system would require a high-speed downstream channel (normally AS0) to carry the digital video and a lower-speed upstream control channel for selection and control. The downstream channel would normally need to be a fixed rate and often a multiple of 1.536 Mbps, the data payload rate for a T1 or DS1 signal. A second bidirectional channel may also have been desired to carry data.

The digital video would most probably use the interleaved path because the added delay would not degrade the user's experience watching the video and the added protection from impulse noise would reduce imperfections in the video playback. The control and data channels would likely use the fast path as they would be somewhat interactive and added delay would be perceivable and undesirable.

Considering the descriptions in the above paragraph, an ADSL system would need to use the following functions:

- Both the fast and interleaved channels
- Stuffing and robbing on at least AS0 as its rate must be locked to the video server's rate
- AOC and EOC channels

As it became apparent that the main application for ADSL would be Internet and other data source access, not video on demand, the requirements for an ADSL system changed. For simple Internet access, an ADSL system would normally need a high-speed downstream channel (usually AS0) and a lower-speed upstream channel (usually LS0). In this case, neither channel would need to be locked to an external system clock. Thus no stuffing or robbing would be necessary. In addition, only one of the paths would be necessary. These new applications, with reduced requirements, prompted reduced overhead modes to be defined for ADSL. The reduced overhead modes further reduced ADSL overhead by either 32 kbps or 64 kbps. This reduction is significant for the upstream channel that typically runs at far lower rates than the downstream channel. Two schemes of reduced overhead are defined. Whether or not reduced overhead is used and, if so, which type of reduced overhead is used are decided during training.

### *Reduced Overhead Mode 1*

The first type of reduced overhead mode eliminates the AEX and LEX bytes from both the fast and interleaved paths. The bytes are eliminated whether or not the paths carry AS or LS bearer channels. In this mode, byte stuffing and byte deleting are not allowed, and all overhead is carried in the fast and sync bytes. Note that the fast and sync bytes are always present in this mode. In this mode, the fast byte carries the superframe CRC for the fast path, the indicator bits, and the EOC channel. The sync byte carries the CRC for the interleaved path and the AOC channel.

### *Reduced Overhead Mode 2*

Reduced overhead mode 2 can be used when bearer channels exist in only one of the paths. Such a condition is called single latency operation (as opposed to dual latency, which would be the case when both paths are being used). In this mode, not only are the AEX and LEX bytes deleted from both paths, but the fast and sync bytes are combined.

Note that the combined byte must carry the CRC byte for each superframe, the indicator bits, and the AOC and EOC channels. To differentiate the EOC and AOC channel information, AOC information is sent in frames 4 and 5, 8 and 9, 12 and 13, ..., and 67 and 68. EOC information is sent in frames 2 and 3, 6 and 7, 10 and 11, ..., and 65 and 66 with the exception of bytes 34 and 35 (which carry indicator bits). Table 7.8 shows a comparison of the overhead bytes for each frame in a superframe for normal framing as well as both reduced overhead framing modes.

Table 7.8 Comparison of Fast and Sync Bytes for Different Overhead Modes

| | **Full Overhead** | | **Reduced Mode 1** | | **Reduced Mode 2** | |
|---|---|---|---|---|---|---|
| **Frame Number** | **Fast Byte** | **Sync Byte** | **Fast Byte** | **Sync Byte** | **Fast Byte** | **Sync Byte** |
| 0 | fast CRC | interleaved CRC | fast CRC | interleaved CRC | fast CRC | interleaved CRC |
| 1 | ib0-7 | sync or AOC | ib0-7 | AOC | ib0-7 | ib0-7 |
| 34 | ib8-15 | sync or AOC | ib8-15 | AOC | ib8-15 | ib8-15 |
| 35 | ib16-23 | sync or AOC | ib16-23 | AOC | ib16-23 | ib16-23 |
| others | sync or EOC | sync or AOC | EOC | AOC | AOC/EOC | AOC/EOC |

**Note**

In Reduced Mode 2, either the fast or sync byte is present, but not both.

## Physical Layer Training

In many instances, I have referred to training and initialization in the text for the point in time at which the transmitter and receiver decide on certain operating parameters, options, or filter values. This section describes the training process in more detail.

ADSL training and initialization is broken down into four main stages: activation and acknowledgment, transceiver training, channel analysis, and exchange. These stages are defined in the subsequent sections.

### Activation and Acknowledgment

During the activation and acknowledgment stage, the ATU-R and the ATU-C are turned on and perform an initial handshake. All signals transmitted during this time are single tones at one of the subcarrier's frequencies. While initially identifying themselves, the ATU-R and ATU-C also decide which side will perform loop timing. The ATU-C decides what will be loop timed. Thus the ATU-R must always have the capability to be loop timed. Generally, loop timing at the ATU-R is desired because implementation of

ATU-Cs, which are normally in cards in a shelf, is simplified if all ATU-Cs run off a single clock instead of having multiple timing-recovery circuits.

### *Transceiver Training*

Phase 2 of training and initialization is called transceiver training. During this time, several wideband signals are sent between the ATU-R and ATU-C. The wideband signals allow each unit to calculate the upstream and downstream received power spectral density and to adjust the *automatic gain control* (AGC) at each receiver prior to the analog-to-digital conversion. Also, the wideband signals can be used to train equalizers in each receiver. A quiet period is also available during this stage of training during which a transceiver can train its echo canceler. The signal sent during this time is implementation specific. It does not have to be specified, because the echo filter is trained in the same modem generating the training signal; therefore, the unit at the other end does not need to do anything during this time. Note that even if an echo-canceled mode is not being used, some echo cancellation may still be necessary to reduce the out-of-band power generated by the unit.

### *Channel Analysis*

During the third training phase—channel analysis—options, capabilities, and configuration information are exchanged between the ATU-R and the ATU-C. Four main messages are sent during this time. The ATU-C sends two messages called C-Rates1 and C-MSG1. Likewise, the ATU-R sends two messages called R-Rates1 and R-MSG1.

The C-Rates1 message sends four options from the ATU-C to the ATU-R. Each option consists of a proposed number of bytes each bearer channel will carry per frame in both the upstream and downstream direction. Recall that the number of bytes per bearer channel in each frame is directly proportional to the bit rate of that bearer channel (the bit rate=(number of bytes)(4000). A total of 20 proposed bearer channel rates are sent for each option. The first 10 rates include 4 rates for the downstream AS bearers, 3 for the downstream LS bearers, and 3 for the upstream LS bearers for the fast path. The second 10 rates are the same as the first but for the interleaved path. Generally, many of these will be set to zero, and if a rate is not zero for one of the paths, it must be zero for the other path. Of the four options, one option will eventually be selected unless a rate adaptive startup is negotiated. You will see that in the case of a rate adaptive startup, more options will be provided later during initialization. Also contained in each option is the number of FEC bytes per codeword in both buffers of the upstream and downstream directions and the number of symbols per codeword on the interleaved channel in both directions.

**Note**

Although most of the channel characteristics are known at this point, configuration information of the ATU-R and ATU-C is not yet known (for example, whether trellis coding is supported). Thus, rate options are exchanged, rather than absolute rate settings.

**Note**

Originally, rate adaptation was not a startup option. When video on demand was a main application of ADSL, the four rates could correspond to different possible digital video interface rates (generally multiples of 1.536 Mbps). Providing options allowed the transceivers to choose the highest rate that could be supported.

C-MSG1 is a message sent from the ATU-C to the ATU-R, describing characteristics and capabilities of the ATU-C. This 48-bit message includes the following information:

- Minimum required SNR margin (usually set to 6 dB)
- Vendor identification
- T1.413 specification issue number that the modem corresponds to
- Modem version number
- Trellis-coding support indicator
- Echo-cancellation support indicator
- Extended exchange support indicator (indicates support for either rate adaptation or rates higher than 8 Mbps)
- NTR support indicator
- Indication of whether or not a reduced overhead mode is supported
- Transmit PSD level during initialization
- Maximum number of bits supported per subcarrier

The information exchanged in C-MSG1 lists all information necessary to ensure that the ATU-R knows all of the ATU-C's capabilities.

R-Rates1 is very similar to C-Rates1 in that four options are sent listing possible rate scenarios. In this case, however, only upstream options are listed, and in fact, the options are copied directly from C-Rates1 (thus they are redundant). Only six bearer channel byte counts per option are sent, three for the LS channels in the upstream fast path and three for the LS channels in the upstream interleaved path. FEC redundancy bytes and codeword size are also echoed for each option.

R-MSG1 contains configuration and support information for the ATU-R. This 48-bit message consists of many of the same parameters as C-MSG1, excluding the SNR margin, NTR support indication, transmit PSD level during initialization, reduced overhead support, and extended initialization support.

Wideband tones are also sent during the channel analysis phase to further refine equalizer settings and analyze the channel. These tones are similar to those sent during the transceiver training phase except that the former include a cyclic prefix, whereas the signals sent during transceiver training did not.

## *Exchange*

The final phase of training is called the exchange phase. Originally, the exchange phase consisted of the ATU-R and ATU-C deciding which upstream and downstream options transmitted in C-Rates1 and R-Rates1 (if any) would be used. The exchange phase was expanded when rate adaptive capability was added to ADSL. If rate adaptation is supported on the loop, before choosing any options, the exchange phase first proposes a new set of options more optimized to the loop's characteristics. The first message exchanged in this phase in the case of rate adaptation is R-MSG-RA. R-MSG-RA is sent by the ATU-R and is an 80-bit message. R-MSG-RA indicates the maximum number of bits that the downstream channel can support (given the channel characteristics) with the margin specified in C-MSG1. Also sent in R-MSG-RA are the assumptions used to calculate $b_{max}$. The purpose of R-MSG-RA is to indicate to the ATU-C more information about the downstream channel so that the ATU-C can later propose a more optimal set of rate options. Information sent during R-MSG-RA consists of the following:

- The number of FEC check bytes assumed for the $b_{max}$ calculation
- The number of data payload bytes in an FEC codeword assumed for the $b_{max}$ calculation
- The number of tones carrying payload assumed for the $b_{max}$ calculation
- The estimated average loop attenuation
- The estimated coding gain assumed for the $b_{max}$ calculation
- The performance margin calculated for $b_{max}$
- The total number of bits per DMT symbol that can be supported ($b_{max}$)

After sending R-MSG-RA, the ATU-R sends R-Rates-RA. R-Rates-RA consists of eight bits and selects the highest proposed downstream rate from C-Rates1 that can be supported given the channel characteristics measured. R-MSG-RA can also indicate that no option from C-Rates1 can be supported or that it wants to wait to receive four more options before making a decision.

The ATU-C responds to R-MSG-RA and R-Rates-RA with C-Rate-RA and C-MSG-RA. C-Rates-RA is very similar to C-Rates-1 in that four upstream and downstream options are proposed. These new options are more closely related to the optimum rates that the channel can support. C-MSG-RA follows C-Rates-1. It is a 48-bit message containing the required SNR margin for the downstream channel.

When rate adaptation is not used, the messages described above are not transmitted. Thus the rest of the exchange phase described in the following paragraphs is the same, regardless of whether rate adaptation is used.

The ATU-R needs to send three more formatted messages to the ATU-C to complete the selection of downstream parameters. The first two messages are R-MSG2 and R-Rates2. The information in R-MSG2 is similar to the information in R-MSG-RA. The exact items sent in this message are the following:

- The estimated loop attenuation
- The performance margin for the option that will be selected in R-Rates2
- The maximum number of bits that the downstream direction could support given the performance margin dictated by the ATU-C

R-Rates2 selects the option from R-Rates-RA or R-Rates1 that lists the highest downstream rates that can be supported given the performance margin specified. The information in R-MSG2 is based on the option specified in R-Rates2. Note that the option chosen corresponds to an option proposed in C-Rates1 if rate adaptation is not used and corresponds to an option proposed in C-Rates-RA if rate adaptation is used.

The ATU-C responds to R-MSG2 and R-Rates2 with C-MSG2 and C-Rates2. C-MSG2 contains the same information as R-MSG2 contains but for the option selected for the upstream channel. C-Rates2 specifies the final decision on the upstream option and downstream option to be implemented. The ATU-C may change the option selected in R-Rates2. The ATU-R will abide by the ATU-C's decision.

After the upstream and downstream options are selected and known by the ATU-R and ATU-C, one more message in each direction is necessary before a transition is made into steady-state operation. The first message is called C-B&G and is transmitted from the ATU-C to the ATU-R. This message transmits the number of bits ($b_i$'s) and the fine gain adjustments ($g_i$'s) to be used on each subcarrier in the upstream direction. The final message from the ATU-R to the ATU-C is called R-B&G and specifies the number of bits and the fine gain adjustment to be used on each subcarrier in the downstream direction. After this final exchange, a transition is made into steady-state operation. This state is called Showtime.

It may seem that the training and initialization phases of ADSL are not completely optimized especially when rate adaptation is used. The reason for this impression is that the original training specification did not include rate adaptation (and thus R-MSG-RA, C-MSG-RA, R-Rates-RA, and C-Rates-RA were not defined). The rate adaptation procedure was meant not only to provide a mechanism for adaptation but also to maintain backward compatibility to modems that were designed to the original specification and do not support this feature. Most ADSL modems designed for Internet access include rate adaptation capabilities.

### *ATM over ADSL*

When an ADSL system is designed to carry ATM cells only in the bearer channels, it may take on added functionality. For example, the ADSL modem will be required to insert idle ATM cells when no payload cells are available to send. In addition, it will be required to scramble and unscramble the cell payload at the transmitter and receiver, respectively. One other functionality that may be necessary is called a *network timing reference* (NTR). An NTR is an 8 kHz clock signal present at the ATU-C that must be reproduced at the ATU-R. Four indicator bits (ib20–ib23) are used to pass information necessary to construct such a clock at the ATU-R. These bits are called ntr0, ntr1, ntr2, and ntr3.

To properly reconstruct the 8 kHz NTR at the ATU-R, the ATU-C generates an 8 kHz clock called a *local timing reference* (LTR) by dividing its 2.208 MHz sampling clock by 276. In every superframe, the phase difference between the NTR and LTR is coded into the indicator bits and sent to the ATU-R. The phase offset is measured in cycles of the 2.208 MHz clock. The bits ntr3 through ntr0 are coded as a 2's complement signed integer and can take on the value of -8 to +7. The ATU-R can use these bits to reconstruct the NTR clock. The ATU-R already has the LTR clock because the 2.208 MHz clocks at each end should already be locked to one another.

## *CAP/QAM ADSL Systems*

Though DMT is normally considered the standard type of ADSL, early types of systems using CAP and QAM were important to the development of DSL. Additionally, some ADSL deployments continue to be based around these technologies. The following sections discuss CAP- and QAM-based implementations. Though the systems are distinctly different, some designs can emulate either a CAP implementation or a QAM implementation based on parameters passed between the modems during initialization.

> **Note**
>
> Also, CAP/QAM techniques are sometimes used in other DSL implementations, including certain non-2B1Q versions of HDSL and proposals for VDSL.

## *Modulation and Demodulation Blocks*

Figure 7.13 shows the block diagrams of a CAP transmitter and a QAM transmitter.

Figure 7.13 Block diagrams of (a) a CAP transmitter and (b) a QAM transmitter.

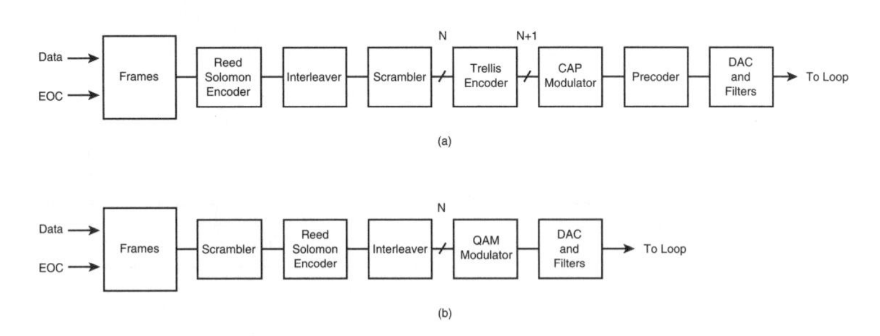


Likewise, Figure 7.14 shows the block diagrams for CAP and QAM receivers.

Figure 7.14 Block diagrams of (a) a CAP receiver and (b) a QAM receiver.

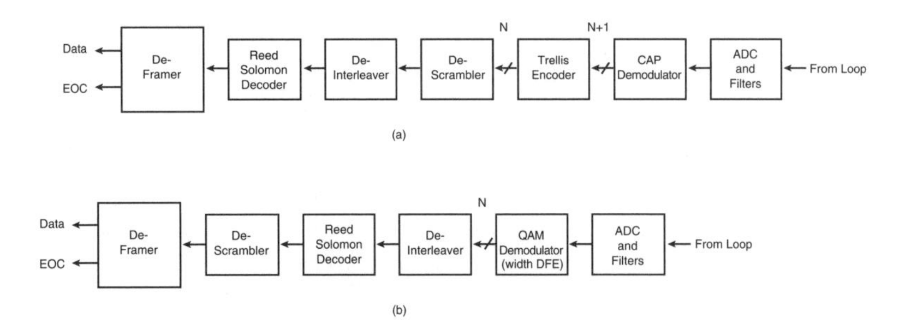


For both modulation technologies illustrated in Figure 7.13 and Figure 7.14, the basic block diagrams of the upstream and downstream units are identical. (However, the implementations are not identical. As with DMT-based ADSL, the upstream and downstream channels use different frequency bands and have different speeds.) Note that many of the same blocks are used in these transmitters and receivers as were used in the DMT version. As you will see shortly however, the implementation of the blocks is quite different. At the time of publication, the specific implementation of CAP and QAM systems is somewhat under development. Consequently, the discussion of the basics of the block implementation is less rigorous than the discussion of DMT-based ADSL systems.

## *Framing Block*

Different types of CAP and QAM framing blocks have been implemented in ADSL modems. In addition, several framing blocks are proposed as possible implementation standards. Two of these methods, one for CAP and one for QAM, are shown in Figures 7.15 and 7.16.

Figure 7.15 A common framing structure for CAP ADSL.

| | | |
|---|---|---|
| Synchronization Word 7 bits | | FEBE 1 bit |
| DATA 424 bytes | | |
| Dying Gasp 1 bit | Reserved 7 bits | |
| EOC 1 byte | | |
| Growth 1 byte | | |
| NTR 1 byte | | |
| CRC-6 6 bits | RAI 1 bit | Rcvd 1 bit |

Figure 7.16 A common framing structure for QAM ADSL.

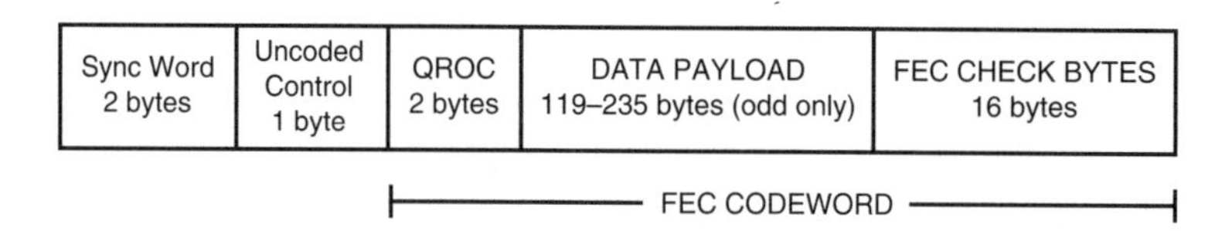


The frame shown in Figure 7.15 is 432 bytes long and variable in time of duration depending on the baud rate and bit rate of the CAP signal. This frame structure is in contrast to DMT frames that were a variable number of bytes long but fixed in time duration. Also, no superframe structure exists, and no sync frame exists. Instead, each frame starts with a synchronization field. Of specific note in the CAP frame structure are the byte reserved for an EOC channel, the room reserved for growth, and the NTR byte. Table 7.9 describes the fields in a CAP ADSL frame.

Table 7.9 Fields in a CAP ADSL Frame

| Field | Length | Description |
|---|---|---|
| Sync Word | 7 bits | 7-bit Barker code used to delineate frames. MSB to LSB, the code is 1110010. |
| FEBE | 1 bit | Indicates that one or more frames received since the last transmitted frame have had errors. |
| Data | 424 bytes | Data payload. |
| Dying Gasp | 1 bit | Set to one by the ATU-R if a loss of power is detected. |
| Reserved | 6 bits | Reserved for future definition. |
| EOC | 1 byte | Transmits EOC information. The EOC channel for CAP is a clear channel, not register based like that for DMT. |
| Growth | 3 bytes | Reserved for future definition. |
| NTR | 1 byte | Reserved for the future when a network timing reference is expected to be defined. |
| CRC-6 | 6 bits | Contains a CRC checksum for the current frame. |
| RAI | 1 bit | Remote Alarm Indication. Indicates that the local receiver is out of frame and cannot achieve frame delineation. |
| RSVD | 1 bit | Reserved for future definition. |

The frame structure shown for QAM-based systems in Figure 7.l6 is based on Reed-Solomon frame size. This type of frame is variable in both length and time duration. Of specific note in Figure 7.16 is the QAM RADSL Operations Channel (QROC). Depending on the first 2 bits of this two-byte field, the QROC carries either 14 bits of EOC information or two bytes of QROC op codes and data used for training and adaptation. Table 7.10 describes the fields of the QAM frame.

Table 7.10 Fields in a QAM ADSL Frame

| Field | Length | Description |
|---|---|---|
| Sync Word | 2 bytes | Delineates frames. It is given by 0x3F0C. |
| Uncoded Control | 1 byte | Reserved for future use. |
| QROC | 2 bytes | Field that transmits EOC messages end to end, or QROC messages that control transmitter and receiver parameters. |
| Payload | 119–235 bytes | Data payload. |
| FEC Redundancy | 16 bytes | Reed-Solomon check bytes. |

As a final note, several variations of CAP and QAM frames are possible. Some contain room for more logical channels, and others are geared toward systems not using the FEC blocks. It is not yet clear which, if any, of these framing structures will become most popular for ADSL implementations.

## *Scrambler*

CAP scrambling is a little more involved when compared to the scrambling method used in DMT. During initialization, simple self-synchronizing scramblers are used. In the upstream direction, the generator polynomial is given by Eqtn. 7.15.

Eqtn. 7.15

$$G(x) = 1 + x^{18} + x^{23}$$

In the downstream direction, the scrambling generator polynomial is given by Eqtn. 7.16.

Eqtn. 7.16

$$G(x) = 1 + x^{5} + x^{23}$$

During normal operation, the scramblers are locked. Figure 7.17 shows a locked scrambler and a locked descrambler.

Figure 7.17 An example of (a) a locked scrambler and (b) a locked descrambler.

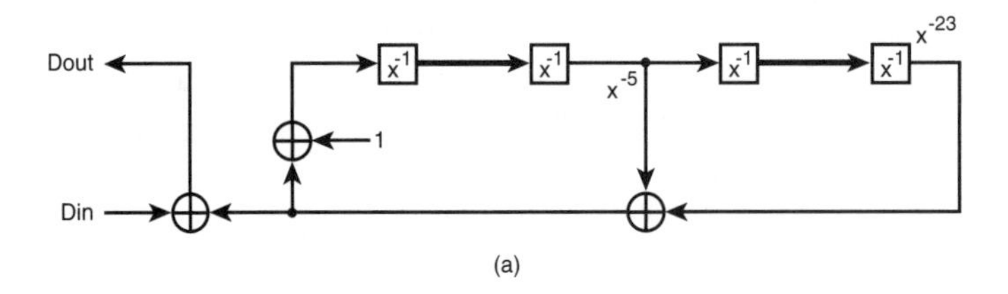


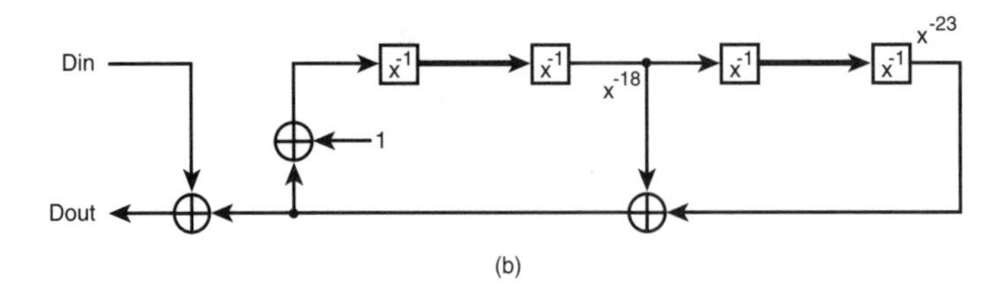

Locking must occur when the input data to the scrambler is all ones. Note also for a CAP implementation that the scrambling is done after the FEC block. This approach is different from both the DMT system and the QAM system.

A QAM ADSL implementation does not use different types of scrambling during training and normal operation. Instead, the upstream and downstream scramblers are simple self-synchronizing scramblers with generator polynomials identical to those used in the upstream and downstream directions for CAP ADSL.

## *FEC Block*

Both CAP and QAM ADSL systems employ Reed-Solomon coding and decoding; however, the number of codewords used may differ. For both systems, the coding is done with bytes as the basic symbols. Thus all arithmetic is done in GF(256). For CAP, the default codeword size is 68 bytes, 4 of which are check bytes. It is possible to request a variable-size codeword that is specified if supported by the peer modem. The number of check bytes continues to remain fixed, and only the number of data bytes changes. Recall that four check bytes can correct a burst error of 2 bytes in length.

QAM ADSL systems provide 16 check bytes able to correct a burst error 8 bytes in length. The number of data bytes per codeword can be between 119 and 235, but it must be odd. An option also exists to bypass the FEC block and send data uncoded.

## *Interleaving/De-interleaving Blocks*

CAP uses a version of a convolutional interleaver. The interleaving depth is variable; however, a common value suggested is an interleave depth of 4 RS codewords. QAM systems also use a variable-length interleaver with possible interleaved depths of 2, 4, 8, or 16 RS codewords. A depth of 1 is also allowed, essentially turning off interleaving. The QROC field is included in the RS codeword; however, the synchronization word and uncoded control octet are not included.

## *Trellis Encoder*

Trellis encoding is defined for CAP ADSL, but not for the QAM implementations. The encoder is an eight-state, two-dimensional code. The two LSBs of each CAP symbol are the inputs to the encoder. A single coded bit is the trellis encoder output. The encoder is shown in Figure 7.18.

Figure 7.18 Block diagram of a trellis encoder used in CAP ADSL.

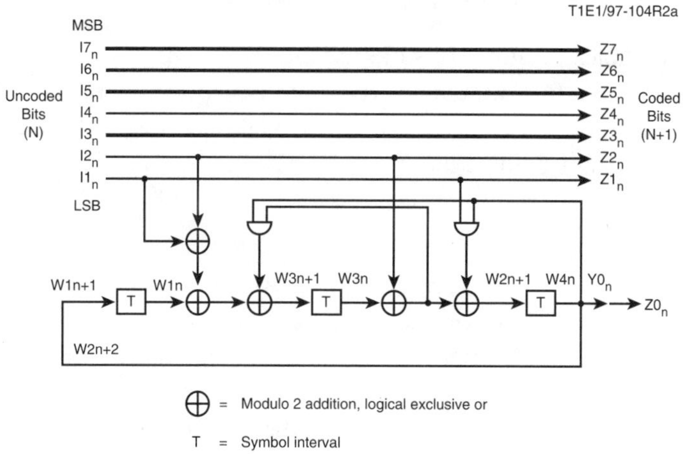

The N bits on the left side are uncoded payload bits from a single CAP symbol. The N + 1 bits on the right side are the uncoded bits as well as a single coded bit. Note that N depends on the size of the constellation being used in the CAP modulator. For example, for 128-CAP, N + 1 must be equal to seven, and thus N is equal to six. Because of the trellis-encoding structure, the smallest constellation allowable is 8-CAP.

## *CAP Modulator*

CAP modulation was discussed in Chapter 6. For ADSL, a CAP modulator may have a constellation size that is any power of two between 8 and 256. Thus the input to the modulator block from the trellis-encoding block must be between three bits (for 8-CAP) and eight bits (for 256-CAP). A CAP modulator also contains a Tomlinson precoder section. As discussed in the preceding chapter, the precoder is in lieu of a DFE in the CAP receiver. After constellation encoding, the complex encoder output is passed to the precoder. Training for the precoder occurs during initialization. The receiver in the upstream direction must contain at least 3 complex precoder filter taps. The receiver in the downstream direction must contain at least 16 complex taps. The actual number of taps is passed between the endpoints before training begins.

FDM is used to separate upstream and downstream channels for CAP implementations. No echo-cancellation functionality is defined. Figure 7.19 shows the placement of the upstream and downstream channels.

Figure 7.19 Upstream and downstream frequency regions for a CAP ADSL system.

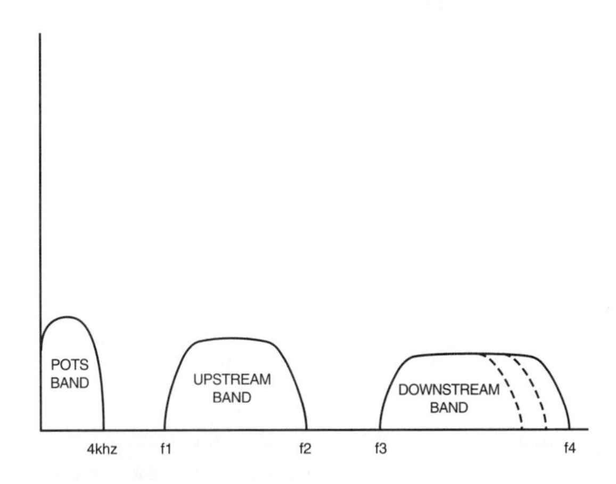


Parameters for Figure 7.19 are shown in Table 7.11 for the upstream direction and in Table 7.12 for the downstream direction.

Table 7.11 Placement of CAP Upstream Spectrum

| Symbol Rate (kBaud) | Start Freq. f1 (kHz) | Stop Freq. f2 (kHz) | Center Freq. (kHz) |
|---|---|---|---|
| 85 | 35 | 132.75 | 83.9 |
| 136 | 35 | 191.4 | 113.2 |

Table 7.12 Placement of CAP Downstream Spectrum

| Symbol Rate (kBaud) | Start Freq. f1 (kHz) | Stop Freq. f2 (kHz) | Center Freq. (kHz) |
|---|---|---|---|
| 136 | 240 | 396.4 | 318.2 |
| 340 | 240 | 631 | 435.5 |
| 680 | 240 | 1022 | 631 |
| 952 | 240 | 1344.8 | 787.4 |
| 1088 | 240 | 1491.2 | 865.6 |

Note that the upstream and downstream channels always start at a fixed frequency. Square root raised cosine filters are used to shape the upstream and downstream bands. These filters typically have a parameter of 15 percent excess bandwidth.

Because of the different symbol rates and the different constellation sizes, CAP transmission can support a variety of data rates. The constellation size and data rate for each direction, and thus the upstream and downstream data rates, are negotiated during training.

## *QAM Modulator*

An ADSL QAM modulator supports constellation sizes consisting of all powers of two between 4 and 256 inclusive. Note that the only constellation size supported by QAM and not by CAP is the four-point constellation. QAM can support this constellation because QAM does not do trellis encoding as is done in a CAP transmitter. The QAM modulator defined for ADSL uses a quadrant-differential encoding technique for constellation encoding. Depending on the constellation size, N bits are passed to the modulator from the interleaver block per symbol. The two MSBs of the N bits along with the quadrant of the preceding symbol are to determine which quadrant the encoded point will be in. For all but 4-QAM, the rest of the N bits are used to select a point from the quadrant. For the 4-QAM system, only one point per quadrant exists, so further encoding is not necessary.

QAM ADSL also uses FDM to separate the upstream and downstream bands. The baud rates defined for the upstream channel include 20 kBaud, 40 kBaud, 84 kBaud, 100 kBaud, 120 kBaud, and 136 kBaud. The baud rates defined for the downstream channel include 40 kBaud, 126 kBaud, 160 kBaud, 252 kBaud, 336 kBaud, 504 kBaud, 806.4 kBaud, and 1008 kBaud. The frequency ranges for the upstream and downstream channels are more flexible than those for CAP with negotiation during initialization necessary to pick the center frequency (and thus the frequency range of each signal). The upstream channel is generally contained in the 30 kHz to 138 kHz range and the downstream channel above 138 kHz.

## *CAP Demodulator*

A CAP demodulator in an ADSL system contains a matched filter as well as the FFF part of a decision feedback equalizer system. Remember that the feedback filter is not necessary because the transmitter employs a precoder. The CAP demodulator performs detection on a symbol-by-symbol basis, decoding at the baud rate of the system. The CAP demodulator decodes symbols using a constellation identical to that used by the CAP modulator. The output of the constellation encoder contains N + 1 bits that are sent to the trellis decoder.

## *QAM Demodulator*

A QAM demodulator normally consists of a matched filter and a complete decision feedback equalizer. Both the feed-forward filter and the feedback filter are complex. Detection is performed on a symbol-by-symbol basis. Constellation decoding again involves preceding symbol quadrant locations because of the differential quadrant encoding implemented by the QAM modulator. At the symbol rate, the demodulator passes N bits to the de-interleaver block where N depends on the size of the constellation used.

### *Final Notes on CAP and QAM ADSL*

Specifications for CAP- and QAM-based ADSL are undergoing the standardization process. The process was somewhat delayed due to the differences between proposed implementations. Recall from Chapter 6 that the actual CAP and QAM line-coding waveforms are very similar and sometimes identical. Had the rest of the transmitter and receiver parameters, including framing, FEC, equalization, and frequency spectrums been the same, it might be possible for CAP and QAM peer modems to communicate properly. Interoperability is much more difficult because, as seen in this chapter, these parameters are much different in the two techniques.

## *Summary*

This chapter dealt with DMT, CAP, and QAM implementations of ADSL. The blocks necessary to implement DMT transmitters and receivers were discussed. Detailed frame and superframe structures were explained along with how the bits from these frames map into the tones used by DMT. Several DMT options including echo cancellation and reduced overhead were discussed. Finally, functions embedded in ADSL to handle ATM were presented. For CAP and QAM, the parameters of the transmitters and receivers were discussed. The differences between the two approaches were pointed out.

## *Exercises*

1. Calculate the line rate (excluding the bandwidth needed for the sync symbol) needed to support a downstream DMT ADSL AS0 channel with a data rate of 4.8 Mbps assuming that a 16-byte Reed-Solomon FEC scheme is used and no other logical channels are used. Assume full overhead mode.
2. Repeat exercise 1 assuming that the following channels are used:
   - S0 on the fast path with a data rate of 1.6 Mbps using eight FEC check bytes
   - LS0 on the fast path with a data rate of 128 kbps using four FEC check bytes
   - AS on the interleaved path with a data rate of 640 kbps—assume that the number of symbols per Reed-Solomon codeword is one.

3. Repeat exercise 2 assuming that the number of symbols per Reed-Solomon codeword is eight.

4. Draw the downstream data frame structure for a DMT ADSL system using only AS0 on the fast path. Assume that four Reed-Solomon check bytes per frame are used and that the total line rate is 6.4 Mbps (does not include the bandwidth needed for the sync symbol).

5. For a line rate of 1.6 Mbps, calculate the maximum downstream data rate available for channel AS0. Assume that it uses the interleaved path with four symbols per FEC codeword and 16 check bytes. Assume that full overhead is used. Repeat this exercise for each of the reduced overhead modes.

6. Assume a simplified DMT system existed consisting of four bins carrying 2, 8, 12, and 2 bits, respectively. Using the tone-ordering algorithm, show how a fast path buffer with 8 bits and an interleaved buffer with 16 bits would map onto these tones.

7. Calculate the power output in a single DMT bit assuming a nominal output of -40 dBm/Hz. Find the total power in the upstream and downstream directions.

8. Find the SNR necessary on a subcarrier to support N bits per DMT symbol for N = 2, 3, ...16.

9. Based on the information from exercise 8, develop a table of SNR, Bits Supported, and Fine Gain Adjust Value.

10. For a CSA #6 Loop, find the maximum $b_i$ value for all DMT subcarriers.

## *Endnotes*

1. "T1.413 Issue 2 Letter Ballot Document," ANSI Document T1E1.4/98-007R4, June 12, 1998.

2. "Draft Physical Layer Specification for CAP/QAM-Based Rate Adaptive Digital Subscriber Line (RADSL)," ANSI Document T1E1.4/97-104R2a, October 15, 1997.

3. European Telecommunications Standards Institute, *Transmission and Multiplexing; Asymmetrical Digital Subscriber Line (ADSL); Requirements and Performance*, November 1996.

4. J. Tellado and J. Cioffi, "PAR Reduction in Multicarrier Transmission Systems," ANSI Document T1E1.4/97-367, December 8, 1997.

5. Aradhana Narula and Frank Kschischang, "Unified Framework for PAR Control in Multitone Systems," ANSI Document T1E1.4/98-183, June 1, 1998.

6. Jose Tellado and John M. Cioffi, "PAR Reduction with Minimal or Zero Bandwidth Loss and Low Complexity," ANSI Document T1E1.4/98-173, June 1, 1998.

CHAPTER 8

# *ADSL in WAN Networks*

In this chapter:

- An overview of common protocols found in LANs and WANs relevant to networks utilizing ADSL access
- Several candidate architectures for WANs that use ADSL as an access technology
- A discussion of options and issues for ATU-R connectivity to the network

Until now, ADSL and other DSL technologies have been covered at the *physical layer*. Sometimes the physical layer is also called *layer 1*. Both refer to the first layer in the seven-layer OSI protocol stack shown in Figure 8.1.

Figure 8.1 The seven-layer OSI protocol model; each layer is independent and passes data to and from neighboring layers.

| | |
|---|---|
| Application Layer | Layer 7 |
| Presentation Layer | |
| Session Layer | |
| Transport Layer | |
| Network Layer | |
| Data Link Layer | |
| Physical Layer | Layer 1 |

The physical layer is characterized by the type of transmission used, including the parameters of the transmitter and receiver blocks discussed in Chapter 7, "ADSL Modulation

Specifics." These parameters include framing structure, scrambling, FEC method, and modulation type.

When looking at a network as a whole, the layers above the physical layer play a large role in how the network is set up, provisioned, and managed as well as how data is sent over the network. ADSL, when used in the access link of a network, should not affect the network's operation and in most cases the underlying line code should be irrelevant.

**Note**

In some cases, the actual access speed and asymmetry, but not the line code, of the DSL might affect the design of the higher layer protocols.

This chapter discusses protocols used in networks that ADSL is used as an access medium. Several of the more common types of architectures being designed and deployed are covered along with the protocols. Many of these architectures assume that the Internet standard TCP/IP protocols are used somewhere in the protocol stack. Before jumping directly into the architectures, several sections discuss the basic features of the various protocols. (This presentation is not a complete and thorough treatment of each protocol—in many cases a book on each subject would be necessary for complete treatment.)

## *Protocol Basics*

The first two protocols discussed are *Ethernet* and *Internet Protocol* (IP). These are the most common protocols in LANs. IP is also the protocol used by the public Internet. Figure 8.2 shows how IP and Ethernet fit into the seven-layer OSI model. Additionally, TCP is shown in this figure as an example wherever a Web browser is being used. The common physical layer of 10BaseT is used in the example. After discussing TCP/IP and Ethernet, asynchronous transfer mode (ATM), Point-to-Point Protocol (PPP), tunneling, and other protocols are discussed.

Figure 8.2 Mapping a typical TCP/IP implementation into the seven-layer OSI protocol model.

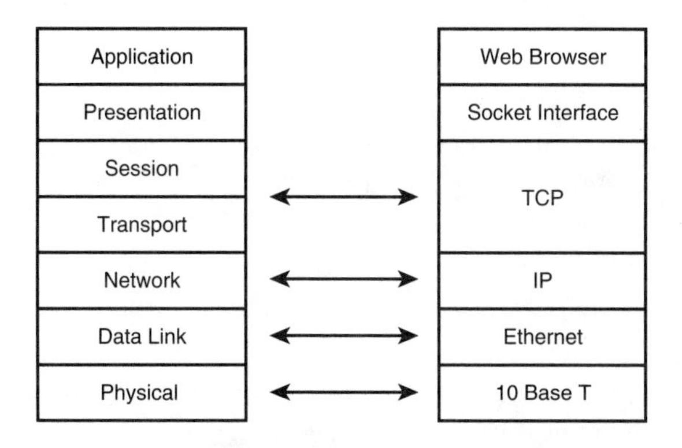


## *Ethernet*

Ethernet is a layer 2 protocol. It is commonly subdivided into a *multiple access control* (MAC) sublayer and a *logical link control* (LLC) sublayer. Figure 8.3 shows two Ethernet frame formats carrying IP traffic. The first is consistent with IEEE 802.2/802.3, and the second with RFC 894.[1,2] RFC 894 encapsulation is more commonly used.

Figure 8.3 Ethernet frame formats according to (a) IEEE 802.2/802.3 and (b) RFC 894.

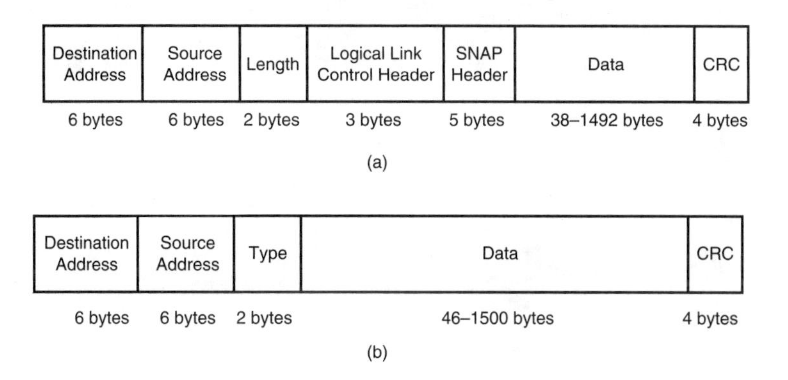


The Ethernet frame using RFC 894 encapsulation contains four fields in addition to the IP data field. A *Source Address field* and a *Destination Address field* identify the Ethernet port sending (source) and receiving (destination) the Ethernet packet. The six-byte addresses used for the source and destination address fields are also called *Ethernet addresses* and sometimes *hardware addresses.* Typically a manufacturer of Ethernet cards preconfigures the Ethernet address for each card (and each card or port has a different Ethernet address). Typically, Ethernet addresses do not have local significance. That is, ports on the same network have Ethernet addresses with no common portion among the six address bytes. This scheme is in contrast to IP addresses on the same network that have a common portion.

**Note**

Some Ethernet cards allow a user to change some or all of the six bytes.

The two-byte Type field in part b of Figure 8.3 identifies the type of layer 3 protocol being encapsulated in the frame. If the public Internet is being accessed, the protocol is IP and the type value assigned for IP encapsulation is 0x0800. It is possible to encapsulate other types of protocols in an Ethernet frame as well, resulting in a different type value. The Type field is sometimes called the *ethertype*.

The final field in the Ethernet frame in part b of Figure 8.3 is the CRC field. This four-byte field covers the entire frame (except for the CRC itself). Note that this CRC has no relationship to the CRC discussed for ADSL (which is purely a physical layer entity). The CRC field determines whether the layer 2 packet was completely and successfully received from the physical layer.

## *IP*

The IP is a layer 3, or network layer, protocol. Figure 8.4 shows the IP packet format.

Figure 8.4 A breakdown of the fields in an IP header.

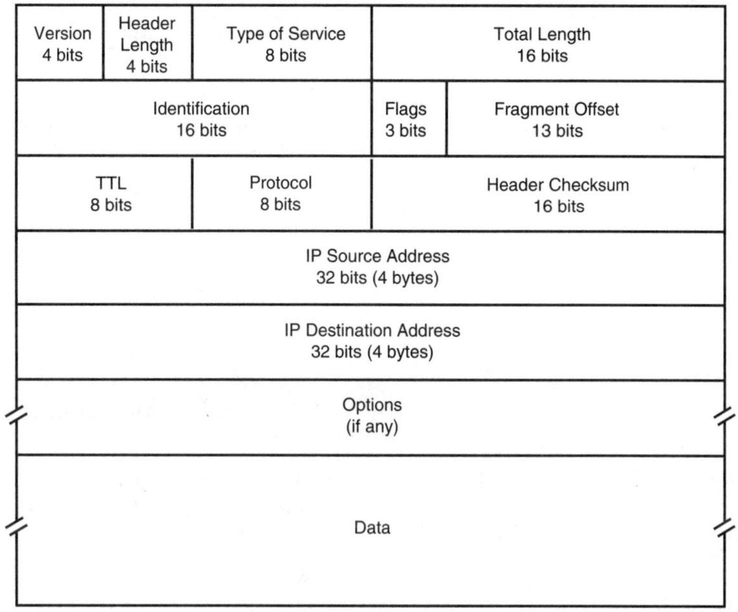

<table>
<tr><td>Version<br>4 bits</td><td>Header<br>Length<br>4 bits</td><td>Type of Service<br>8 bits</td><td colspan="2">Total Length<br>16 bits</td></tr>
<tr><td colspan="3">Identification<br>16 bits</td><td>Flags<br>3 bits</td><td>Fragment Offset<br>13 bits</td></tr>
<tr><td colspan="2">TTL<br>8 bits</td><td>Protocol<br>8 bits</td><td colspan="2">Header Checksum<br>16 bits</td></tr>
<tr><td colspan="5">IP Source Address<br>32 bits (4 bytes)</td></tr>
<tr><td colspan="5">IP Destination Address<br>32 bits (4 bytes)</td></tr>
<tr><td colspan="5">Options<br>(if any)</td></tr>
<tr><td colspan="5">Data</td></tr>
</table>

Table 8.1 describes the many fields in an IP packet. Of these fields, the two most important to our discussion are the *Source IP Address* field and the *Destination IP Address* field.

Table 8.1 IP Packet Fields

| Field Name | Length | Description |
|---|---|---|
| Version | 4 bits | The IP protocol version number (currently 4). |
| Header Length | 4 bits | The number of 32-bit words in the header. |
| Type of Service (TOS) | 8 bits | Contains four fields specifying the precedence (usually ignored), delay requirement (high or low), throughput maximization (high or low), and a reliability parameter. |
| Total Length | 16 bits | Specifies the number of bytes in the IP packet, including the header. |
| Identification | 16 bits | Modulo 16 counter that identifies an IP packet. |

*continues*

Table 8.1 Continued

| Field Name | Length | Description |
|---|---|---|
| Flags | 3 bits | The DF bit signifies whether the IP datagram may be fragmented, and the MF bit indicates whether the packet is the final fragment. One of these three bits is not used and always set to zero. |
| Fragment Offset | 13 bits | If the data is a fragment, this bit indicates where it belongs in the entire datagram. The unit of measurement is 64-bit units. |
| Time to Live (TTL) | 8 bits | Indicates how many router hops this packet may pass through before being discarded. Normally, routers decrement this value. The packet is discarded when this value hits zero, keeping a packet from eternally existing in a network. |
| Protocol | 8 bits | Identifies the protocol that follows the IP header (TCP, UDP, IGMP, and so on). |
| Header Checksum | 16 bits | A checksum to ensure the reliability of the header. |
| Source IP address | 32 bits | The IP address of the host that sent the IP packet. |
| Destination IP Address | 32 bits | The address of the host for which this packet is intended. |
| Options | Variable | May set routing or security restrictions on the packet. Options are usually not used. |
| Data | Variable | The IP data payload. |

Each of the fields documented in the preceding table is four bytes (32 bits) long. A network administrator normally assigns four-byte IP addresses to each host (computer) on a network. All computers on a LAN typically have a common portion of their IP addresses that identifies the network followed by a unique portion representing each individual host.

Consider the example in Figure 8.5.

Figure 8.5 An example of two subnets with multiple hosts connected by a router.

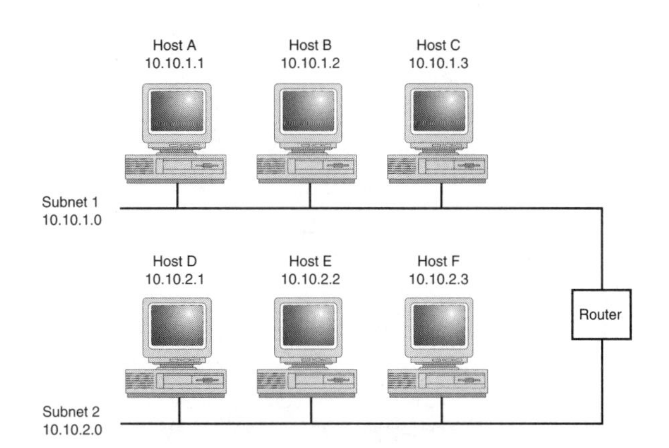


This figure shows two *subnets*, or logical networks: 10.10.1.0 and 10.10.2.0. The first item to note is the method used to write addresses. This method is called *dotted-decimal notation*. Each decimal number represents one of the four bytes making up the IP address. Because each number represents eight bits, the number may be any positive integer between 0 and 255. In this example, each subnet has three hosts. Hosts on the same subnet have the same first three bytes. However, the common part of a network address (representing the subnet) may differ by three bytes depending on the network.

## Note

Associated with a network is a 32-bit subnet mask containing 1s for the N MSBs and 0s for the 32-N LSBs (N may be any number between 1 and 31). A bitwise AND function performed on the mask and IP address from a network yields the network part of the IP address. A bitwise NAND function yields the host portion of the IP address.

You can find very detailed and informative discussions of IP in *TCP/IP Illustrated, Volume 1: The Protocols*; *TCP/IP Illustrated, Volume 2: The Implementation*; *and TCP/IP: Architectures, Protocols and Implementation.*[3,4,5]

IP networks can be viewed as a number of connected LANs. Switches, bridges, and routers are devices that connect hosts together on Ethernet and IP networks. Switching and bridging are layer 2 functions, and routing is a layer 3 function. Figure 8.6 shows a typical bridging application.

Figure 8.6 A bridging application using two half-bridges and modems to connect hosts on the same subnet.

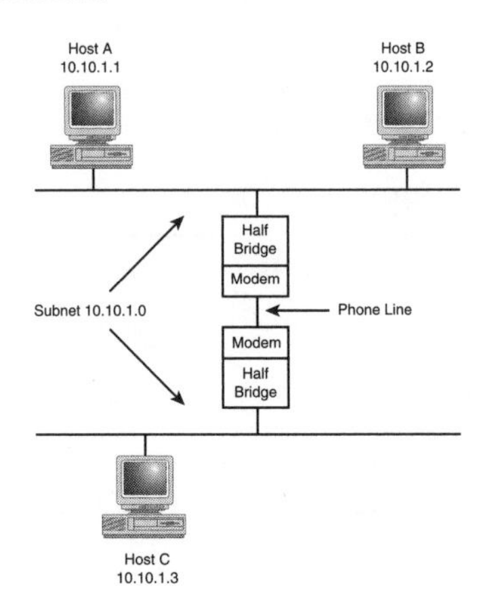


The two half-bridges in Figure 8.6 make up a bridge connecting two Ethernet segments. Ethernet packets destined for ports on the other Ethernet segment are transparently sent over the bridge. To function properly, the bridge must know which Ethernet addresses are on each segment. This information can be manually configured in the bridge but is more often learned by the half-bridges. To learn the location of different Ethernet ports, each half-bridge monitors the Source Ethernet Addresses of packets on its Ethernet segment. All source addresses are entered into a table. If a half bridge sees a packet with a Destination Ethernet Address not in its table, that host must be on the other side of the bridge and the Ethernet frame will be forwarded. A bridge does not change the contents of a packet, including the Ethernet source and destination addresses. Bridges allow two Ethernet segments to appear to be connected. Segments separated by a bridge are typically part of the same IP subnet.

An Ethernet switch connects to multiple Ethernet segments and simply switches packets between segments depending on the destination Ethernet address. Switches are convenient for separating segments to reduce traffic on each segment. Segments connected by an Ethernet switch normally are part of the same IP subnet. A switching table kept by the Ethernet switch keeps a mapping between the Ethernet Destination address and the port to which the frame should be switched. Figure 8.7 shows multiple Ethernet segments connected by a switch and the corresponding switching table.

Figure 8.7 Multiple Ethernet segments connected by a switch.

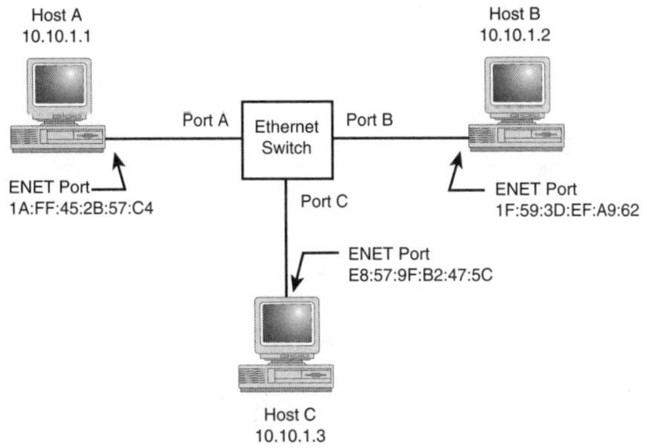

Host A
10.10.1.1

Host B
10.10.1.2

Port A
Ethernet Switch
Port B

ENET Port
1A:FF:45:2B:57:C4

Port C

ENET Port
1F:59:3D:EF:A9:62

ENET Port
E8:57:9F:B2:47:5C

Host C
10.10.1.3

| Input Port | Ethernet Destination Address | Output Port |
|---|---|---|
| A | 1F:59:3D:EF:A9:62 | B |
| A | E8:57:9F:B2:47:5C | C |
| B | 1A:FF:45:2B:57:C4 | A |
| B | EF:57:9F:B2:47:5C | C |
| C | 1A:FF:45:2B:57:C4 | A |
| C | 1F:59:3D:EF:A9:62 | B |

A router is a layer 3, or network layer, device that transfers IP packets, based on the destination IP address. Routers connect different local IP networks. Figure 8.8 shows a router connecting two networks. Overlaid on Figure 8.8 is an example of a packet being sent from host A to host B.

The following events transpire to make the transfer in Figure 8.8 happen:

- Host A has an IP packet to send to host B. The destination IP address of the packet identifies host B and the source IP address of the packet identifies host A.
- Host A knows that host B is not on the same subnet. Host A thus sends the packet to host A's default router. The source and destination IP addresses of the packet still identify host A and host B respectively. The source Ethernet address identifies host A's Ethernet port, and the destination Ethernet address identifies the router's Ethernet port.
- The router receives the packet, recognizing the destination Ethernet address as its own. The router's IP layer recognizes that the destination IP address does not belong to the router. The router then looks into a routing table and identifies an output port where the IP packet (or IP datagram) should be sent to reach host B.

Figure 8.8 An example of routing between several subnets; the addresses show the IP and Ethernet addresses present in packets on each side of the router when routed from host A to host B.

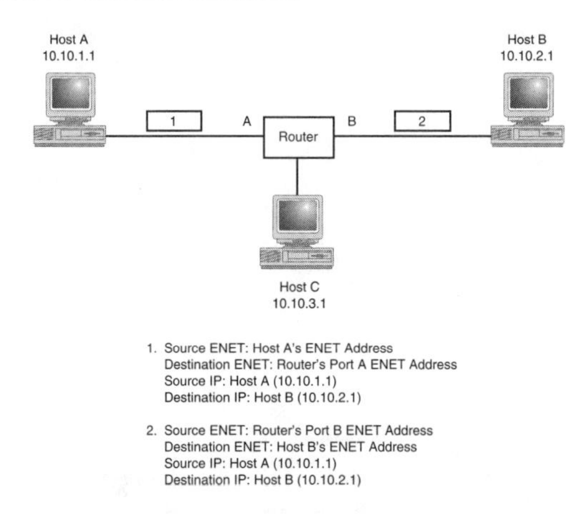


**Note**

In many cases, the destination host may not be directly connected to the first router. In such a case, the router forwards the packet to another router. Eventually, barring any errors, the host will be reached.

- The router does not modify the IP part of the packet. It reformats the Ethernet part of the message to include the router's Ethernet port on subnet B as the source Ethernet address and the Ethernet address of host B as the destination Ethernet address. The newly formatted Ethernet packet is then sent on to port B's Ethernet segment.
- Host B's Ethernet port recognizes an incoming packet with its Ethernet address in the Destination Ethernet Address field. Recognizing that this packet is for itself, host B's Ethernet port forwards the IP portion of the packet to its IP layer. The IP layer recognizes its own IP address in the Destination IP Address field and knows that the IP packet is for host B. The packet will then be sent further up host B's protocol stack for handling.

Note from this example the difference in usage of IP and Ethernet addresses. If a host in California needs to send an IP packet to a host in New York via the Internet, the sending host must know the IP address of the host in New York, but does not need to know the Ethernet address. (Most probably, the IP address of the destination host would be found using a *Domain Name Server* [DNS].) Public IP addresses on the Internet are administered and maintained by several organizations. Understanding the difference between IP addresses and Ethernet addresses, as well as the difference between layer 2 and layer 3 devices, is necessary to really understand the networking issues discussed later in this chapter.

## *TCP*

The IP and Ethernet protocols do not ensure that a packet sent to a destination actually arrives at the destination. A channel is not set up, and the receiving host does not send an acknowledgment. For this reason, IP is called a *connectionless* protocol. TCP, on the other hand, is a *connection-oriented* protocol. When data is to be transferred to another host, TCP uses IP to contact the destination and tell the host at the destination that it wants to set up a *session*; a session is basically a connection made between two hosts and acknowledged by both endpoints. When a session is set up, TCP sees to it that data is transferred between the endpoints and that the receiving host sends acknowledgment packets. If an acknowledgment is not received for a certain packet, the TCP protocol resends the packet.

## *ATM*

*Asynchronous transfer mode* (ATM) is newer than the IP and Ethernet protocols. It is sometimes referred to as *cell transfer*, whereas IP and Ethernet are thought of as packet transfer technologies. ATM is a layer 2 protocol. Figure 8.9 shows a protocol stack of ATM carrying IP.

Figure 8.9 A protocol stack with IP running over ATM.

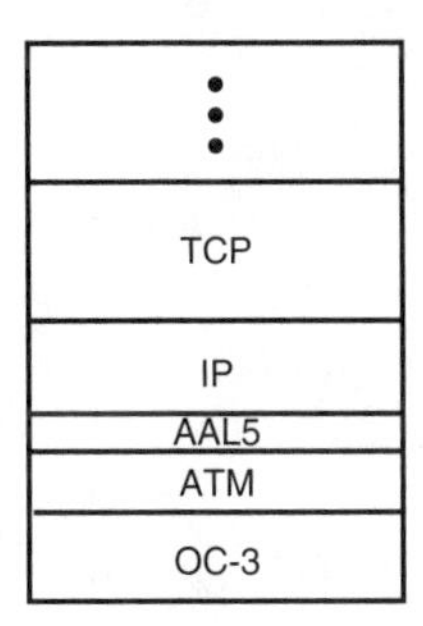


Here, an OC-3 (155 Mbit/s optical fiber) is assumed as the physical layer. The *ATM adaptation layer 5* (AAL5) layer shown adapts the higher-layer protocol (IP) so that it will fit nicely into ATM cells.

In ATM, instead of variable-length packets, fixed-length cells are created and sent. The cell length is 53 bytes, 48 of which are payload bytes and 5 of which make up a header. Figure 8.10 shows the format of an ATM cell. Of particular interest in the ATM cell header are the *virtual path identifier* (VPI) and *virtual channel identifier* (VCI) values. ATM switches use the VPI and VCI values to distinguish local connections and direct cells from an input to an output. The fixed, short size of ATM cells allows switches to run faster than most IP routers.

Figure 8.10 Format of an ATM cell at a user-to-network interface.

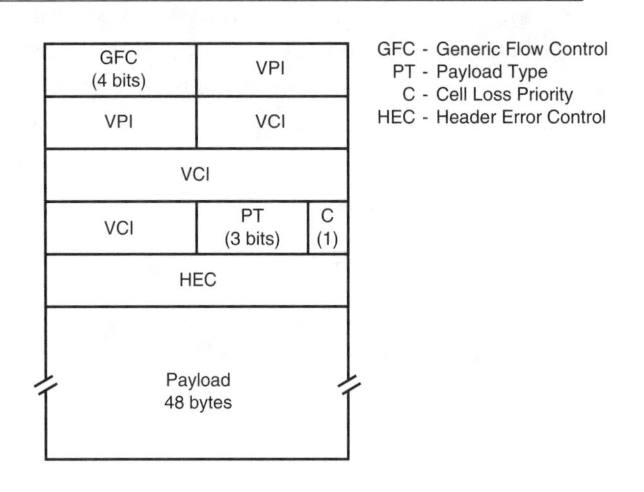


ATM is a connection-oriented protocol. A connection is set up between the ATM endpoints before payload data is transmitted. A connection either can be provisioned into the equipment by an operator or can be dynamically set up using signaling. The former type of connection is called a *permanent virtual circuit* (PVC), and the latter is a *switched virtual circuit* (SVC). PVCs exist until an administrator unprovisions the connection. SVCs exist for the duration of the ATM session and are set up and torn down by the signaling protocol. ATM SVC setup is very analogous to dialing a telephone in that an end station's ATM address is specified and sent to the network. ATM adds the capability of specifying how much bandwidth and what type of traffic will flow through the configured path. It is said that a *Quality of Service* (QoS) is requested for the call. All switches through which the virtual circuit will pass have the ability to deny the connection if they cannot support the requested QoS. Supporting QoS is a major advantage of ATM over a protocol such as Ethernet. Some disadvantages of ATM are the high cost compared to Ethernet and the relative amount of deployed equipment with respect to Ethernet (especially equipment supporting SVCs). This chapter discusses cell architectures, packet architectures, and mixed architectures for networks including ADSL as an access mechanism.

For detailed discussions of ATM, refer to *ATM: Foundations for Broadband Networks* and *Asynchronous Transfer Mode: ATM Architecture and Implementation.*[6,7]

# *PPP*

PPP is used on many dial-up modem connections. It is specified in RFC 1661.[8] PPP is a layer 2 protocol and provides a simple method to encapsulate any protocol inside. PPP consists of a link-negotiation phase called *Link Control Protocol* (LCP) and a network negotiation phase called *Network Control Protocol* (NCP).

PPP adds several bytes of overhead to a packet. The format of a PPP packet is shown in Figure 8.11.

Figure 8.11 The format of a PPP packet.

| flag 0x7E | Address 0xFF | Control 0x03 | Protocol | Data | CRC | flag 0x7E |
|---|---|---|---|---|---|---|
| 1 byte | 1 byte | 1 byte | 2 bytes | | 2 bytes | 1 byte |

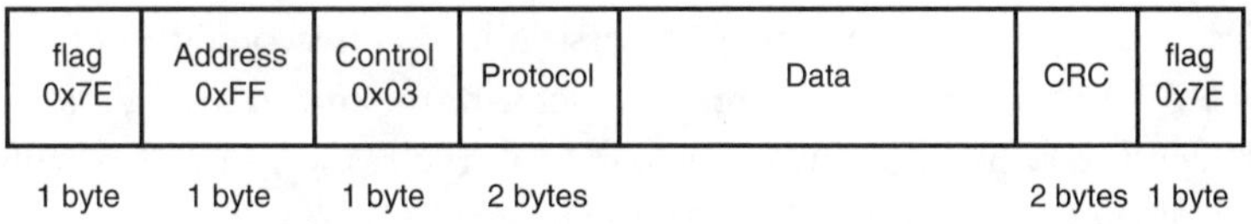

The type of information in the Data field is specified by the value in the Protocol field. Thus two bytes are used to indicate what type of protocol is being encapsulated by PPP. The value most relevant for ADSL access is the protocol value for the IP protocol, 0x0021.

When a link is first brought up, LCP negotiation is performed. During LCP, many different data-link parameters are negotiated. Some of these include the maximum PPP packet size, the authentication protocol to be used, and protocol and header compression options. LCP packets are identified by the protocol field value of 0xC021. These packets are sent when the link is first being made.

After LCP negotiation, NCP negotiation occurs. For IP, the NCP negotiation used is called *Internet Protocol Control Protocol* (IPCP), which is defined in RFC 1332.[9] IPCP allows an IP address to be assigned to an end station. NCP packets are identified by the protocol field value of 0x8021.

Two important authentication protocols can be used with PPP. They are Point-to-Point Authentication Protocol (PAP) and Challenge Authentication Protocol (CHAP). PAP consists simply of one of the endpoints (normally the dial-up user) sending a username and a password to the other end (normally the network being accessed). CHAP consists of the one end of the connection (normally the network end) sending a challenge to the other end and that end replying to the challenge.

PAP or CHAP is normally run between the LCP and NCP negotiations. When used with IP, PPP provides a powerful method to accomplish the following:

- Configure a link so that packets can be reliably sent between the endpoints (LCP)
- Make sure, with reasonable certainty, that the user accessing the network is an authorized user (PAP or CHAP)
- After having confidence in the user's identity, assign that user an address on the accessed network (NCP)

When all of the negotiation and authentication activities are complete, the connection is ready to send user data between the two endpoints.

## *Tunneling*

PPP provides an effective way to manage a session between two endpoints. However, one drawback of PPP is that it has no addressing mechanism, and so a single logical connection existing between the two endpoints participating in the session can carry only one PPP session. Because of the lack of addressing mechanism, PPP does not allow a group of PPP sessions to be aggregated at an access node and sent over the same channel to a common destination. Tunneling methods do allow this however.

A *tunnel* simply refers to a single logical channel over which sessions that normally do not share a logical channel are sent. Three of the more common tunneling methods are *Point-to-Point Tunneling Protocol* (PPTP), *Layer Two Forwarding* (L2F), and *Layer Two Tunneling Protocol* (L2TP). L2TP is a superset of PPTP and L2F developed by proponents of both of these protocols and is intended to replace them.

L2TP is designed to multiplex multiple PPP sessions over a common channel. Essentially, this is done by assigning a call number to each PPP session at one end of the tunnel before multiplexing so that the sessions can be differentiated at the other end of the tunnel. Figure 8.12 shows an example of a number of PPP sessions between subscribers and a service provider.

In part a of Figure 8.12, PPP aggregation does not occur. In this situation, each session requires a separate physical channel between each subscriber and a service provider. Part b of Figure 8.12 shows an example in which PPP sessions are aggregated, but tunneling is not used. In this case, separate logical channels between the aggregation point and the service provider are necessary to keep the PPP sessions identifiable. Part c of Figure 8.12 shows aggregation with tunneling. In this case, only one logical connection is needed between the aggregation point (commonly called the *L2TP access concentrator* or LAC) and the other tunnel endpoint (commonly called the *L2TP network server* or LNS) at the service provider.

Figure 8.12 Comparison of access systems (a) without aggregation, (b) with aggregation but without tunneling requiring, and (c) with aggregation and tunneling.

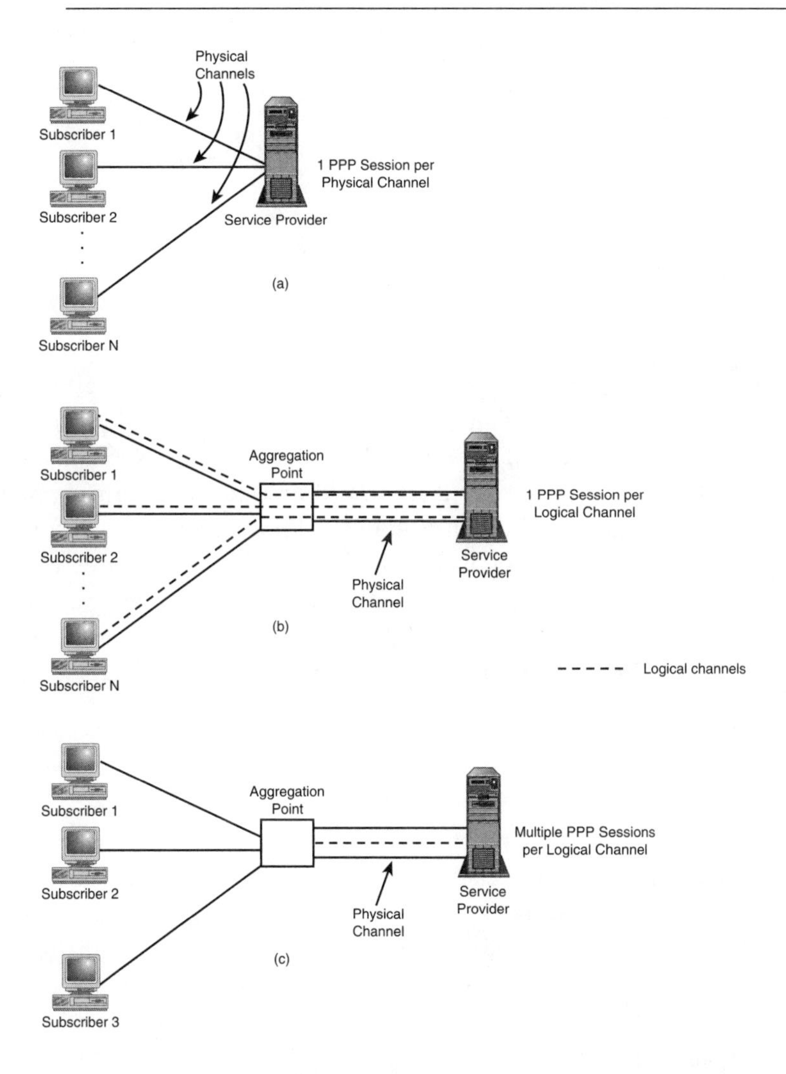


## Note

For example, in an ATM network, each PPP session would have a unique VPI/VCI assigned to it. In a frame relay network, each PPP session would have a unique data link connection identifier (DLCI) assigned to it.

The value of L2TP is illustrated in Figure 8.13.

Figure 8.13 An access network using L2TP tunneling between access points and a service provider.

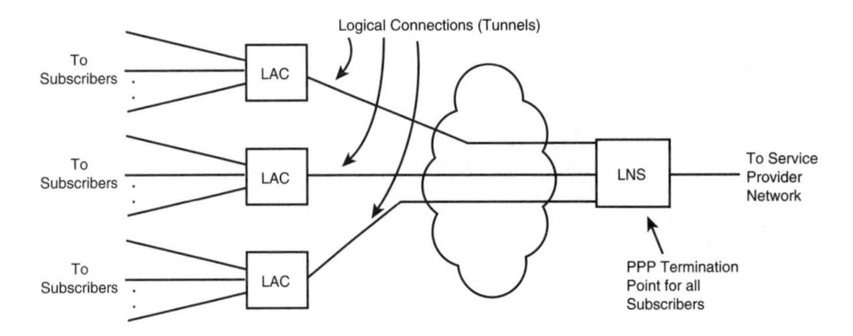


In this figure, access to some network (the Internet, for example) is located at a central point (sometimes called a *point of presence,* or POP). Access nodes (LACs) are located at various points away from the LNS. Because PPP provides methods to authenticate users (from which billing can be derived) and assign addresses to users, terminating the PPP sessions at a common point (the LNS), instead of at all the access points, is convenient. On the other hand, provisioning a logical connection for every subscriber between all the LACs and the LNS is usually not convenient. L2TP solves this issue by allowing a single logical connection between each LAC and the LNS to transport the PPP sessions. You can find more detailed information on L2TP in the Internet Draft, "Layer Two Tunneling Protocol (L2TP)."[10]

## End-to-End Players

A firm understanding of networks with ADSL access involves understanding the different players that take part in making the access possible and providing services. These include end users or subscribers, network access providers (NAPs), network transport providers (NTPs), and network service providers (NSPs). The next few sections discuss these players and where they fit into the end-to-end picture. Figure 8.14 illustrates the different players. In many cases, the NAP, NTP, and NSP may not be separate companies; a single company may do any or all of these functions.

Figure 8.14 Illustration of the end-to-end players relevant to a DSL access network.

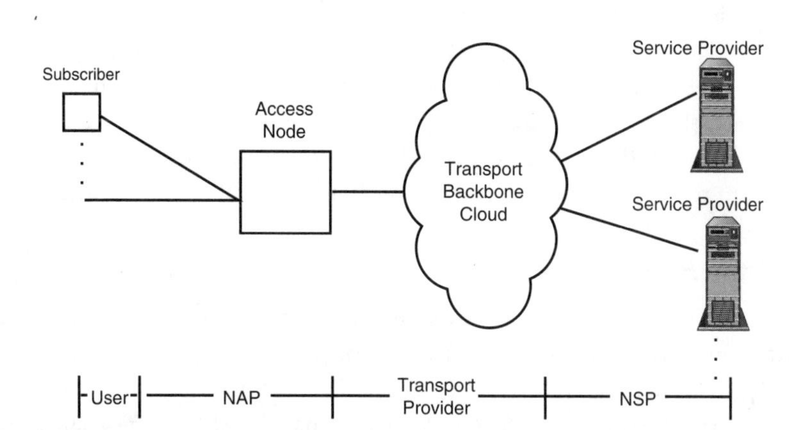


## *Users*

Users or subscribers are one endpoint of an end-to-end connection. Users include residential subscribers wanting Internet access, small business offices wanting connectivity to other offices, and telecommuters. A user might have a full-time dedicated connection to a single location or may want to connect to various places at various times (connect to the office during the day and to the Internet in the evening).

## *Network Access Providers*

NAPs normally own and operate the equipment that terminates the physical ADSL link and aggregates traffic. Traffic from a NAP may go to one or more NSPs. NAPs may terminate more than simply ADSL, including other DSL flavors or even voice-band modem access.

## *Network Transport Providers*

A *network transport provider* moves traffic between NAPs and NSPs. The transport provider typically owns the transport cloud shown in Figure 8.14. (Other arrangements are possible as well. For example, the transport provider may lease some of the cloud.) Note that the cloud is typically made up of various types of interconnected switches and routers. A transport provider might carry traffic from various NAPs or NSPs. In addition, the transport provider might carry traffic totally unrelated to ADSL networks, including long-distance voice service or private, dedicated connections between corporate locations.

### Network Service Providers

NSPs are typically an endpoint of a PPP session originating from a user. Examples of NSPs are Internet service providers, corporate networks providing access to corporate telecommuters, or banking institutions providing online banking functions. In many cases, NSPs provide users access to some type of network (the Internet or a corporate LAN). The NSP frequently assigns IP addresses to users via IPCP, authenticates users to make sure that they are allowed to access the network, and keeps track of billing and usage information.

### ILECs and CLECs

Incumbent local exchange carriers (ILECs) and competitive local exchange carriers (CLECs) are companies that can be NAPs, transport providers, and NSPs. Traditionally, ILECs are Regional Bell Operation Companies (RBOCs). Competitive carriers have emerged in the 1990s since the Telecom Act ruled that the ILECs must fairly lease their copper facilities to competitors to open up the market of providing basic services. ILECs generally own central offices, the bulk of the copper in their area's network, and the bulk of the equipment in the central office. CLECs typically lease space from ILECs in a central office, lease copper on which to provide service, and in some cases, lease equipment and service from the ILEC.

**Note**

Often, central office space is literally a 10-foot by 10-foot cage to which the CLEC receives a key. Copper pairs are run to the cage for the CLEC by the ILEC, and all of the CLEC's equipment must be stationed inside the cage.

## Architectures for DSL Networks

Many different architectural options exist for networks involving DSL access. Depending on the application of the network, some architectures may work better than others. For example, a virtual LAN network in which some traffic might be meant for all users would have different requirements than a dial-up network meant to privately connect subscribers to the corporate networks or the Internet. The following sections discuss some of the more common architecture approaches. In many cases, these architectures might coexist (at least in part) and hybrid approaches may be implemented.

### Bridging Architectures

Earlier types of ADSL networks were based on bridging across the loop with routing in the network backbone. Such a network is shown in Figure 8.15.

Figure 8.15 A bridging architecture for ADSL access.

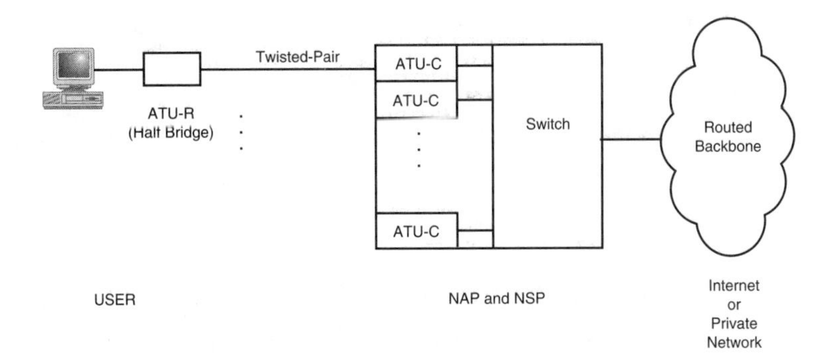


In this figure, the end users are connected to the network through external modems with Ethernet interfaces. (The ATU-R and ATU-C are considered half-bridges.) It is also possible to form a switching backbone; however, the provisioning needs and scalability of the network might be affected because the switches would require very large switch tables. In addition, MAC addresses can normally not be assigned to end stations because they are commonly permanently configured during the ports' manufacturing.

**Note**

Even if the MAC address were not factory configured, no common protocols exist to assign addresses.

In this type of network, the NAP typically is the NSP allowing a user to access a network. That network might be the Internet, a private network with gateways to the Internet, or other locations such as corporate networks.

## *Upstream Traffic*

A packet generated by an end user's computer will be seen by the ATU-R. The ATU-R half-bridge can learn what Ethernet addresses exist on its Ethernet segment by monitoring and tabulating the source MAC address of all packets that it does not generate. It will bridge to the ATU-C any packet received on its local Ethernet having an Ethernet Destination Address field that does not match an entry in its address table. At the ATU-C, the Ethernet packets bridged from the ATU-R would be aggregated by either a hub or an Ethernet switch and sent to a router. The router is at the edge of an IP routed backbone connecting the access node to various gateways including, for example, the Internet and corporate networks.

### *Downstream Traffic*

Traffic destined for a subscriber is routed through the backbone to the proper access node. The access node router must route the packet to the proper port interfacing the switch. Note that the access router must insert the end user's Ethernet address into the destination Ethernet field of the packet before sending the packet to the switch. The switch can then switch the packet to the appropriate ATU-C to be bridged over to the user's ATU-R and Ethernet segment.

### *Architecture Evaluation*

Ethernet bridging over the loop combined with routing in the backbone offers a very simple architectural option for applications where the NAP and NSP are the same entity. In this case, a true transport provider does not necessarily exist. This type of network may have security and scaling limitations, depending on the implementation. Handling of broadcast messages as well as address assignment also raises security issues discussed in the second bullet below. Although some of the problems can be overcome with special implementations, the solutions engender other problems. Among the issues that arise with this type of network are the following:

- Secure address management—If PPP is not used (and thus PPP authentication methods are not used), dynamic address assignment can be troublesome, as most dynamic address assignment protocols do not include authentication. Addresses can be permanently assigned to users; however, this task would be difficult and might consume too many IP addresses. (Normally, an NSP has fewer IP addresses to assign than it has users, as all users are usually not active simultaneously.)
- Broadcast packet handling—Ethernet bridges and switches can sometimes be configured to block broadcast packets. However, some broadcasts must be passed between the NAP router and the end PC's (address resolution protocol (ARP) packets, for example). This process opens the door for packets from one private LAN to be present on another private LAN.
- Session establishment—In this type of network, it can be difficult to monitor the actual network usage by a user. Such information is often useful for billing purposes.
- Flexibility—This type of network is meant for the joint NAP/NSP arrangement. If the NSP is located at the other side of the backbone, more problems of security and scalability arise. Certain relationships between the NSP and NAP can alleviate some of these problems; however, these are not always practical in real life.

## *ATM End-to-End Architectures*

A more popular architecture for ADSL access systems is the ATM end-to-end architecture. Several variations of this architecture have appeared, including those based on true PVCs, those based on SVCs, and some hybrid approaches. PVC- and SVC-based networks are discussed in the next few sections. Although different types of protocol models have been implemented, the most popular involve PPP. For this reason, models involving PPP are the primary focus. ATM architectures are discussed in the ADSL Forum contributions, "TR-002: ATM over ADSL Recommendations" and "Enhancements to Core Networks Architectures for ADSL Access."[11,12]

### *PVC-Based Models*

The block diagram in Figure 8.16 shows a network with PVCs end-to-end. The figure shows several subscribers along with several NSPs. The NAP, transport provider, and all NSPs are treated as independent entities.

Figure 8.16 An end-to-end PVC access architecture with PPP over ATM.

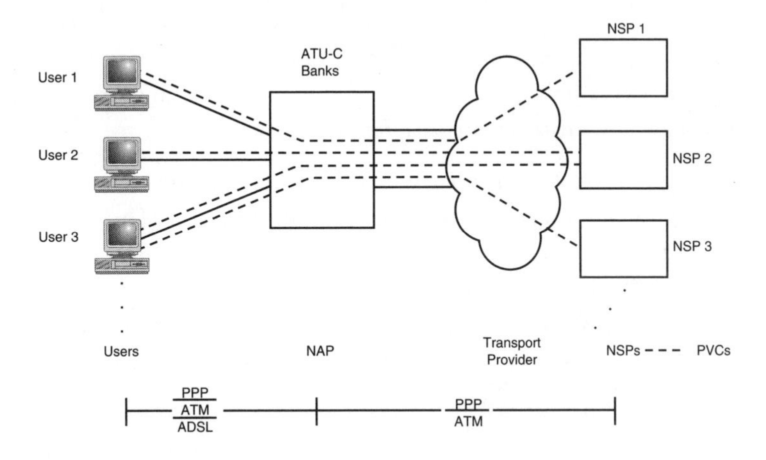


Figure 8.16 also shows the protocol model for the connection. It is important to note that both ATM and PPP are present end-to-end. Over the loop, these protocols are carried over ADSL and in the transport backbone over whatever physical layer is being used. (Often this is DS3 or OC-3.) Typically, the NSP to which traffic is forwarded is a gateway to another network. This network could be a private corporate network, the Internet, a banking institution's network, or perhaps some type of entertainment provider.

The model assumes neither a specific type of network at the user location (the premise network) nor a specific type of NSP network. It should be noted that to the access and transport domains, the protocols in the premise network and NSP do not matter, as they operate at the ATM layer only and do not interact with the higher-layer protocols. A good portion of the time, the higher-layer protocol above ATM will be IP. Much of this discussion is based on that assumption.

**Note**

The protocols in the NSP network and the premise network must be compatible with one another, or one of the two entities must have the proper protocol conversion equipment to ensure proper operation.

Shown in Figure 8.16 are the PVCs provisioned for this sample network. A PVC is provisioned between user 1 and NSP 1, between user 2 and NSP 2, and one each between user 3 and NSPs 2 and 3. These PVCs would normally be provisioned when a user signs up for service with a specific NSP. If user 1 decides to terminate service with NSP 1 in favor of NSP 2, the existing PVC between user 1 and NSP 1 must be removed and a new PVC provisioned between user 1 and NSP 2. Note that user 3 has a PVC provisioned to both NSP 2 and NSP 3 and can access service from either entity. The key idea being illustrated here is that to access more than one NSP, multiple PVCs must be provisioned end-to-end. In Figure 8.16, user 1 can access only NSP 1 and cannot access any of the others. Likewise, user 2 can access only NSP 2. In an end-to-end PVC environment, dynamically selecting an NSP, also known as *service selection*, is not possible.

Note in Figure 8.16 that PPP is terminated at the user and at the NSP gateways. Further, the NAP and transport providers play no part in the PPP session. Thus the NSP can control all PPP negotiation parameters, can implement the authentication method and verification, and can issue IP addresses without intervention or knowledge on the part of the NAP or transport providers. This feature is important to the practical business practices of the NSP (an NSP normally would not want to share its private password databases with another party).

Because the end-to-end PVC for each user is provisioned at subscription time and remains available until the user wants to change service providers (which in many cases could be years), no information short of counting actual user cells can be used to determine the actual user's usage of the service. Thus ATM alone provides limited options for usage tracking. An advantage of using PPP over ATM is that PPP can help determine the amount of time a user is actually accessing the NSP. Entries can be made in a database detailing when a user logs on to an NSP's network as well as when a user logs off. This information opens up more options on how billing policies can be implemented (flat monthly fees versus usage-time fees, for example).

In all, an ATM PVC implementation has many benefits including the capability to provide a guaranteed bit rate through the access and transport domains for every user. A drawback of this type of access network is the coordination between NSPs, transport providers, and NAPs and provisioning needs necessary to connect all users to the desired NSPs (one PVC per user per destination). This problem is amplified as the number of users grows and as users change from one NSP to another (often called user churn). Another drawback of an end-to-end PVC architecture is the lack of service selection. If a user wants to access more than one NSP, multiple PVCs must be provisioned for the user, further complicating the provisioning problem.

## *SVC-Based Model*

An SVC model overcomes many of the issues associated with a PVC model. Figure 8.17 shows an end-to-end ATM SVC architecture.

Figure 8.17 An end-to-end SVC access system with PPP over ATM.

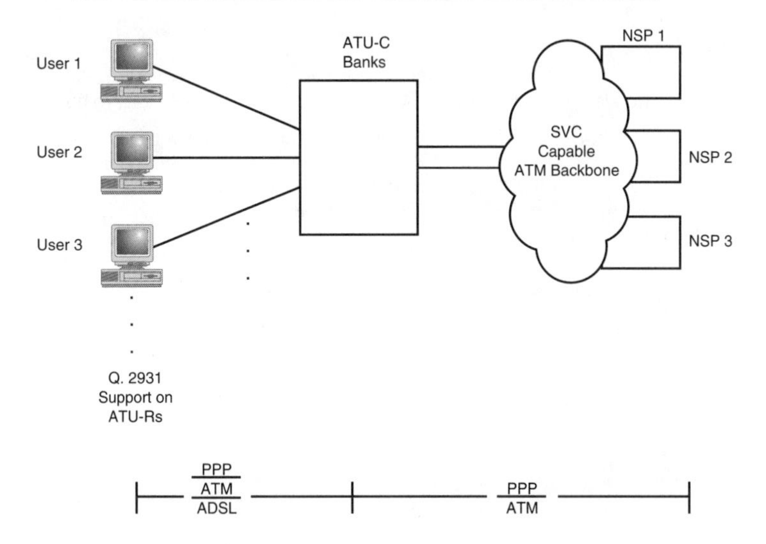


The basic diagram is the same as that shown in Figure 8.16 except that the transport backbone is now SVC capable. Therefore, PVCs do not have to be provisioned at subscription time. Instead, virtual circuits are dynamically created when a subscriber wishes to access an NSP (begin a session) and are torn down when a subscriber ends the session. Service selection is done by a signaling specification known as Q.2931.[13] This signaling method transmits all the information necessary to indicate a desired destination as well as the bandwidth characteristics of the session. Assuming that an adequate virtual circuit can be set up between the user and the desired NSP, the connection will be made. Service selection in an SVC network is sometimes simply referred to as dialing, as it mimics the functionality of dialing in the telephone system.

Note that the data protocol stacks in Figure 8.17 did not change from those in Figure 8.16. Although PPP is no longer necessary to delineate sessions (because call starts and stops can be recorded), it is still convenient for authentication and address assignment. Thus PPP end-to-end would remain in SVC access networks.

The major drawback of an SVC ADSL access network is that the access node, backbone, and all endpoints must be ATM based and support SVCs. For many different reasons, this requirement is not widely met. For example, some transport networks are based on technologies other than ATM. Also, many existing NSPs do not have ATM-based gateways and for business reasons are not motivated to upgrade their gateways to ATM. Although an end-to-end SVC network is very enticing, true deployment based on it will have to wait until basic upgrades are made to the network.

## *Tunneling Architectures*

In an attempt to bridge the gap between PVC networks and SVC networks, tunneling architectures have been proposed. Figure 8.18 shows a network employing tunneling.

Figure 8.18 A tunneling access system with PPP end-to-end.

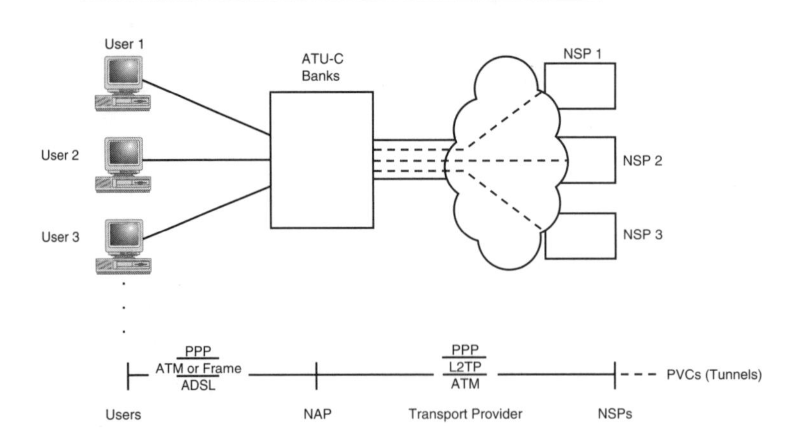


The notable characteristic of this network is that L2TP is present in the transport backbone between the NAP and the NSP. In addition, a single PVC is provisioned between the NAP and each NSP. Each PVC represents a single tunnel. All users' traffic passing through the NAP in Figure 8.18 destined for NSP 1 would be sent on the tunnel between the

NAP and NSP 1. Thus, as compared to a PVC end-to-end network, a single PVC can carry traffic from multiple users. Likewise, all traffic between NSP 2 and the NAP is sent on the tunnel between the two. PPP is again present as an end-to-end protocol and provides the methods for authentication, address assignment, and session delineation. Over the actual access loop, both cell (ATM) and frame (FUNI) implementations are possible.

Users in a tunneling network must use a dialing method to access an NSP. Two dialing methods are possible. The first uses Q.2931 signaling. In this case, the signaling is terminated at the NAP, and the NAP forwards the user's traffic onto the appropriate NSP. The desired destination will be known after decoding the Q.2931 messages. This NAP model differs from the use of Q.2931 in the SVC end-to-end network in which the switches as well as the NSP endpoint participate in the session set up. Upon receiving a Q.2931 request for a connection to an NSP, the NAP (LAC in L2TP terms) initiates a call setup with the NSP (LNS in L2TP terms). Essentially, this call setup assigns a call ID to the user's PPP session so that the session can be uniquely identified at each end of the tunnel.

A second possible method of service selection in a tunneling environment is to encode the desired destination name into the authentication name sent by the user. When a user initiates a PPP session, the LCP negotiation is done by the LAC, not the LNS. A call ID is requested and received for the user after the LCP phase. PPP is "temporarily terminated" at the LAC, and then the PPP endpoint is handed off to the LNS. In addition to LCP negotiation, authentication information may be received by the LAC. Authentication information often contains a user name. If, as a rule, the username is formatted as *`user_name@nsp.com`*, where *`user_name`* identifies the user and *`nsp.com`* identifies the desired NSP, this field can have the dual purpose of identifying the user as well as serving as a dialing string. For example, if the author wanted to access NSP 1, the username field might be `Dennis.Rauschmayer@NSP1.com`. The value in using this type of dialing method is its simplicity compared to supporting Q.2931.

Tunneling networks with some type of service selection method mimic an SVC network without needing ubiquitous support of SVC signaling. Though provisioning in the backbone is still necessary, the number of PVCs needing provisioning scales with the number of NSPs, instead of with the number of subscribers as in the PVC end-to-end case. In addition, churn of PVCs is less likely. Because L2TP is independent of the protocol over which it runs, deployment in backbone networks consisting of technologies other than ATM is possible as well as in hybrid networks. These factors have made tunneling architectures a practical compromise for xDSL access networks.

Tunneling networks have some less attractive features as well. Compared to ATM end-to-end systems, a guaranteed bandwidth can not be maintained for individual subscribers (even if the backbone is all ATM) because multiple subscribers share the same PVC. This situation can be somewhat overcome by offering differently loaded tunnels between NAPs and NSPs. For example, a regular and a "gold" tunnel where a premium rate is charged for the "gold" tunnel, but the ratio of bandwidth to subscribers is higher than for the regular tunnel. However, the true QoS offered by ATM cannot be matched. Also, L2TP requires that PPP packets be reassembled at the NAP before being placed into the tunnel. This step is inconvenient if cells are used over the loop.

## *Premise Network Considerations*

In the architectures discussed in the previous sections, no mention was made of considerations in the premise network. Several popular options are common in the premise network, including an external ATU-R connected by Ethernet to a computer (or computer network), an external ATU-R connected though an ATM-25 port to a computer, and an internal ATU-R, sometimes called a *network interface card* (NIC).

When a NIC or an ATM-25 connection is used, much of the protocol processing including PPP can be implemented on the host PC. This approach greatly simplifies the ATU-R design as it essentially has only physical layer functionality. Address assignment to the user is straightforward as well because an IP address assigned though PPP is expected. Figure 8.19 shows premise network configurations.

Figure 8.19 Premise network configurations where (a) the ATU-R is connected via an ATM 25 interface and (b) the ATU-R is inside the PC; in both cases, ATM and PPP are terminated inside the PC.

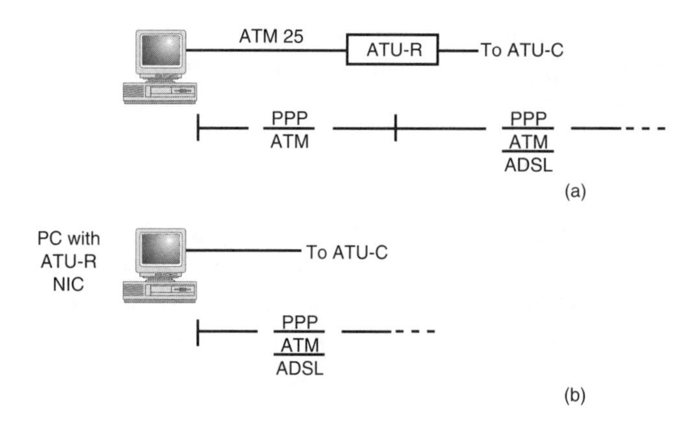


When an ATU-R is connected to a user's computer or computer network via Ethernet, address handling and PPP termination are a bit more involve. This case is illustrated in Figure 8.20.

Figure 8.20 An example of a premise network in which the ATU-R is connected to the PC via Ethernet; in this case, the PPP termination point is in the ATU-R and not the PC.

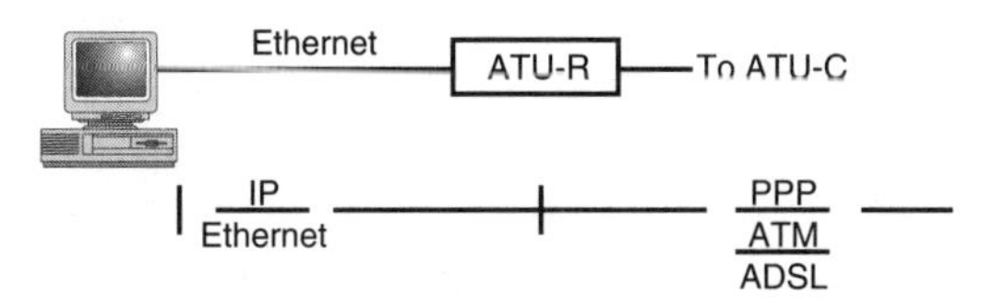


Because Ethernet ports and drivers are usually not equipped to handle PPP, the PPP endpoint is typically found in the ATU-R instead of in the PC. Also, the PC will typically already have an IP address assigned to its Ethernet port before being connected to the NSP. This IP address is sometimes called the *private* IP address. This situation raises a problem because the NSP normally requires the user to take on an IP address consistent with the NSP's address plan. One method to solve this problem is *network address translation* (NAT). In this case, the ATU-R obtains an IP address on behalf of the PC though IPCP. When traffic from the PC destined for the NSP passes through the ATU-R, it substitutes the obtained IP address for the private IP address in the source IP address field. For traffic passing in the other direction, the ATU-R substitutes the private IP address for the obtained IP address in the destination IP address field. When this step is properly done (which includes recalculating IP and TCP checksum values), the operation is transparent to both the PC and the NSP. NAT can run into problems with certain protocols such as file transfer protocol (FTP). Some datagrams sent using this protocol embed the IP address of the user into the packet in ASCII form. For FTP to work through an NAT device, FTP packets must be recognized, decoded, and adjusted to reflect the translated IP address. In addition, some IP security methods do not work with NAT. NAT is discussed in further detail in RFC 1631, "The IP Network Address Translator (NAT)."[14]

Several other methods exist to combat the issues of an Ethernet attached ATU-R. One method involves creating a tunnel through the premise network and extending the PPP session all the way to the PC so that protocol processing can be done locally. Other protocols are also being proposed to handle this situation. No one method of dealing with the premise network dominates, and various methods will continue to exist with the important quality that they all look the same to the NAP.

## *Summary*

The first part of the chapter reviewed the basics of networks and protocols. Then the discussion turned to the way in which these protocols fit together in WANs incorporating ADSL. Several types of architectures, including bridging architectures, ATM end-to-end architectures, and tunneling architectures, were considered, and their advantages and disadvantages were examined. The role of PPP in access networks was discussed along with the many benefits it provides. The chapter concluded with a discussion of issues and options in the premise network. Different common implementations of an ATU-R were described, along with the address-assignment and protocol-processing requirements of each.

## *Exercises*

1. Find RFC1660 on the Internet. Explain the LCP parameters available for negotiation.
2. ARP and Ping commands are commonly available on workstations. Use the Ping command to verify connectivity to various hosts. View the ARP table after each Ping. What type of information is given in the ARP table?
3. Find the IETF specification for L2TP on the Internet. Draw the frame structure for an L2TP packet.
4. Draw the L2TP frame structure of an L2TP packet carrying an IP frame. Do the same for an Ethernet frame.
5. Assume that 10,000 ADSL users are accessing 20 NSPs through a transport provider. If 50 access nodes exist and each user accesses only one NSP, how many PVCs are necessary in an end-to-end PVC model? How many are necessary if each user accesses two NSPs? Four NSPs?
6. Repeat exercise 5 assuming a tunneling model is employed.
7. What is the maximum number of sessions that can be multiplexed over an L2TP tunnel?

## *Endnotes*

1. W. Stallings, *Handbook of Computer Communications Standards, Volume 2: Local Network Standards.* New York, NY: Macmillan, 1990.

2. C. Hornig, "Standard for Transmission of IP Datagrams over Ethernet Networks," RFC 894, 1984.
3. W. Richard Stevens, *TCP/IP Illustrated Volume 1: The Protocols.* Reading, MA: Addison-Wesley, 1994.
4. Gary Wright and W. Richard Stevens, *TCP/IP Illustrated Volume 2: The Implementation.* Reading, MA: Addison-Wesley, 1995.
5. S. Feit, *TCP/IP: Architectures, Protocols and Implementation.* Washington, DC: McGraw Hill, 1993.
6. U. Black, *ATM: Foundations for Broadband Networks*. New York: Prentice Hall, 1995.
7. J. Martin, et al., *Asynchronous Transfer Mode: ATM Architecture and Implementation.* New York: Prentice Hall, 1996.
8. W. Simpson, "The Point-to-Point Protocol (PPP)," RFC 1661. July 1994.
9. G. Mcgregor, "The PPP Internet Protocol Control Protocol (IPCP)," RFC 1332. May 1992.
10. Kory Hamzeh, et al., "Layer Two Tunneling Protocol (L2TP)," Internet Draft draft-ietf-pppext-l2tp-01.txt. October 1996.
11. ADSL Forum, "TR-002: ATM over ADSL Recommendations." March 1997.
12. J. Wojewoda, et al., "Enhancements to Core Networks Architectures for ADSL Access," ADSL Forum Contribution 97-072. June 1998.
13. ITU-T, "Recommendation Q.2931: B-ISDN Application Protocols for Access Signaling." February 1995.
14. K. Egevang and P. Francis, "The IP Network Address Translator (NAT)," RFC 1631. May 1994.

CHAPTER 9

# VDSL Overview

In this chapter:

- Basic deployment locations of VDSL
- Requirements for VDSL systems
- Proposed modulation methods for VDSL
- VDSL in WAN networks

ADSL satisfies the need to send up to 8 Mbps or so over twisted pairs. However, there is still a need to send more data at even higher speeds. VDSL fills this void by delivering up to 50 Mbps. Unlike ADSL, VDSL offers both symmetric and asymmetric configurations. Typically, VDSL uses far more bandwidth than ADSL uses and has a much shorter reach due to the higher bit rates supported. This chapter covers the basic goals of this relatively new technology and presents some candidate implementations. As with ADSL, some implementations promote DMT, and others use CAP/QAM techniques. The chapter also includes a brief discussion of how WAN networks might implement VDSL.

## VDSL Deployment Locations

Like ADSL, VDSL can provide frequency separation from baseband, leaving room for Plain Old Telephone Service (POTS). (As you will see later, in addition to POTS, other services such as ISDN can also be run on the same twisted pair as VDSL.) Thus each end of a twisted pair carrying VDSL and POTS uses a splitter to separate the two signals. At the subscriber's location, the modem terminating VDSL is called the *VTU-R*. The very high speeds of VDSL suggest that some sort of high-speed optical network will feed a bank of centralized VDSL modems. This central point might be a CO of some remote optical node. For this reason, the peer modem to the VTU-R modem is called the VTU-O (as opposed to VTU-C, which seems to follow from the ADSL notation), denoting an optical feeder.

**Note**

However, other methods of feeding VDSL banks are certainly possible. For example, VDSL can be used as a means to multiplex and carry multiple T1 streams over a short distance.

Figure 9.1 shows an example of VDSL range and reach from a central deployment location. If the location is a CO, the actual area to which VDSL could be deployed, or the coverage area, would be much smaller than the area served by the CO (typically extending well beyond 12 kft). Limiting deployment to this percentage of the overall serving area severely limits the business potential of VDSL.

Figure 9.1 Range and reach of VDSL from a central location.

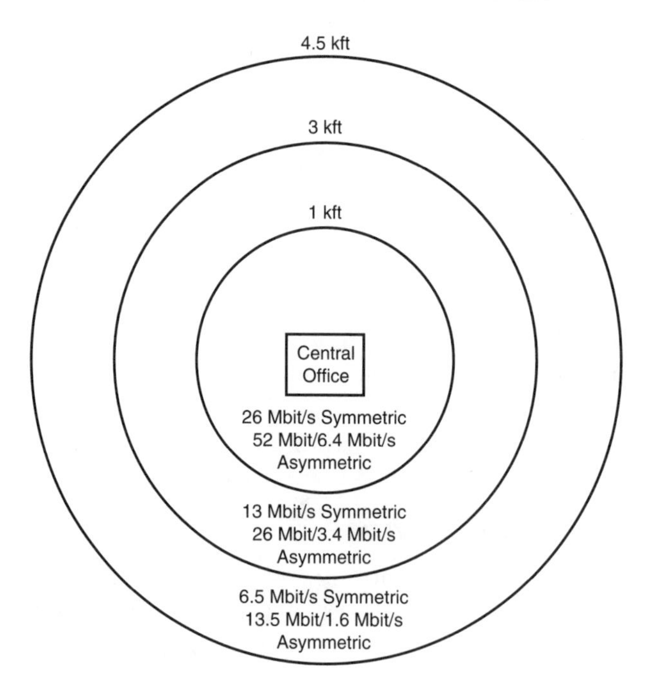


Instead, VDSL deployment will likely take place from the CO as well as from *optical network units* (ONUs). These nodes typically are deployed in, and serve, neighborhoods, industrial parks, and other areas with high telecommunications traffic patterns. ONUs are sometimes used in context when discussing technologies such as fiber to the curb (FTTC), fiber in the loop (FITL), and fiber to the neighborhood (FTTN). The common feature of these technologies is that high-capacity fiber is deployed to an ONU that serves a small area. The media connecting subscribers to the ONU can be coaxial cable, some sort of wireless link, or more probably, a twisted pair. The combination of a high-capacity link to a serving node, the small area served by each node, and the popularity of twisted pairs connecting subscribers to the serving node make a network utilizing ONUs a perfect fit for a technology like VDSL.

The example in Figure 9.2 shows an area with VDSL deployment using ONUs to serve more remote areas and VDSL deployment from the CO (CO) to serve less remote areas. In the figure, simple optical links serve each ONU. In reality, optical rings or some other type of optical distribution might be more practical.

Figure 9.2 Coverage area of VDSL when deployed from remote ONUs.

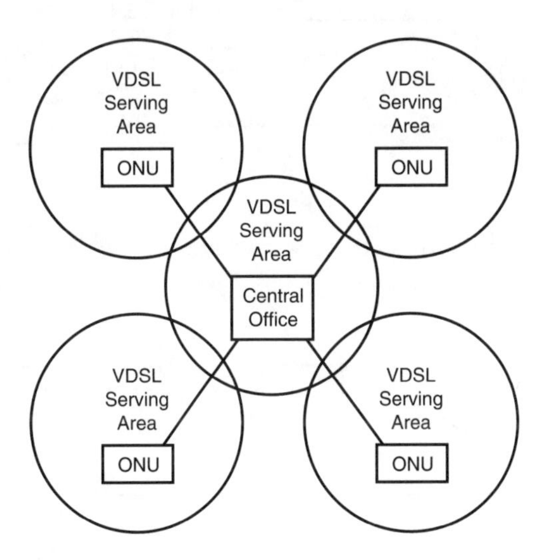


The total optical bandwidth available to an ONU is usually not greater than the sum of potential bandwidths of all ONU users. For example, if an ONU serves 20 subscribers and each has a 50 Mbps VDSL link, the total potential bandwidth for that ONU is 1 Gbps, much more than the ONU would normally be fed. The ratio of the optical bandwidth available to the ONU to the potential aggregate bandwidth of all users is called the *oversubscription ratio*. Oversubscription ratios are engineered so that performance remains reasonable for all users.

## *Crosstalk Issues*

As with FDM ADSL, VDSL does not have self-NEXT. The trend toward technologies with no self-NEXT, or at least with limited self-NEXT, has evolved because of the increased performance achievable and, in some cases, the design simplifications possible with such implementations. Modulation schemes for VDSL and how they avoid self-NEXT is discussed later in this chapter.

Many of the crosstalk analyses in previous chapters are applicable to VDSL. In addition, the short deployment range of VDSL poses several new crosstalk possibilities. Consider the two deployment configurations in Figure 9.3 and Figure 9.4.

**Figure 9.3** Crosstalk scenario with VDSL and another DSL technology in the CO mix.

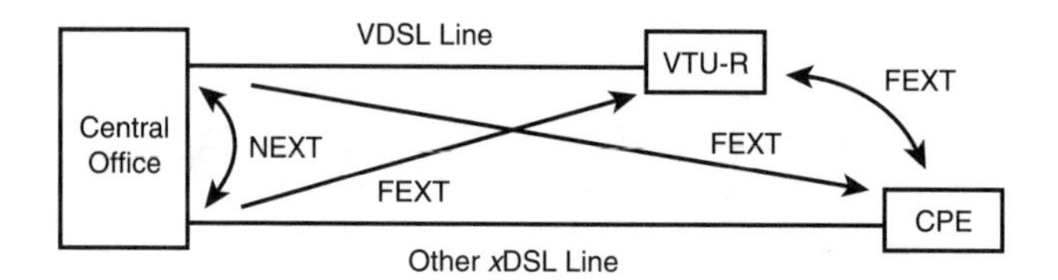


**Figure 9.4** Crosstalk scenario with VDSL and another DSL technology in the customer premise (CP) mix.

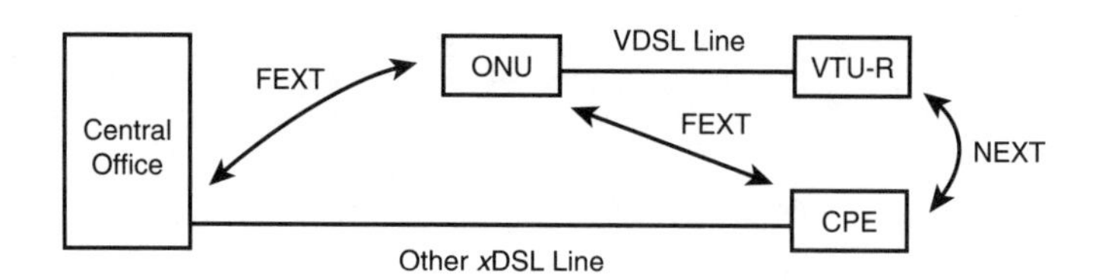


Figure 9.3 is sometimes called CO mix.[1] In Figure 9.3, VDSL and another xDSL technology are both provided from the CO and share a binder group between the CO and some point at which the VDSL is removed. (The VDSL might be terminated at this point, or it might be routed into a different binder group.) The various types of NEXT and FEXT that can exist are also shown in Figure 9.3. The FEXT between the two endpoints is normally not a factor because the distance is usually large enough to attenuate the disturbing signal. However, other FEXT paths may exist as well. Whether these FEXT signals will be significant or not depends on the lengths of the two channels as well as the frequencies used by the upstream and downstream channels.

In the configuration shown in Figure 9.4, the VDSL signal is terminated at a remote ONU and the other xDSL signal at the CO. At one point, the twisted pair segments carrying the signals share a common binder. This configuration is sometimes called the CP mix.[2] Note that it is not necessary for the VTU-R and the other xDSL CPE modem to be located in the same building, but the twisted pairs carrying them must share a binder group for some length near the respective endpoints. Figure 9.4 also shows the possible significant NEXT and FEXT paths for the CP mix configuration. The significance of the FEXT paths depends on the location of the ONU as well as the frequencies used by the channels. Note that the endpoints of the VDSL circuit can contribute either NEXT or FEXT into the endpoints of the other xDSL technology, depending on the deployment configuration. This flexibility adds a new degree of freedom in analyzing crosstalk.

Of particular interest is the coexistence of ADSL and VDSL. In the CO mix configuration, ADSL and VDSL are both provided from the CO and share a binder group between the CO and the point at which the VDSL is removed. (The VDSL might be terminated at this point, or it might be routed into a different binder group.) For no NEXT to exist between these services at the CO, the following two conditions must exist:

- The downstream VDSL signal must not overlap the upstream ADSL signal (the VDSL downstream signal must start higher than 138 kHz).
- The downstream ADSL signal must not overlap the upstream VDSL signal (the VDSL upstream signal must not overlap 138 kHz to 1.1 MHz).

In addition, the downstream ADSL signal may cause a reasonable level of FEXT into the downstream VDSL signal since the distance between the ADSL transmitter at the CO and the VTU-R receiver is small. Likewise, FEXT from the VDSL signal may pose a problem to the upstream ADSL signal.

For no NEXT to exist between the ADSL and VDSL in the CP mix configuration, the upstream and downstream frequency spectrums of the respective technologies must not overlap. Additionally, VDSL FEXT into the downstream ADSL signal might occur, and ADSL FEXT can occur into VDSL depending on the ONU location. Owing to the short loop lengths supported by VDSL, the FEXT, though less severe than NEXT, might have a detrimental impact on the ADSL downstream.

Because of the need to eliminate NEXT between ADSL and VDSL systems and the possibility that FEXT may indeed cause performance degradation, the spectrum allotted to VDSL falls above that used by either the upstream or downstream ADSL channels when ADSL and VDSL are both present. In addition, when dealing with other technologies, the deployment configuration of VDSL must be taken into account to decide which points contribute NEXT or FEXT to other points. The specific frequency allocations for VDSL are discussed later in this chapter (after other influencing factors are discussed).

## *RFI Issues*

VDSL receivers must deal with the issue of radio frequency interference (RFI). Included in RFI issues are *ingress* and *egress*. The cause of RFI ingress is in-band radio waves from nearby antennae incident upon a twisted pair carrying the VDSL. An amateur radio antenna is a good example of an RFI ingress disturber. Figure 9.5 illustrates this basic situation.

Figure 9.5 RFI ingress into VDSL due to a local transmitter.

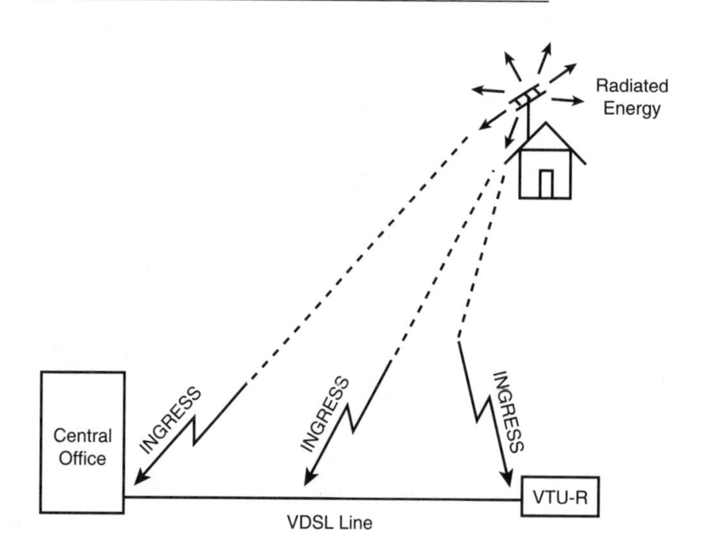


The factors influencing the amount of ingress include the power output of the antenna, the distance between the antenna and the twisted pair, the relative orientation and shielding properties of the binder group, and the balance of the twisted pair itself. Normally, the RFI ingress excites each wire in the twisted pair, thus creating a longitudinal ingress signal on the pair. Because the balance of the twisted pair is not ideal (normally it is 30 dB to 35 dB for high frequencies), some of the ingress leaks into the differential signal.

RFI ingress signals are often very narrow in bandwidth when compared to the VDSL signal. This characteristic is helpful, as the ingress will affect only a small portion of the usable bandwidth.

Another issue is that the ingress signal may be very large compared to the received VDSL signal. In this case, the analog front end of the receiver must be carefully designed so as not to saturate. In addition, some action must be taken so that the analog-to-digital converter (ADC) performs with the proper precision. The issue here is that a large ingress signal will swamp the converter, causing it to waste bits of precision on this unusable signal instead of on the received VDSL signal. Because the VDSL signal is quantized with less precision, the quantization noise, or imprecision due to the conversion process, increases, effectively decreasing the achievable bit rate on the channel. The ADC must have enough range to handle the ingress signal as well as enough precision to adequately quantize the VDSL received signal. Even after quantization, further RFI cancellation may be necessary in the digital domain as documented in Jeong and Yoo's article, "An Efficient Digital RFI Canceler for DMT-Based VDSL," and Wiese and Bingham's article, "Digital Radio Frequency Cancellation for DMT VDSL."[3,4]

To facilitate the design of the ADC, the analog front end of a VDSL receiver might employ circuitry to reduce RFI ingress. Recall that ingress is normally present both longitudinally and differentially on a twisted pair. Figure 9.6 shows a method to use the longitudinal ingress signal to reduce the metallic ingress signal.

Figure 9.6 A method to reduce ingress that uses a longitudinal signal.

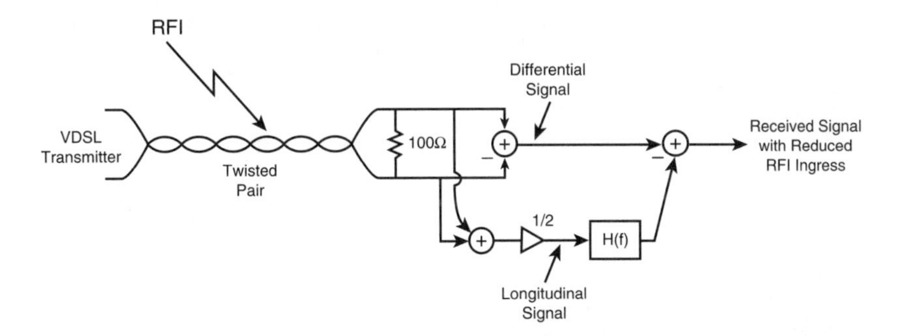


The filter tries to match the unbalance of the twisted pair that is essentially the mechanism by which the ingress reaches the differential signal. If the filter matched the unbalance perfectly, the signal at the output of the summation would consist only of the received VDSL signal (as well as crosstalk and background noise). Figure 9.7 illustrates this concept via a block diagram.

Figure 9.7 A specific implementation of an RFI ingress cancellation circuit.

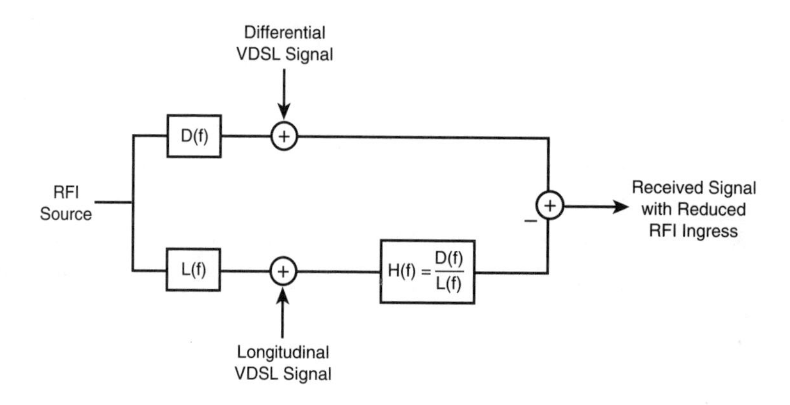


A specific implementation is described in Cioffi, et al.'s article, "Analog RF Cancellation with SDMT."[5] Ingress reduction on the order of 35 dB can be achieved by this method. Note that this would be equivalent to gaining more than five bits of precision in the ADC.

RFI egress is also a concern with VDSL. Egress is illustrated in Figure 9.8.

Figure 9.8 An example of RFI egress.

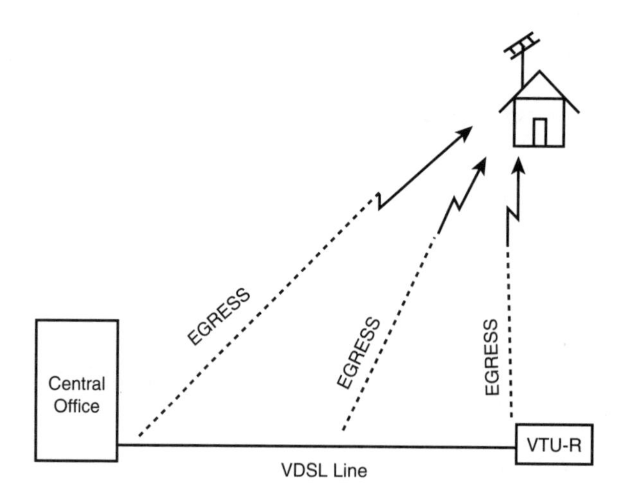


Here the VDSL signal is radiated from the twisted pair and can disturb the signal received by local antennae if these received signals overlap the VDSL spectrum. To combat this problem, the VDSL transmit power in frequency regions reserved for wireless or radio services must be lowered (these regions were discussed in Chapter 4, "Power Spectral Densities and Crosstalk Models"). Normally, 20 dB reductions in these areas will be adequate to mitigate problems from VDSL RFI egress.

# VDSL Requirements

This section discusses the more fundamental requirements put forth for VDSL systems. These requirements are captured in the technical reports "Very High Speed Digital Subscriber Line System Requirements"[6] and "Transmission and Multiplexing : Access Transmission Systems on Metallic Access Cables; Very High Speed Digital Subscriber Line (VDSL); Part I: Functional Requirements."[7] It should be noted that at the time of publication, the requirements were not completely static. Also, some aspects of the candidate implementations in this chapter may not completely satisfy all requirements in the aforementioned reports.

## Rate and Reach Goals

In contrast to ADSL, VDSL technology provides either symmetric or asymmetric data rates to a subscriber. The reason that both need to be supported is that some VDSL applications are best served by an asymmetric connection, whereas others are best served by a symmetric connection. Video on demand (VoD), for example, is an application that benefits from an asymmetric connection. Another application, ATM LAN extension, might better be served by a symmetric connection. In addition, applications that do not even exist might turn out to be application drivers for VDSL. Defining symmetric and asymmetric modes of operation up front does not limit the types of application that might one day use VDSL.

**Note**

This situation developed with ADSL. The initial application driver was VoD; however, Internet/LAN access later became an even stronger application driver.

Table 9.1 shows asymmetric data rate goals for VDSL along with the approximate reach goals for each configuration. How the rates are provisioned (hard configuration as compared to automatically adapting to the loop) is yet to be determined. At the very low end of the asymmetric rates (6.5 Mbps/ 0.8 Mbps), VDSL approaches the performance limits of ADSL. Though it might seem logical to define even lower rates for VDSL and not bother with ADSL, this approach would most probably cause modems to be more complex than necessary and would impact cost effectiveness.

Table 9.1 Asymmetric Data Rate Goals for VDSL

| Length Classification | Downstream Rate (Mbps) | Upstream Rate (Mbps) | Reach Goal (kft) |
|---|---|---|---|
| Short | 52 | 6.4 | 1 |
| Medium | 26 | 3.2 | 3 |
| Long | 13 | 1.6 | 4.5 |
| Long | 6.5 | 0.8 | 6 |

Table 9.2 shows rate and reach goals for symmetric VDSL operation. Many of these rates are primarily intended to support ATM access systems (for example, 26 Mbps would support 25.6 Mbps ATM adapters along with some overhead). It is expected that some of the symmetric as well as asymmetric rate and reach goals will churn and evolve as VDSL goes through the standardization process. However, Tables 9.1 and 9.2 provide good insight into the probable operating ranges of VDSL.

Table 9.2 Rate and Reach Goals for Symmetric VDSL Operation

| Length Classification | Downstream Rate (Mbps) | Upstream Rate (Mbps) | Reach Goal (kft) |
|---|---|---|---|
| Short | 26 | 26 | 1 |
| Medium | 13 | 13 | 3 |
| Long | 6.5 | 6.5 | 4.5 |

## *Frequency Allocation*

The frequency ranges available for VDSL to use are limited by the need for compatibility with other DSL technologies as well as RFI concerns discussed previously. The base power spectral density (PSD) mask for VDSL appears in Figure 9.9.

Figure 9.9 The basic PSD requirements of a VDSL transmitter.

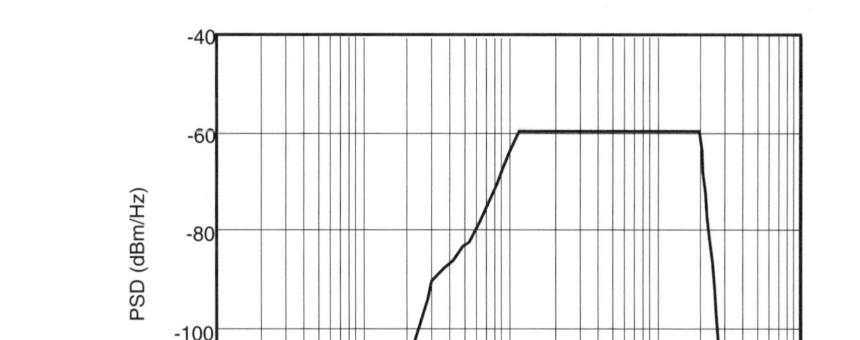


This basic template has a PSD of –60 dBm/Hz from 1.1 MHz to 20 MHz. This range would normally be considered the usable part of the spectrum, and the rest would be considered sidebands. In the ADSL band, the PSD rolls off so as to minimize the crosstalk degradation into adjacent ADSL lines. A POTS or ISDN link could be run on the same line as a VDSL system using this mask, given that a splitter was used to separate the signals.

The mask shown in Figure 9.9 may cause excess egress and interfere with local antennas. In this case, a notching option can be applied resulting in the mask shown in Figure 9.10. Here bands known to be prone to causing egress into existing radio/wireless services are notched down to a level of –80 dBm/Hz (essentially reducing the egress power in the notched areas by a factor of 10). The notched frequency regions can still transmit information if the received SNRs across the regions are favorable.

**Note**

Of course, if egress is a problem that causes a band to be notched, ingress may likely exist in the same band as well (the local antennas may transmit as well as receive), further degrading the SNR and causing the notched bands to be useless for transmitting data.

Figure 9.10 A VDSL PSD with notches to reduce RFI egress.

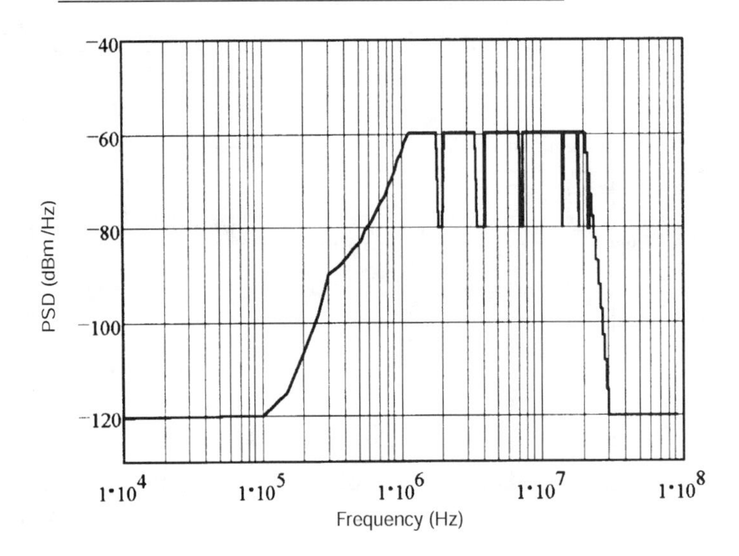


A power boost option has also been proposed for VDSL. When ADSL compatibility is not necessary (perhaps because a service provider did not deploy ADSL), this option would not only boost the VDSL level in the 1.1 MHz to 20 MHz region but also allow the usable end to stretch down to about 300 kHz. Below 300 kHz, the PSD would roll off and eventually intersect with that shown in Figure 9.9 so that ISDN and POTS could be supported.

It may take trial deployments of VDSL to determine which spectral mask properties are more desirable. Real-life deployment may also flag new problems needing to be overcome. In addition, VDSL spectrum requirements may depend on territory. In any case, Figure 9.9 provides at least a starting point to understanding VDSL PSD development.

One final note on the VDSL PSDs discussed is that the masks do not assume anything about symmetric versus asymmetric operation nor upstream and downstream channels. As I discussed previously, implementations are free to use the frequencies within the defined mask in any way they want, although self-NEXT avoidance will likely be necessary for performance reasons. The following sections discuss different proposed implementations of VDSL. You will see that several methods are employed to abide by the PSD requirements and, at the same time, avoid self-NEXT.

# *Modulation Methods for VDSL*

Several different modulation methods have been developed to satisfy the requirements of VDSL. At the time of this writing, VDSL was still very early in the standardization process, and many details of the proposals are yet to be specified (for example, framing content is not yet known). Though many proposals currently exist for VDSL, most fall into two categories: a time-domain duplexing (TDD) category and a frequency-domain duplexing (FDD) category. Different modulation methods within each category exist as well. The ideal method by which standardization can be achieved is to choose a modulation method and fill in all of the implementation details necessary for a standards document. In reality, one or more of the methods are likely to become standardized (or at least implemented by service providers), mainly because of the lack of consensus or compromise on behalf of the standardization body members. The following sections describe some of the candidate implementations.

## *TDD Approaches*

TDD is a half-duplex approach for sending signals between two endpoints. Basically, one modem transmits while its peer modem only receives. After a set amount of time, the process reverses, and the first modem only receives while its peer transmits. This basic concept is shown in Figure 9.11.

Figure 9.11 An example of TDD.

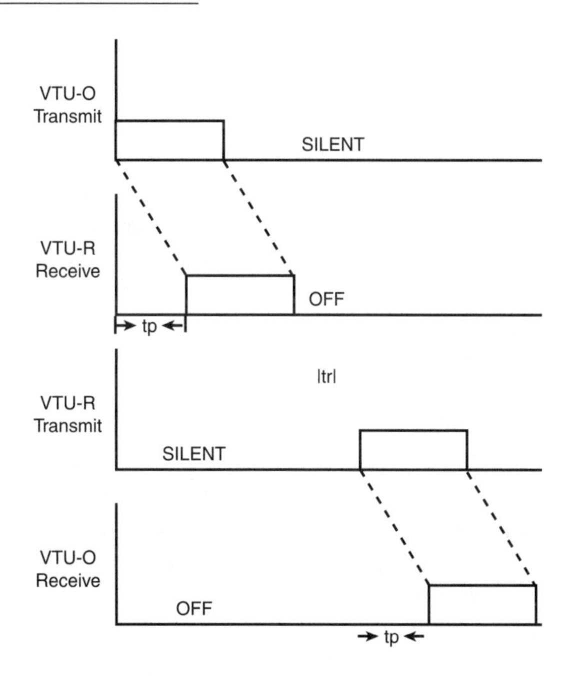

Sometimes this method is known as "ping-pong" for obvious reasons. In each direction, the entire usable frequency band can be used to send information, and thus no separate upstream and downstream PSD need be defined.

The most popular TDD approach proposed for VDSL is the synchronous DMT approach (SDMT). Essentially, this technology employs many of the same fundamental blocks used in ADSL along with the capability to gate the output of the transmitter. (Though many of the same blocks are used, the parameters, such as FFT/IFFT size, will be different.) Figure 9.12 shows a block diagram of an SDMT transmitter and receiver.

Figure 9.12 Block diagram of an SDMT transmitter and receiver.

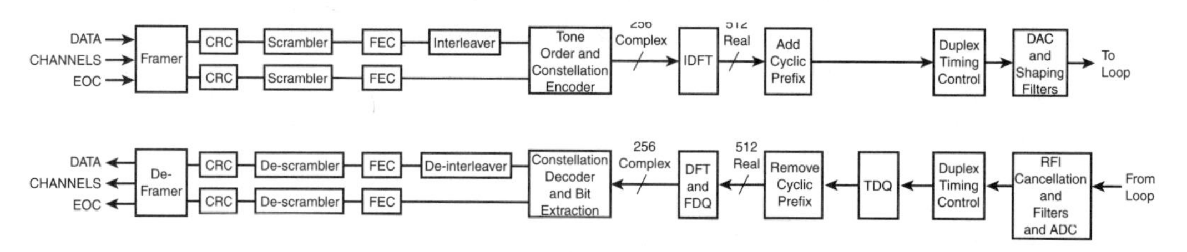


Figure 9.12 is very similar to the DMT-based ADSL diagram discussed in Chapter 7, "ADSL Modulation Specifics." The actual implementation of these blocks will no doubt undergo changes in the early stages of VDSL. A discussion of the blocks in order to explain the main concepts of SDMT follows.

## *Non-Modulation-Specific Blocks*

Some of the blocks in Figure 9.12 should be well understood by now, and implementation details are not necessary for an understanding of SDMT. These blocks include the scrambler and descrambler, the FEC blocks, and the interleaver and de-interleaver blocks. In addition, block implementation is not yet static. Some of the common implementations seen for other DSL systems, including self-synchronizing scramblers and descramblers, Reed-Solomon FEC blocks in GF(256), and convolutional interleaving, are likely implementation candidates.

To a lesser extent, framing and tone ordering/constellation encoding are not modulation-specific blocks. Of course, the tone-ordering and constellation-encoding algorithms must know the number of subchannnels in the modulation path. For SDMT, the number of subchannels supported is 256. As with DMT-based ADSL, some of these subchannels must not be used so that the SDMT output spectrum conforms to the VDSL PSD mask. Because SDMT is a half-duplex modulation scheme, each transceiver will use 256 tones, as opposed to DMT ADSL, where upstream and downstream transmitters used a different number of tones.

## *SDMT Modulator*

The SDMT modulator accepts a vector of 256 complex values from the constellation encoder. Each of these values represents a constellation point for a subchannel or equivalently, the cosine wave magnitude (real part) and sine wave magnitude (imaginary part) of the time-domain waveform produced for a subchannel. The complex conjugates of the 256 complex values are then appended to the vector so that the new vector has complex conjugate symmetry. An IFFT is performed, as given by Eqtn. 9.1.

Eqtn. 9.1

$$x_k = \sum_{i=0}^{511} Z_i e^{\frac{j\pi ki}{256}} \quad \text{for k = 0..511}$$

Here the vector Z represents the vector with complex conjugate symmetry. The output vector of the IFFT contains 512 time-domain values (all real). This IFFT transform is identical to that done in a downstream DMT ADSL transmitter.

## *Cyclic Prefix Block*

The cyclic prefix block adds the last L sample points of the 512-point time-domain vector to the beginning of the vector. L can be any integer less than or equal to 64. The new time-domain vector will be given by (x511–L,x511–L+1,x511–L+2,Ö,x511,x0,x1,Ö,x511). Adding a cyclic prefix makes a channel appear to be circular so that equalization can be done in the frequency domain. The 512 time-domain samples along with the L sample cyclic prefix represent one SDMT symbol. The choice of L is further discussed in the next section.

**Note**

If the channel's impulse response is longer than the cyclic prefix, the channel will not look circular. A time-domain filter in the receiver must be used in this case.

## *Superframe Structure*

SDMT uses a superframe structure with duration of 500 μs. Thus the superframe frequency is 2 kHz. A superframe consists of a number of downstream symbols followed by a quiet period and then a number of upstream symbols followed by a quiet period. Figure 9.13 illustrates an SDMT frame structure.

Figure 9.13 Sample SDMT frame structure.

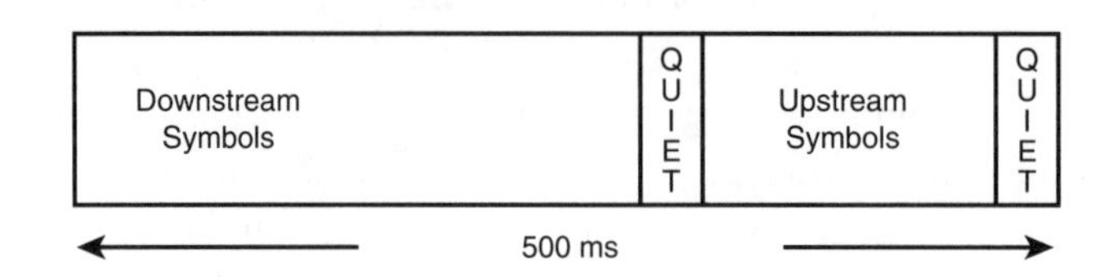


By varying the number of downstream symbols sent and the number of upstream symbols sent, the average upstream and downstream bit rates as well as the bit rate ratio can be controlled.

Consider the examples shown in Figure 9.14. The first shows an 8:1 downstream to upstream ratio, the second a 2:1 ratio, and the third a 1:1 ratio (symmetric example).

Figure 9.14 Possible implementations of an SDMT frame for different upstream and downstream bit-rate ratios.

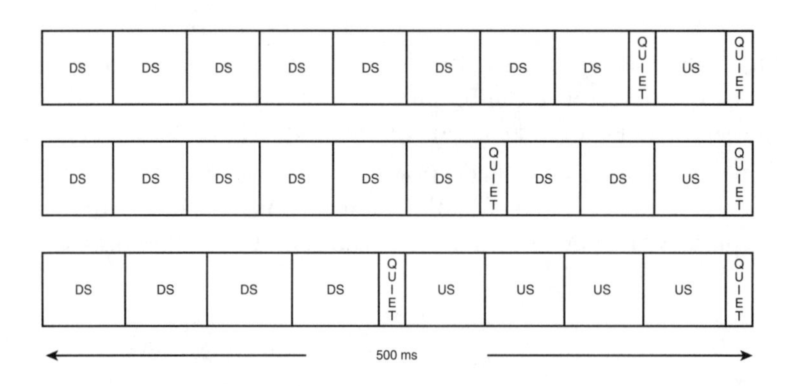


Self-NEXT avoidance is relatively straightforward when lines sharing a binder group use the same symmetry ratio. In such a case, the frame structure of all lines must be aligned so that all VTU-Os transmit simultaneously and likewise receive simultaneously. If no VTU-O transmits while another VTU-O receives, then self-NEXT does not exist. Aligning the superframes of all VDSL lines in a binder group is not trivial, but is certainly possible. For the 2:1 ratio case, self-NEXT elimination is illustrated in Figure 9.15.

Figure 9.15 Self-Next avoidance in an SDMT system.

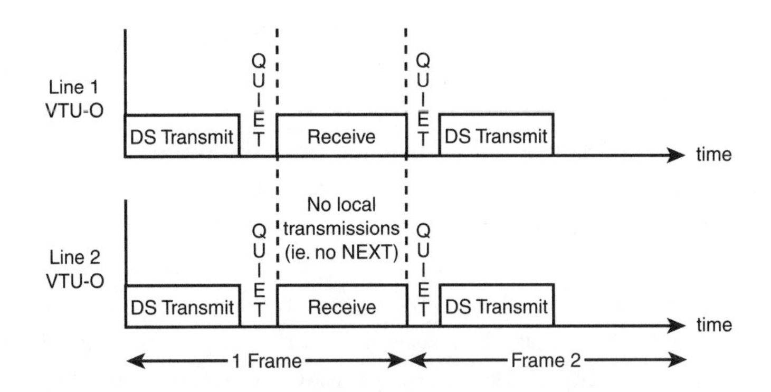


## Note

A proposal on how to align the superframes of all VDSL lines in a binder group includes using a common network clock for synchronization, using stuff and delete methods,[8] and even using GPS signals.[9]

When mixing different data rate ratio configurations within a binder group, complete self-NEXT elimination is not possible. However, frames can be arranged to reduce the occurrence of self-NEXT. See Figure 9.16 for an example of limited self-NEXT due to frame alignment of a symmetric and an asymmetric VDSL mix.

Figure 9.16 Two SDMT frames with different upstream-to-downstream ratios aligned to limit the amount of mutual NEXT.

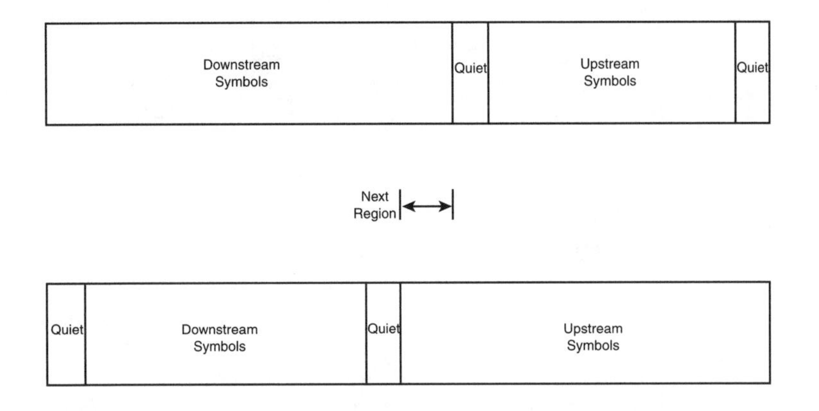

Though it is not possible with this scheme to perfectly eliminate self-NEXT, the self-NEXT levels are predictable and line rates can be adjusted accordingly when self-NEXT is present.

## *Sampling Considerations*

Up to now, the number of samples in a frame has been discussed, as well as the length of a superframe. Not discussed yet is the duration of a sample (or the inverse of that being the sample rate). Following from the sample rate is the number of frames in a superframe, the highest frequency used in the system, and the width of each frequency bin. Remember that all blocks up to the output of the modulator have functionality that does not depend on the output sample rate. Of course, the reason is that the length of a superframe is fixed (500 μs); the higher number of frames per superframe implies that the blocks leading up to the modulator must perform their functions more quickly.

Three sample rates are proposed for SDMT: 11.04 MHz, 22.08 MHz, and 44.16 MHz. For these rates, Table 9.3 shows the maximum operating frequency, bin width, number of samples per superframe, and approximate number of frames per superframe. When more frames exist per superframe, the net data rate in each direction can be higher. Thus short lines supporting the extremes of the required VDSL rates use a sample rate of 44.16 MHz.

Table 9.3 Maximum Operating Frequency, Bin Width, Number of Samples per Superframe, and Approximate Number of Frames per Superframe

| Sampling Frequency (MHz) | Maximum Frequency (MHz) (approximate) | Bin Width (kHz) | Samples per Superframe | Frames per Superframe |
|---|---|---|---|---|
| 11.04 | 5.52 | 21.5625 | 5520 | 10 |
| 22.08 | 11.04 | 43.125 | 11040 | 21 |
| 44.16 | 22.08 | 86.25 | 22080 | 43 |

For the lowest sample rate, the total number of samples per superframe is 5520. Given that each frame contains 512 base samples along with a cyclic prefix of between 0 and 64 samples, the maximum possible number of frames per superframe (assuming a 40 sample cyclic prefix) is 10, leaving no samples to be used for the quiet period (30.4 us). This scenario is not very practical, as some line turnaround time will be necessary for proper half-duplex operation.

The length of the quiet time can be increased by reducing the number of samples used in the cyclic prefix (which may increase the front-end filtering needs of the receiver) or by decreasing the number of frames per superframe to nine. The latter method would leave, using the maximum length cyclic prefix, a minimum of 336 samples for the quiet times, or approximately 30.4 us. This time would be divided between the two required quiet periods (though it might not be evenly divided). For the higher sample rates, similar calculations can be done to determine the lengths of the cyclic prefixes and quiet times. It is intended that many of the parameters including the sample rate, the number of frames per superframe, and length of the cyclic prefix and quiet times be negotiated during startup.

### *RFI Cancellation Circuitry*

Generally, the front end of an SDMT receiver contains an RFI cancellation block. As discussed earlier, this block is necessary to combat ingress from local RFI sources. Normally, some type of cancellation is necessary before the analog-to-digital conversion process to adequately digitize the received signal with minimal quantization noise.

### *Receive Blocks*

Much of the functionality in the receive path of an SDMT modem follows from the discussion of the receive path of a DMT modem. Time-domain equalization is done to shorten the channel impulse response to less than the length of the cyclic prefix. An FFT is done on a block of 512 received samples, yielding a complex vector with complex conjugate symmetry. Frequency-domain equalization is performed on the lower 256 values of the FFT output, each complex value representing the received component from a single bin. Constellation decoding and error correction is done before the data is demultiplexed by a framer.

### *Final Words on SDMT*

As a candidate for VDSL, SDMT has several advantages. First, the use of TDD allows for easy control over the upstream and downstream data rate ratio. Second, because SDMT is half-duplex, a hybrid is not needed on VDSL modems, and the sharing of digital circuitry between the transmitter and receiver (the FFT/IFFT circuitry for example) is possible, allowing for complexity reduction. Third, the nature of multicarrier modulation allows for flexibility in dealing with narrowband interference from RFI and can easily reduce egress in a band by simply reducing the power on a tone.

One issue related to SDMT is the requirement to properly synchronize frames for VDSL lines in a common binder group. If all lines emanate from a single ONU, this synchronization might be feasible by using a common clock. However, when independent ONUs feed VDSL into the same binder group, achieving frame synchronization is much more difficult.

## *FDD Approaches*

FDD approaches to transmission are similar to those discussed for ADSL. This approach uses full-duplex transmission and separates upstream and downstream transmissions by using different frequency bands for each. CAP/QAM-based techniques lend themselves nicely to an FDD approach, as the basic spectrum can be positioned anywhere, depending on the hilbert pair filter responses in the case of CAP and center frequency in the case of QAM. A second method discussed here is an FDD method utilizing DMT. This method is sometimes known as Zipper.

**Note**

FDD approaches to transmission include CAP/QAM-based ADSL and FDM ADSL. Echo-canceled ADSL is partially an FDD approach, but not completely because some of the upstream and downstream bands overlap.

### *CAP/QAM-Based FDD Systems*

CAP and QAM technologies offer another method of implementing VDSL. The basic blocks in the modulation system are similar to those used in CAP/QAM ADSL implementations. For VDSL, the proposed CAP/QAM implementation can run in symmetric mode as well as asymmetric mode. In asymmetric mode, the upstream channel's frequency range starts at about 1.12 MHz and can go up as high as about 2.12 MHz. Different symbol rates are defined as well as different constellation sizes to control the upstream data rate. The downstream channel begins at around 2 MHz and can range up to about 18 MHz. Again, different baud rates and constellation sizes can be used to provide different downstream data rates. Self-NEXT is avoided simply by using different frequency bands for transmission in the two directions.

Symmetrical service is handled in a very interesting way by CAP/QAM-based VDSL. For symmetric operation, two upstream channels exist: one from 1.12 MHz to 1.6 MHz, and one from 11.30 MHz extending up to 19.8 MHz. The downstream channel sits between the two upstream channels and ranges between 2.4 MHz and up to 10.7 MHz. Again no self-NEXT exists because the upstream and downstream channels do not overlap in frequency.

Figure 9.17 shows the various spectrums utilized by CAP/QAM VDSL for symmetric and asymmetric service.

Figure 9.17 Upstream and downstream spectra for CAP/QAM-based VDSL for (a) asymmetric rates and (b) symmetric rates.

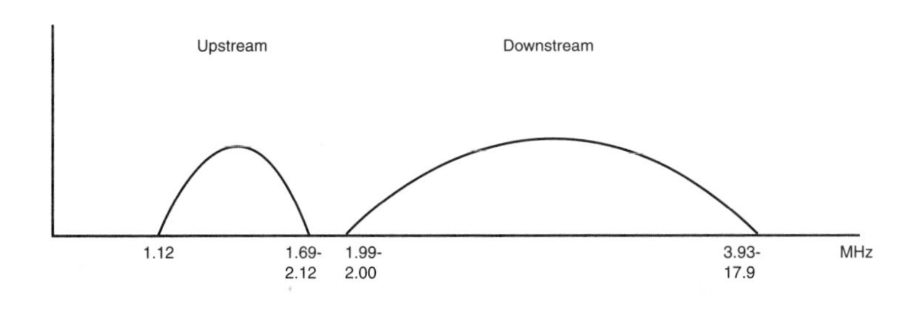


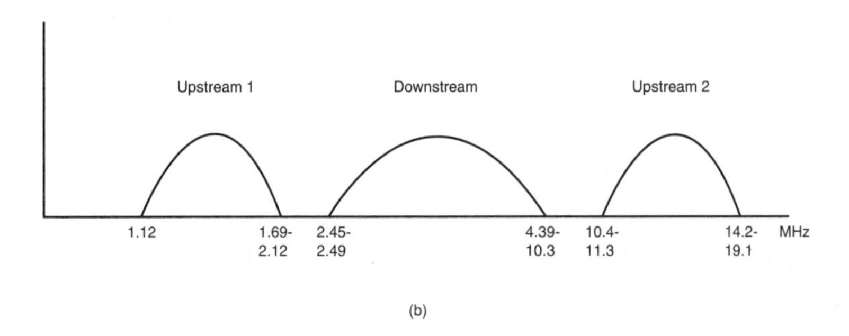


The figure shows different downstream baud rate options (and thus frequency ranges). When mixing symmetric and asymmetric services, in some cases self-NEXT will occur. Note that the highest baud rate downstream carrier overlaps the higher upstream regions in the symmetric frequency plot, which would cause a problem. However, some of the other downstream symmetric options would not, as they end below the higher upstream channel. As CAP/QAM VDSL modems mature, some of the frequencies may change; however, the basic implementation strategy is likely to remain the same.

RF egress control is a bit more difficult for CAP/QAM-based modems. Recall that SDMT-based modems could simply lower the power in a subcarrier at frequencies where power reduction was necessary. CAP/QAM modems must use time-domain filters to do this. Digital filters can normally be implemented for this task, but their flexibility with respect to changes in the power reduced frequencies and the amount of power reduction is somewhat reduced.

### *DMT-Based FDD Systems*

A second FDD method for VDSL uses DMT and is a relatively new modulation scheme named Zipper. Zipper is a time-synchronized version of DMT using 2,048 subchannels going up to 11.04 MHz. Each subchannel has a width of about 5.309 kHz. A subchannel in Zipper can be used for upstream transmission or for downstream transmission, but not for both. If all lines in a binder use a similar bin assignment, then no self-NEXT will be present. (The frames must be time aligned as well so that crosstalk is orthogonal to the received signal.) In addition, subchannel usage can be chosen in a way that minimizes NEXT interaction with other services. Zipper has many of the same benefits as SDMT with regard to ingress and egress handling flexibility. Though a relatively new addition to VDSL implementation, this modulation method may develop into a powerful tool for high-speed access.

## *VDSL in WAN Networks*

The speeds that VDSL supports make it suitable for many types of applications. Many applications that exist today may use VDSL as a transport mechanism; some applications yet to be developed may also use VDSL. A good example of VDSL usage is the full-service network described later in the chapter. The following sections describe video services and data services that utilize VDSL.

### *Video Services*

The high-rate options of VDSL make it a very good access technology for VoD. Figure 9.18 shows a system using VDSL in this application.

Figure 9.18 VDSL used in a network providing video services.

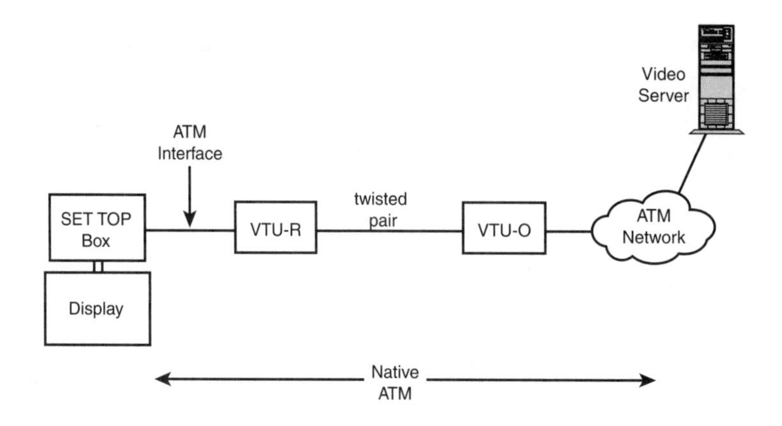

The video server in this example sends digital video over ATM through access equipment to the VDSL loop. At the customer end, some type of video codec converts the digital video to the proper signal for the display screen. The digital video would likely be compressed, and the codec would perform decompression. It is also likely that PPP would not be used on this connection, as it might not be the most efficient protocol for carrying high-speed video. Another control channel might use PPP for authentication and billing.

### *Data Services*

Many data service possibilities exist for VDSL. In the near future, VDSL might be overkill for residential Internet access and Web surfing. However, back-hauling data traffic between locations using twisted pairs might be more reasonable. VDSL could also be used to connect larger offices and corporations to data networks in lieu of fiber connections. As time goes on, more and more opportunities are likely to exist for VDSL to carry data.

### *The Full-Service Network*

Because of the high bit rates supported by VDSL, it is often thought of as an access mechanism to a full-service network (FSN). This type of network is intended to serve all the communication needs of a subscriber, including voice, video, and data applications. Such an all encompassing network could replace today's telephone system and cable system, and it could add extra features such as video telephony. A full-service access network (FSAN), consisting primarily of telephone and service providers, is working to piece together existing standards where available and stimulate new standards to define a FSN.

## *Summary*

This chapter focused on VDSL. The basic needs that VDSL is intended to fulfill were discussed along with several candidate VDSL implementations. The approach differences of these implementations were highlighted, along with some pros and cons of each. Finally, VDSL network applications were discussed. The true direction and application drivers for VDSL are yet to be seen, but any technology promising the delivery speeds of VDSL has great potential.

## *Endnotes*

1. K. S. Jacobsen, "Spectral Compatibility of ADSL and VDSL, Part I: The Impact of VDSL on ADSL Performance," ANSI Contribution T1E1.4/97-404, December 1997.

2. K. S. Jacobsen, "Spectral Compatibility of ADSL and VDSL, Part I: The Impact of ADSL on VDSL Performance," ANSI Contribution T1E1.4/98-035, March 1998.

3. Byung-Jang Jeong and Kyung-Hyun Yoo, "An Efficient Digital RFI Canceler for DMT-Based VDSL," ANSI Contribution T1E1.4/97-422, December 1997.

4. Brian Wiese and John Bingham, "Digital Radio Frequency Cancellation for DMT VDSL," ANSI Contribution T1E1.4/97-460, December 1997.

5. J. Cioffi, et al., "Analog RF Cancellation with SDMT," ANSI Contribution T1E1.4/96-084, April 1996.

6. "Very High Speed Digital Subscriber Line System Requirements," Revision 13, ANSI Contribution T1E1.4/98-043, March 1998.

7. K. Foster, "Transmission and Multiplexing : Access Transmission Systems on Metallic Access Cables; Very High Speed Digital Subscriber Line (VDSL); Part I: Functional Requirements," ETSI TM6 Draft Technical Report, DTS/TM-06003-1, V0.0.7 February 1998.

8. J. Cioffi, "Add/Delete SDMT Solution for Unsynchronized VDSL Lines," ANSI Contribution T1E1.4/96-247, September 1996.

9. Sven-Rune Olofsson, et al., "Synchronizing Optical Networks Using GPS," ANSI Contribution T1E1.4/97-283, September 1997.

# Index

## C

# E

# M



# N



# O

## U



## V



## W-Z